NEUROMETHODS

Series Editor
Wolfgang Walz
University of Saskatchewan
Saskatoon, SK, Canada

For further volumes:
http://www.springer.com/series/7657

Neuromethods publishes cutting-edge methods and protocols in all areas of neuroscience as well as translational neurological and mental research. Each volume in the series offers tested laboratory protocols, step by step methods for reproducible lab experiments and addresses methodological controversies and pitfalls in order to aid neuroscientists in experimentation. *Neuromethods* focuses on traditional and emerging topics with wide ranging implications to brain function, such as electrophysiology, neuroimaging, behavioral analysis, genomics, neurodegeneration, translational research and clinical trials. *Neuromethods* provides investigators and trainees with highly useful compendiums of key strategies and approaches for successful research in animal and human brain function including translational "bench to bedside" approaches to mental and neurological diseases.

General Anesthesia Research

Edited by

Marco Cascella

Istituto Nazionale Tumori - IRCCS - Fondazione Pascale, Napoli, Italy

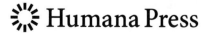 Humana Press

Editor
Marco Cascella
Istituto Nazionale
Tumori - IRCCS - Fondazione Pascale
Napoli, Italy

ISSN 0893-2336   ISSN 1940-6045 (electronic)
Neuromethods
ISBN 978-1-4939-9893-7   ISBN 978-1-4939-9891-3 (eBook)
https://doi.org/10.1007/978-1-4939-9891-3

This Humana imprint is published by the registered company Springer Science+Business Media, LLC, part of Springer
Nature
The registered company address is: 233 Spring Street, New York, NY 10013, U.S.A.

Preface to the Series

Experimental life sciences have two basic foundations: concepts and tools. The *Neuromethods* series focuses on the tools and techniques unique to the investigation of the nervous system and excitable cells. It will not, however, shortchange the concept side of things as care has been taken to integrate these tools within the context of the concepts and questions under investigation. In this way, the series is unique in that it not only collects protocols but also includes theoretical background information and critiques which led to the methods and their development. Thus, it gives the reader a better understanding of the origin of the techniques and their potential future development. The *Neuromethods* publishing program strikes a balance between recent and exciting developments like those concerning new animal models of disease, imaging, *in vivo* methods, and more established techniques, including, for example, immunocytochemistry and electrophysiological technologies. New trainees in neurosciences still need a sound footing in these older methods in order to apply a critical approach to their results.

Under the guidance of its founders, Alan Boulton and Glen Baker, the *Neuromethods* series has been a success since its first volume published through Humana Press in 1985. The series continues to flourish through many changes over the years. It is now published under the umbrella of Springer Protocols. While methods involving brain research have changed a lot since the series started, the publishing environment and technology have changed even more radically. Neuromethods has the distinct layout and style of the Springer Protocols program, designed specifically for readability and ease of reference in a laboratory setting.

The careful application of methods is potentially the most important step in the process of scientific inquiry. In the past, new methodologies led the way in developing new disciplines in the biological and medical sciences. For example, Physiology emerged out of Anatomy in the 19th century by harnessing new methods based on the newly discovered phenomenon of electricity. Nowadays, the relationships between disciplines and methods are more complex. Methods are now widely shared between disciplines and research areas. New developments in electronic publishing make it possible for scientists that encounter new methods to quickly find sources of information electronically. The design of individual volumes and chapters in this series takes this new access technology into account. SpringerProtocols makes it possible to download single protocols separately. In addition, Springer makes its print-on-demand technology available globally. A print copy can therefore be acquired quickly and for a competitive price anywhere in the world.

Saskatoon, SK, Canada   *Wolfgang Walz*

Preface

Topics of Research in General Anesthesia: State of the Art and Perspectives

The anesthetics used in everyday clinical practice are reassuringly safe. They provide us with a portfolio of rapidly acting and titratable agents. The technique used for the administration of anesthesia and monitoring of vital signs has also improved dramatically during these last decades. There is also a rapid development in surgery, surgical technique, and patients with complex and severe disease that are scheduled for surgery, along with a growing number of elderly and fragile patients having surgery. Thus, the demands for safe and effective anesthesia, securing not only a safe and effective intraoperative course but a rapid recovery following more complex surgery in the elderly and fragile patient, have become of outmost importance.

Stable intraoperative course with minimal deviation in vital signs, maintaining homeostasis with secure and adequate depth of anesthesia, avoiding any risk for unintentional light anesthesia, and subsequent risk for awareness with recall must be acknowledged. Avoiding too deep anesthesia and subsequent cardiovascular depression and prolonged recovery should likewise be avoided.

Goal-directed anesthesia, titrating anesthetics to each patients' unique needs, and balancing the surgical stress are common requests nowadays. Choosing an anesthetic agent may also impact quality of recovery and long-term outcomes. Combining drug choice and goal-directed drug delivery may indeed have an important impact on patient quality of recovery and potentially long-term prognosis. The potential risk for neurocognitive side effects during recovery has become a major concern in the handling of the elderly and fragile patients. Optimizing preoperative, prehabilitation, and fine-tuning anesthesia is important to understand in order to minimize the risk for postoperative cognitive side effects, postoperative delirium, and postoperative cognitive dysfunction. The effects on the cognitive function, both in the neonate and in the elderly, are a matter of debate. Extensive research is devoted to study the impact of anesthetics on the central nervous system, i.e., what can specific drugs impact and what impact do specific drugs have on the neuro-inflammatory response when associated to surgery and anesthesia. Further studies translating the basic research into the effect of surgery/anesthesia on the developing brain and in the elderly with signs and symptoms of dementia are urgently needed. Anesthetic techniques and their impacts on the prognosis in patients with cancer disease are also discussed. There are data suggesting that anesthetic choice may impact the risk for metastasis and cancer reoccurrence. There is also extensive research looking at the effects of anesthetics on cancer cells and systems involved in cancer progression. Preclinical research, as well as prospective randomized clinical trials, is ongoing.

We are currently at an exciting time where anesthesia has evolved from being tasked to only take care of patients during the surgical procedure to perioperative medicine specialty, supporting the patient throughout the preoperative preparation and postoperative period. Quality of recovery, avoidance of neurocognitive side effects and having long-term outcomes, and the development of chronic pain and disease progress are becoming the new objective for anesthesiologists.

Stockholm, Sweden   *Jan G. Jakobsson*

Contents

About the Editor

MARCO CASCELLA MD currently works in the Department of Supportive Care at the National Cancer Institute, Istituto Nazionale Tumori IRCCS Fondazione "G. Pascale," Naples, Italy, where he has responsibility for research in anesthesia and pain medicine. He is also annual professor of Physiology on the Degree Course in Biomedical Laboratory Techniques at the Faculty of Medicine, University of Naples "Federico II," and of Anesthesia and Pneumology on the Degree Course of School of Nursing at the same university. He is an active member of the scientific board on cancer pain and palliative care of the Italian Society of Anesthesia and an active researcher with special interests in the mechanisms of general anesthesia, anesthesia-related neurocognitive phenomena, antibiotic therapy in critically illness patients, and pain medicine. His work also includes preclinical investigations, with particular focus on nutraceuticals as promising therapeutics in neurodegenerative disease as well as on the role of opioids in the processes of carcinogenesis. He is head and co-investigator of different research projects and serves as editor for research projects sponsored by the Italian Ministry of Health. He has been chairmen and speaker at numerous conferences and conventions and has published over 100 scientific publications, including peer-reviewed manuscripts, books, and book chapters.

Contributors

NAGOTH JOSEPH AMRUTHRAJ • *S.S.D. Sperimentazione Animale, Istituto Nazionale Tumori - IRCCS - Fondazione G. Pascale, Naples, Italy*

CLAUDIO ARRA • *S.S.D. Sperimentazione Animale, Istituto Nazionale Tumori - IRCCS - Fondazione G. Pascale, Naples, Italy*

THEODOROS ASLANIDIS • *Intensive Care Unit, St. Paul General Hospital, Thessaloniki, Greece*

DANIELA BALDASSARRE • *Department of Medicine and Surgery, University of Salerno, Salerno, Italy*

ANTONIO BARBIERI • *S.S.D. Sperimentazione Animale, Istituto Nazionale Tumori - IRCCS - Fondazione G. Pascale, Naples, Italy*

FRANCESCA BIFULCO • *Department of Anesthesia and Pain Medicine, Istituto Nazionale Tumori - IRCCS - Fondazione G. Pascale, Naples, Italy*

SABRINA BIMONTE • *Department of Anesthesia and Pain Medicine, Istituto Nazionale Tumori - IRCCS - Fondazione G. Pascale, Naples, Italy*

GRAZIELA BITER • *Department of Anesthesiology, CHU Saint-Pierre, Université Libre de Bruxelles, Bruxelles, Belgium*

SERENA BOCCELLA • *Division of Pharmacology, Department of Experimental Medicine, University of Campania "L. Vanvitelli", Naples, Italy*

MARCO CASCELLA • *Istituto Nazionale Tumori - IRCCS - Fondazione Pascale, Napoli, Italy*

ARTURO CUOMO • *Department of Anesthesia and Pain Medicine, Istituto Nazionale Tumori - IRCCS - Fondazione G. Pascale, Naples, Italy*

VITO DE NOVELLIS • *Division of Pharmacology, Department of Experimental Medicine, University of Campania "L. Vanvitelli", Naples, Italy*

DIANA DI FRAJA • *Cardiac Anesthesia and Intensive Care Unit, A.O.R.N. "Dei Colli", Monaldi Hospital, Naples, Italy*

RAFFAELA DI NAPOLI • *Department of Anesthesiology, Institut Jules Bordet, Université Libre de Bruxelles, Bruxelles, Belgium*

MARCO FIORE • *Department of Women, Child and General and Specialized Surgery, University of Campania "Luigi Vanvitelli", Naples, Italy*

FRANCESCA GARGANO • *Unit of Anesthesia, Intensive Care and Pain Management, Department of Medicine, University Campus Bio-Medico of Rome, Rome, Italy*

ALDO GIUDICE • *Epidemiology Unit, Istituto Nazionale Tumori, IRCCS—Fondazione G. Pascale, Naples, Italy*

FRANCESCA GUIDA • *Division of Pharmacology, Department of Experimental Medicine, University of Campania "L. Vanvitelli", Naples, Italy*

MAHER KHALIFE • *Department of Anesthesiology, Institut Jules Bordet, Université Libre de Bruxelles, Bruxelles, Belgium*

GAELE LEBEAU • *Department of Psychiatry, AP-HP, Western Paris University Hospital group, Paris, France*

LIVIO LUONGO • *Division of Pharmacology, Department of Experimental Medicine, University of Campania "L. Vanvitelli", Naples, Italy*

SABATINO MAIONE • *Division of Pharmacology, Department of Experimental Medicine, University of Campania "L. Vanvitelli", Naples, Italy*

IDA MARABESE • *Division of Pharmacology, Department of Experimental Medicine, University of Campania "L. Vanvitelli", Naples, Italy*

NADER D. NADER • *Department of Anesthesiology, University at Buffalo, Buffalo, NY, USA; Anesthesiology Sv, VA Western NY Healthcare System, Buffalo, NY, USA; Department of Anesthesiology, Jacobs School of Medicine and Biomedical Science, Buffalo, NY, USA*

FILOMENA OLIVA • *Department of Medicine and Surgery, University of Salerno, Salerno, Italy*

ENZA PALAZZO • *Division of Pharmacology, Department of Experimental Medicine, University of Campania "L. Vanvitelli", Naples, Italy*

CONCETTA PALMIERI • *Department of Anesthesia, Intensive Care and Hyperbaric Medicine, "Santobono-Pausilipon" Children's Hospital, Naples, Italy*

ORNELLA PIAZZA • *Department of Medicine and Surgery, University of Salerno, Salerno, Italy*

GORIZIO PIERETTI • *Department of Plastic Surgery, University of Campania "L. Vanvitelli", Naples, Italy*

ANTONIO PISANO • *Cardiac Anesthesia and Intensive Care Unit, A.O.R.N. "Dei Colli", Monaldi Hospital, Naples, Italy*

MAIKO SATOMOTO • *Department of Anesthesiology, Toho University School of Medicine, Tokyo, Japan*

MARIANTONIETTA SCAFURO • *Department of Woman, Children, General and Specialistic Surgery, University of Campania "L. Vanvitelli", Naples, Italy*

GIULIO SCALA • *Villa Esther Hospital, Avellino, Italy*

GIULIANA SCARPATI • *Department of Medicine and Surgery, University of Salerno, Salerno, Italy*

MATTHEW UMHOLTZ • *Department of Anesthesiology, Brandon Regional Hospital, Brandon, FL, USA*

HELENE VULSER • *Department of Psychiatry, AP-HP, Western Paris University Hospital group, Paris, France; Faculty of Medicine, Paris Descartes University, Sorbonne Paris Cité, Paris, France*

STEFAN WIRZ • *Department of Anesthesiology, Intensive Medicine, Pain/Palliative Medicine, CURA-Hospital Bad Honnef, Bad Honnef, Germany; Center for Pain Medicine, CURA-Hospital Bad Honnef, Bad Honnef, Germany; Center for Weaning, CURA-Hospital Bad Honnef, Bad Honnef, Germany*

Chapter 1

The Challenge of Accidental Awareness During General Anesthesia

Marco Cascella

Abstract

Intraoperative unconsciousness and amnesia are main goals of general anesthesia. Although these objectives are achieved in the vast majority of cases, in very rare circumstances they are not "completely" obtained, or maintained, and in turn, accidental consciousness and subsequent memorization of sensorial information may occur during intended anesthesia. This failure of anesthesia is termed as *general anesthesia awareness* (or *accidental awareness during general anesthesia*, or simply *awareness*). While the incidence of this complication is rare, the clinical features and its potentially devastating psychological sequelae impose a thorough knowledge of the phenomenon.

During anesthesia, the patient may be occasionally cognizant responding to commands or may wake up. This intraoperative awakening, termed as "wakefulness," must be not confused with the awareness. The discriminating element for a proper definition of awareness, indeed, is the concomitant presence of two elements that correspond to higher cognitive functions: consciousness and memory processing of the intraoperative experience.

Although not all wakefulness episodes complete the memory processing (encoding, storing through consolidation, and retrieval), consolidated unexpected experiences can be expressed as explicit-spontaneous, or induced, reports. The pathways of this declarative, or explicit, memory produce, in turn, the *awareness with recall* phenomenon. Alternatively, the intraoperative experience can be processed without requiring conscious memory content and expressed as inexplicable changes in behaviors, or performances, or through the mechanism of priming, in which exposure to one stimulus influences a response to a subsequent stimulus, without conscious guidance, or intention. The processing of information via nondeclarative, or implicit, memory system, leads to the other awareness subtype: the *awareness without explicit recall.*

Referring to the state of the art of research, the aim of this chapter is to dissect the multiple aspects of this anesthesia-induced complication. Although in recent years research has allowed us to understand many aspects of the phenomenon, its complete characterization still seems far away, and several controversies and dark sides remain. In particular, the chapter addresses topics related to definitions and classification, epidemiology, clinical features, risk factors, management, and strategies useful for prevention.

Finally, the interest in the subject is justified as it represents a fascinating matter of investigation which intersects study areas in the contexts of the "general anesthesia research" and neuroscience. Mechanisms of anesthesia, and impact of anesthetics on consciousness and memory, represent an attractive way for studying brain and mind through an "experimental model" (general anesthesia) that is carried out on several tens of millions of patients, every year.

Marco Cascella (ed.), *General Anesthesia Research*, Neuromethods, vol. 150, https://doi.org/10.1007/978-1-4939-9891-3_1,
© Springer Science+Business Media, LLC, part of Springer Nature 2020

Key words General anesthesia awareness, Accidental awareness during general anesthesia, Explicit memory, Anesthesia awareness with recall, Awareness without explicit recall, Memory consolidation, Memory processing, Benzodiazepines

Abbreviations

AAGA   Accidental awareness during general anesthesia
AAWR   Anesthesia awareness with recall
ASA   American Society of Anesthesiologists
AWER   Awareness without explicit recall
BDZs   Benzodiazepines
CIIA   Combination of intravenous and inhaled anesthesia
DoA   Depth of anesthesia
ETAC   End-tidal anesthetic concentration
GA   General anesthesia
GAA   General anesthesia awareness
GABAAR   Gamma-aminobutyric acid A receptor
IA   Intraoperative awakening
ICU   Intensive care unit
IFT   Isolated forearm technique
MAC   Minimal alveolar concentration
NAP5   Fifth National Audit Project from Great Britain
NMBAs   Neuromuscular blocking agents
NMDA N-methyl-D-aspartate
NO   Nitrous oxide
PTSD   Posttraumatic stress disorder
TIVA   Total intravenous anesthesia

1   Introduction

Intraoperative unconsciousness and amnesia, respectively, the incapability of evaluating and processing the information of the environment and the abolished memorization of events that occurred while the patient is under the effect of anesthetics, are the main goals and the foundation of general anesthesia (GA). Although these objectives are achieved in the vast majority of cases, in very rare circumstances they are not "completely" maintained and, in turn, accidental consciousness and subsequent memorization may occur during intended anesthesia. This anesthesia complication, which has been referred to for many years as *awareness*, is generally termed as *general anesthesia awareness* (GAA) or *accidental awareness during general anesthesia* (AAGA). The awareness complication encompasses a spectrum of conditions with different clinical manifestations (Fig. 1).

Awareness during anesthesia is not a new problem because, interestingly, it was observed even at the beginning of the modern

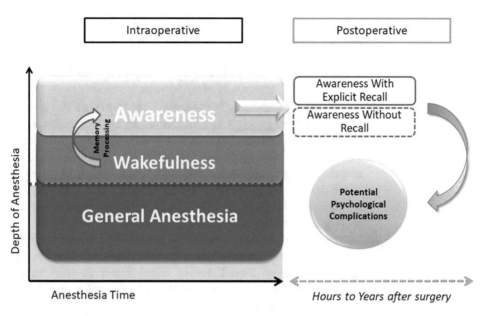

Fig. 1 General anesthesia awareness. Schematic picture and subtypes. During anesthesia, it is not always possible to maintain a state of deep anesthesia (below the red line). Again, patients may be occasionally cognizant responding to commands or may wake up. This intraoperative awakening is termed as "wakefulness." Although not all wakefulness episodes follow the pathways of memory processing (encoding, storing through consolidation, and retrieval), consolidated unexpected experiences can be expressed as explicit-spontaneous or induced report (declarative, or explicit memory), configuring the awareness with recall phenomenon. Alternatively, the intraoperative experience can be processed without requiring conscious memory content (nondeclarative or implicit memory) and expressed as inexplicable changes in behaviors or performances. Both awareness subtypes may be reported (awareness with recall) or expressed (awareness without recall) at the emergence from anesthesia but also at distance from the end of the intervention. Awareness may have psychological sequelae for the patient, including single symptoms (insomnia, depression, anxiety), or grouped into definite syndromes (e.g., post-traumatic stress disorder)

anesthesia. Morton himself described that his anesthetized patients were "half asleep" and experienced pain during the operation [1]. In 1950, Winterbottom described the first well-detailed clinical case in a correspondence to the British Medical Journal entitled "Insufficient anaesthesia". As the patient reported:

> … I woke up in the theatre! … I was awakened by the most excruciating pain in my tummy. I felt as if my whole inside was being pulled out. I wanted to cry out or otherwise indicate my suffering but I couldn't move any part of me. I heard the doctors talking about the gall-bladder and about doing something with it to the small intestine … [2].

Later on, in 1961, Meyer and Blacher [3] firstly illustrated the case of a patient that manifested psychological complications after an episode of awareness. This psychological manifestation was indicated as "traumatic neurosis" because the posttraumatic stress disorder (PTSD) had not yet been characterized by the American Psychiatric Association [4]. The authors titled the report "Traumatic Neurotic Reaction Induced by Succinylcholine Chloride" where they postulated the pathogenetic role of that short-acting depolarizing curare. Undoubtedly, the introduction of the neuro-muscular blocking agents (NMBAs) in anesthesia (1942) has significantly increased the risk and, consequently, the number of GAA cases.

Nowadays, there is "awareness" that although the incidence of this complication is rare, the clinical features and its potentially devastating psychological sequelae justify the high public concern. Furthermore, the awareness-related medicolegal issues are of paramount importance. In this regard, Domino et al. [5] analyzed the database of the Closed Claims Project, which is composed of closed malpractice claims in the USA, and found that approximately 2% of the legal claims against anesthetists concerned awareness.

Because the phenomenon is particularly complex, there is confusion regarding terminology [6] and classification, as well as poor understanding of overall incidence and that in the various settings, clinical significance and potential psychological sequelae, risk factors, and strategies useful for prevention.

Referring to the state of the art of research, the aim of this chapter is to dissect the multiple features of this anesthesia-induced complication. In recent years, research has allowed us to understand many aspects of the phenomenon. Nevertheless, because its complete characterization still seems far away, several controversies and dark sides remain. This topic represents a fascinating matter of investigation which intersects study areas that, in the context of the "general anesthesia research," range from the pharmacodynamics of anesthetics to the mechanisms of GA, the relationship between anesthesia and consciousness, the processes of memorization, and the consequent effort to develop systems and strategies for intraoperative brain monitoring.

2   Definitions and Subtypes

The discriminating element for a proper definition of the GAA is the concomitant presence of two elements that correspond to higher cognitive functions: consciousness and memory. The relationship between consciousness and anesthesia is not so simple to define. The first problem is the exact definition of consciousness, a complex combination of phenomenal experiences with the precise understanding of what they are:. in other words, the link between the brain and the mind. Given its high significance, this matter has been addressed by philosophers, scientists, mathematicians and researchers from different fields; thus, several theories of consciousness have been postulated [7–13] (Table 1). According to a philosophical approach, consciousness represents the state or quality of awareness, or, of being aware of an external object or something within oneself. In medicine, consciousness is assessed by observing

Table 1
Selected consciousness theories (Adapted from [15])

Author(s) (date)	Theory	Brief explanation
Immanuel Kant (Critique of Pure Reason (1781)) and following attempts for explaining the "binding problem"	*Cognitive binding*[a]	Consciousness is the result of binding together in a single unified Percept, different features of an object. There are several levels of binding: neural (convergent binding), systems (assembly binding) and global (synchronous binding) levels
Edelman (1987) [8]	*Neural Darwinism*	The mind and consciousness are purely biological phenomena, arising from complex cellular processes within the brain. Thus, the development of consciousness and intelligence can be explained by the Darwinian theory
Baars (1988) [9]	*Global workspace*[b]	Based on the artificial intelligence concept called the "blackboard." consciousness works like a computer memory, which can call up and retain an experience even after it has passed
Crick and Koch (1990) [10]	*Neural correlates of consciousness*	The neuronal correlates of consciousness constitute the smallest set of neural events and structures sufficient for a given conscious percept or explicit memory. This theory is a structured attempt to explain the binding problem
Thagard and Stewart (2014) [11]	*Semantic pointer competition*	Consciousness results from three mechanisms: Representation by firing patterns in neural populations, binding of representations into more complex representations called semantic pointers, and competition among semantic pointers to capture the most important aspects of an organism's current state
Tononi (2004) [12]	*Integrated information*[b]	Conscious experience represents the integration of a wide variety of information, and this experience is irreducible
Graziano and Kastner (2011) [13]	*The "attention schema" theory*	Specific cortical areas, notably in the superior temporal sulcus and the temporoparietal junction, are used to build the construct of awareness and attribute it to other people. The same cortical machinery is also used to attribute awareness to oneself

[a]For attempts to explain the "biding problem," see ref. 7
[b]Global workspace theory and integrated information theory are not mutually exclusive. The former tries to explain in practical terms whether something is conscious or not, whereas the latter seeks to explain how consciousness works more broadly

and testing the patient's arousal and responsiveness. Based on the degree of the patient's response—assessed by standardized behavior observation scales—consciousness may range from a full alertness and comprehension state to disorientation, delirium, loss of meaningful communication, and finally an unconsciousness state featuring loss of movement in response to painful stimuli [14]. Even in anesthesia, consciousness has a spectrum of conditions extending from a complete waking status to a deep level of anesthesia. Obviously, they are not rigid consciousness/unconsciousness conditions, as consciousness fluctuation during general anesthesia has been also postulated [15].

In nosographic and clinical terms, it is important to distinguish the "true" awareness complication from intraoperative conditions that do not necessarily presuppose consciousness and memory. During GA, in nonparalyzed or partially paralyzed patients and in the case of inadequate pain control, spinal reflex to pain or surgical stimulation may be intact despite a right anesthesia status [16]. Of note, these involuntary movements and reactions such as facial muscles contraction, eye movements, intolerance of orotracheal tube and limb movements as well as sweating and signs of sympathetic activation, show the features of the finalistic activity.

In some circumstances, patients seem to be cognizant responding to commands (*see* the isolated forearm technique, IFT), or may wake up during anesthesia and surgery. This intraoperative awakening (IA), called "wakefulness," is not necessarily associated with postoperative recall or memory of the event. Memory processing encompasses encoding (i.e., acquiring and representing of memory), storage through working memory stage and consolidation (through synaptic and broad cellular events such as transcriptional, translational, posttranslational mechanisms, and feedback and feedforward regulation), and retrieving of information. In the lack of successful memory processing, the sensorial features of the wakefulness fall into oblivion. Thus, because the occurrence of intraoperative wakefulness not necessarily is followed by explicit-spontaneous, or induced, report after emergence, the terms (and phenomena) of wakefulness and awareness must be not confused [17]. The phenomenon of intraoperative wakefulness occurs much more often than we imagine. A recent well-done meta-analysis on IFT concluded that standard GA regimens, with inhaled or intravenous anesthetics, are not able to avoid events of intraoperative consciousness [18]. The discrepancy between intraoperative events of consciousness and lack in postoperative reports is a topic of paramount importance; probably, it is related to the anesthetics functioning and their different effects on consciousness mechanisms and memory processing. From a mechanistic point of view, not only awareness but also wakefulness can be interpreted as examples of anesthesia failure. A fascinating hypothesis for addressing the issue of awakening during GA postulated the "dysanesthesia" phenomenon. This term encompasses states of mind that can arise in the course of anesthesia, featuring an uncoupling of sensation and perceptual experience. In other words, individuals in this state can be aware of events but in a neutral way, and dissociated from the experiences [19]. The impact of memory on this condition is difficult to understand. Nevertheless, wakefulness and awareness must be addressed as distinct phenomena when the difference relies on the role of the "variable" memory.

The GAA phenomenon encompasses two subgroups with profound differences in terms of incidence, clinical features, and mechanism. These subtypes are the *anesthesia awareness with recall* and the *awareness without explicit recall*. The distinctive element lies in the type of underlying memory that has been activated. The awareness with recall allows activation of the declarative, or explicit, memory, whereas the nondeclarative, or implicit, memory is implicated in the genesis of the awareness without conscious content. The neural correlates and mechanisms of implicit memory differ from those of explicit memory. Implicit memory, indeed, is often preserved after brain damage or experimental manipulations that abolish conscious recall. Also, anesthesia has a different impact on the two forms of memory.

2.1 Anesthesia Awareness with Recall

Concerning memory and awareness, patients may be able or not, to report (spontaneously or after questioning) on anesthesia emergence, or subsequently, the type of experience lived during anesthesia. When patients can recall the unexpected experience, it is appropriate to refer to the *anesthesia awareness with recall during general anesthesia* (AAWR) phenomenon. It is a GAA subgroup defined as the unintended experience of intraoperative sensory perceptions and its explicit recall after the end of surgery [20]. Specifically, these unexpected and unwanted experiences can regard hearing sounds, pain, and immobility, sometimes together with extreme anxiety and distress because of the inability to communicate (e.g., due to paralysis induced by NMBAs). The recalling of these experiences can be reported at the emergence from anesthesia, in the recovery room, or several days/months later.

2.1.1 Declarative, or Explicit, Long-Term Memory and Anesthesia

Each episode of AAWR is an example of declarative, or explicit, long-term memory referred to the conscious intentional recollection of factual information, previous experiences, and concepts [21]. These explicit memories ("knowing what") include autobiographical memories from specific events (episodic memory), as well as general facts and information (semantic memory), and are encoded by the hippocampus (especially the episodic memory), entorhinal cortex and perirhinal cortex (within the medial temporal lobe of the brain and transferred from hippocampus), but are consolidated and stored in the temporal cortex. The amygdala is also involved as it attaches emotional significance to memories and, in turn, can modify or form new bits of memory, specifically related to fear and anxiety. Probably, the mechanism of PTSD follows this way [22].

For explaining the discrepancy between IA occurrence and lack of postsurgical recall we must assume that consciousness and memory are separate cognitive processes. Referring to the general anesthetics mechanism, the doses required for unconsciousness are generally higher than those required for amnesia [23]. Moreover, other factors should be considered for explaining the limited recall despite a higher number of IA. Type of sensorial perception, emotional content (and distress) and exposure time (i.e., time of awakening) are significant factors for inducing and completing the consolidation process. Hearing a conversation during surgery creates discomfort for the patient, but the feeling of paralysis is certainly a serious factor of distress. Therefore, the sensations associated with an intense and unpleasant emotional content are definitely more likely to overcome the filters imposed on consolidation and to induce more serious sequelae. The aspect linked to motivation and intensity of sensation is of great importance. Not by chance, in fact, what is most remembered correlates with the most unpleasant experiences, but also with those full of positive contents. Again, the repetitiveness of the experiences (e.g., several episodes of intraoperative wakefulness) is another facilitating factor (memory training). Concerning time of exposure to the event, previous studies indicated that despite short time IA events likely cannot induce memory consolidation, wakefulness phases lasting more than 30 s could increase the risk of GAA [24, 25]. Thus, memory processing under (and after anesthesia) depends on a series of variables that are responsible for encoding and storing the event, but also for the characterization of the related recall.

Acquisitions from the basic and clinical research on memory must necessarily be translated into the study of the GAA phenomena. For example, recent findings suggested that consolidation is not a static and unidirectional process as a stabilized memory can return to a labile state and, in turn, can undergo a subsequent consolidation phase, called reconsolidation [26]. In this view, the recall is an expression of repeated processes of consolidation and reconsolidation and each reconsolidation phase may be altered by several types of interference, including behavioral, psychological, and pharmacological.

2.2 Awareness Without Explicit Recall

The awareness without explicit recall (AWER) represents the other awareness subtype. To understand its assumptions it is necessary to refer to basic concepts of implicit long-term memory processing.

2.2.1 Nondeclarative, or Implicit, or Unconscious Memory

Within the long-term memories, the counterpart of the declarative memory is the nondeclarative one, or implicit memory, or unconscious memory. It is a set of procedural memories ("knowing how") which includes the unconscious memory of skills and how to do things (e.g., the use of objects or movements of the body). These memories are typically acquired through repetition and practice, and are composed of automatic sensorimotor behaviors that are so deeply embedded that we are no longer aware of them. Probably, this type of memory can also influence decision-making [27]. In neuroanatomical terms, implicit memory relies on the putamen, caudate nucleus, motor cortex, and cerebellum involved in motor control. Basal ganglia encode and stores emotions, rewards, habits, movement, and learning with special regard to the elaboration of sequences of motor activity useful, for example, for playing a sport or for playing a musical instrument.

According to Robert Veselis, the implicit memory represents a "hidden" memory, *memories that exist, but that we do not know we possess* [28]. Although we are not aware of their existence, these memories characterize our performance, skills, knowledge, and behavior. They work through the priming mechanism. If memory is intended as a network of nodes representing different pieces of information, the simplest form of learning is the temporary activation of a single node (e.g., one or more words). This element (priming) represents the perceptive trigger (i.e., perceptual priming) because it facilitates subsequent learning (e.g., a whole sentence). Learning notions related to the initial node is known as conceptual priming as it promotes the perception of, or responding with, conceptually related information. Conceptual priming is based on the meaning of a stimulus and is enhanced by semantic tasks. For instance, "patient," will show priming effects on "hospital," as the two words belong to the same category.

Experimentally, through the "positive" priming the subject is first exposed to a brief, or degraded, stimulus (e.g., few words) and after an interval ranging from minutes to months, the participant is then exposed to an incomplete stimulus, such as a word fragment, and is asked to recognize or identify it. If the previous exposure to the prime facilitates the identification of the stimulus and the completion of the task, then the subject must have some implicit memory for it. Thus, the positive priming is a mechanism of the implicit memory in which exposure to one stimulus influences a response to a subsequent stimulus, without conscious guidance, or intention, through perceptual (format of the stimulus), semantic (linguistic meaning of the stimulus), or conceptual (structural meaning of a stimulus) associations. In addition to the positive priming system, the implicit memory works also through "negative" priming. Because this mechanism operates a selection of information, removing "distracting elements" during the memory processing, it represents a strategy by which inhibitory control is applied to cognition [29]. Priming approaches are used to train a subject's memory both in positive and negative ways (i.e., positive and negative primings), for instance in elderly or in those with different degrees of cognitive impairment [30].

2.2.2 Implicit Memory During Anesthesia

During anesthesia, the IFT is a suitable approach for investigating on the implicit memory and for assessing consciousness of the external world (connected consciousness). In particular, consciousness is investigated through behavioral reports rather than the search of explicit postoperative recalls of events. In other words, IFT offers real-time information about the presence of consciousness, by opening a window on consciousness while the patient is "apparently" unconscious [31]. The method is usually performed by putting a cuff on the patient's arm before the administering of the NMBA, so that the patient is able to move the arm in response to a verbal command, if aware. Because subjects need to be aware of their sensory environment to hear the command, a response indicates intraoperative consciousness of sensory stimuli.

In anesthetized patients, the priming-induced memory explains the *awareness without explicit recall* (AWER) phenomenon, which represents the other GAA subtype. As previously explained, priming expresses features of the implicit learning. Because it influences performances (procedural memory) and can affect thoughts, and behaviors, patients who processed implicit memory may express inexplicable changes in performances or behaviors without a conscious memory content.

Of note, while conceptual priming is prevented by adequate anesthesia, perceptual priming (e.g., enhanced word stem completion performance) seems to be preserved even during deep anesthesia [32]. Regardless of the precise mechanism of priming and the impact of anesthesia on implicit memorization, the AWER seems to be frequently detectable. In early studies on IFT during AG for caesarean section, Tunstall and Sheikh [33] found that approximately the half of patients studied responded to a command 2–5 min after the induction of GA, whereas King et al. [34] demonstrated that 97% of patients had a positive response after skin incision although in both investigations none of these women had explicit recall of the episode. Furthermore, Sanders et al. [35] analyzed the data from the huge number of studies on the subject and showed that the incidence of AWER is significantly higher than that of awareness with a conscious recall. However, because these data have been collected on the basis of IFT-related responses (wakefulness), and not all these events undergo memory processing, they are not very indicative of the real extent of the problem. More scientifically accurate data come from IFT-based research performed by priming procedure and investigating on the unconscious memory and its processing. The priming procedure, indeed, indicates that a positive result after completing the testing is suggestive of memory consolidation process under anesthesia without any type of recall [36]. Interestingly, a great number of experiments in the field proved that the mechanism of anesthesia-related memorization and learning is very efficient, even during deep anesthesia [36–38]. For example, in 1992, Jelicic et al. [39] reported that patients under anesthesia who were continuously hearing (via headphones) with statements about common facts of some previously, were postoperatively able of correctly answering more questions, compared with control ($p < 0.005$), and, in turn, demonstrated the activation of preexisting knowledge and priming. In the same experiment, other individuals received "new information" such as the names of nonfamous people. These patients designated more "nonfamous names" as famous (thus falsely attributing fame) than patients in the control group ($p < 0.001$), which demonstrates that information-processing during anesthesia can also take place as unconscious learning. Other investigations proved that post-surgery, subject asked to fill unfinished words were capable of filling the gaps with the words they were hearing intraoperatively without being able to recall hearing them. This process, however, must be separated by the process of intraoperatively positive suggestions which might have the potential to encourage well-being and recovery of patients, as it does not provide acquisition of new information [40].

By summarizing, not all episodes of intraoperative awakening follow the pathways of memory consolidation. On the other side, consolidated unexpected experiences can be expressed as explicit-spontaneous or induced reports. The pathway of the declarative, or explicit memory, configures the AAWR phenomena. Alternatively, the intraoperative experience can be expressed through inexplicable changes in behaviors or performances despite the absence of conscious memory content (nondeclarative or implicit memory). This latter configures the AWER event. The two GAA subtypes represent different neurobiological processes.

2.3 Mechanisms of Anesthetics-Induced Amnesia

The matter of memory formation and storage, as well as the interference of anesthetics (and other drugs during anesthesia) with these mechanisms is extremely complex to dissect. By simplifying, these interferences may affect working and long-term memory processes. Research highlighted that propofol induces anterograde amnesia by interfering on the episodic memory system which operates the memory processing [41]. Of note, Veselis et al. [23] demonstrated that low doses propofol (e.g., during sedation) did not impair encoding or working memory but prevented retention of material in long-term memory. On the contrary, excessive sedation, and a surgical anesthesia, may also impair working memory. Because the range of drug doses which induce memory impairment during sedation is difficult to ascertain, the same authors proved [42], in volunteer participants, that a continuous infusion of propofol (0.9 µg/ml), and midazolam (40 ng/ml) affected long-term, but not working memory. The neural correlates of these actions are not well understood, so far. Although encoding of memory tasks (e.g., verbal inputs) mainly involves the left inferior prefrontal cortex [43], propofol-induced amnesia is not linked to a failure of memory encoding in this cortical area [23]. Again, as demonstrated by Pryor et al. [44], this injected anesthetic did involve the amygdalar activation. Thus, depending on the doses, a propofol regimen could strengthen the amnesia; nevertheless, it does not completely protect against the memorization of any emotional components perceived during an inadequate anesthesia status. On the other hand, Deeprose et al. [45] previously proved that intraoperative stimulation (through headphones) facilitated learning during anesthesia, independently on anesthetic depth. It is a further proof of the efficacy of implicit memory, as well as that implicit memory and explicit memory are differently affected during anesthesia.

Thiopental and methohexital are ultra-short-acting barbiturates used to induce and maintain anesthesia. These drugs have poor amnesic action. An old fascinating study showed that thiopental has mild memory effects compared with propofol and benzodiazepines (BDZs) [46]. Nevertheless, a clinical study showed no clinically significant differences in amnesia compared with propofol [47]. Administration of etomidate is used for rapid sequence intubation and induces amnesia. Concerning mechanisms, Zarnowska et al. [48] demonstrated, in mice, that the amnesic effect of etomidate works through long-term potentiation inhibition and is mediated by the Gamma-Aminobutyric Acid A receptor (GABA$_A$R) that contains the extrasynaptic alpha5 subunit. These data confirm the results of a previous study in which was well demonstrated that GABA$_A$Rs mediate amnesic but not sedative, or hypnotic, effects of etomidate [49]. Taken together, these studies offered the interesting possibility of investigating the anesthetics effects on different cognitive functions (memory and consciousness).

Ketamine is a dissociative anesthetic with several pharmacodynamic properties as it can be administered to induce anesthesia, sedation, and analgesic aims. The cellular mechanisms for its amnesic proprieties are not clear although preclinical studies (in human sympathetic ganglion-like SH-SY5Y cells) postulated that amnesia is due to the inhibition of α4β2 nicotinic acetylcholine receptor [50, 51], which modulates the synaptic release of neurotransmitters in the hippocampus. The N-methyl-D-aspartate (NMDA) receptors are implicated in memory formation. In vivo investigations in rats showed that ketamine induced anterograde and retrograde amnesia via blocking the NMDA receptors, in a dose-dependent manner [52]. Finally, some researchers have emphasized that the glycogen synthase kinase (GSK) 3β/β-catenin signaling—a pathway that regulates transcription of genes important for synaptic plasticity and memory [53]—may play a role in ketamine-induced retrograde amnesia [54].

Inhalational anesthetics of significant clinical interest include volatile anesthetic agents such as halothane, isoflurane, sevoflurane, and desflurane, as well as certain anesthetic gases, such as nitrous oxide (NO) and xenon. These anesthetics may impair learning and memory at concentrations that are subhypnotic and thus lower than the concentration required for anesthesia [55]. Of note, it seems that volatile anesthetics are able to impair the hippocampus-dependent learning [56]. The hippocampal θ-rhythm is a prominent network activity featuring synchronized oscillation at 4–12 Hz. This rhythm is implicated in memory processing through activation of plasticity and inter-regional signal integration [57]. At surgical levels of anesthesia isoflurane, halothane and NO have amnesic proprieties, maybe interfering on this hippocampal activity [58]. Xenon is a rare gas belonging to the noble gases of the periodic table with anesthetic properties mostly due to the non-competitive inhibition of NMDA receptors [59]. According to Haseneder et al. [60], xenon could have a significant amnesic effect. In a study using murine brain slices, they reported inhibition of both NMDA and quisqualate receptors in amygdalan neurons. These findings suggest the role of this gas on the modulation of emotional components of memory.

Compounds included in the class of BDZs present sedative, hypnotic, anxiolytic, anticonvulsant, and muscle relaxant properties by acting on the GABA system at the $GABA_A$ receptor. For these actions, these drugs can be used for sedation before or after surgery, as well as to induce and maintain anesthesia. The BDZs used in clinical anesthesia are the agonists midazolam, diazepam, alprazolam, and lorazepam and the antagonist flumazenil, the latter used in reversing BDZ effects. Midazolam is the most commonly prescribed by the anesthesiologists because of its strong sedative actions and fast recovery time, as well as its water solubility, which reduces pain upon injection. The amnestic effects of the BDZs have been extensively studied [61, 62] (*see* also paragraph 9. Prevention).

3   Incidence. *The Challenge of Awareness Detection*

While GAA is a rare complication, it is difficult to refer to well-established epidemiologic data because the reported incidence strongly depends on the method of detection. For instance, it has been demonstrated a different incidence within the same sample, depending on the methods of postoperative assessment [63]. Thus, the literature offers contradictory data on its real occurrence. Early investigations reported an incidence of AAWR ~1 to 2 per 1000 operations involving GA (0.1–0.2%) [64]. On the other side, data from the fifth National Audit Project from Great Britain (NAP5), evaluated in more than 2.7 million cases, reported an incidence of AAWR of only 1:19,600 (0.005%), that is 20 times less than previously reported [65]. Some authors have criticized the results because of the methodology of data collection (spontaneous complaints/reports of awareness), especially about the absence of structured interviews that may have underestimated the real incidence of awareness [66]. Again, because as stated by Robert Veselis [28], "memory is a behavior" since it requires a behavioral output once the "victim" has decided to report an experience, several factors may influence whether or not patients will decide to, and indeed recall their experiences, either prompted or spontaneously [67].

Subsequently, the SNAP-1 study, designed to evaluate patient-reported outcomes after anesthesia in the UK, confirmed findings from early studies. By using the modified Brice questionnaire the authors found an incidence of 0.12% (1:800) [68]. Other studies, conducted through postoperative interview, confirmed an incidence ranging from 0.1% to 0.2% [69–71]. Thus, we can assume that the complication occurs in about one or two out of every 1000 surgeries.

Despite limitations related to the retrospective analysis and the absence of specific tools for direct awareness detection, in patient underwent surgery for cancer disease and routinely assessed by the team of the psycho oncology service we detected an incidence of 1:10,550 (0.0095%). The issue of awareness occurrence emerged during the psychological evaluation and then was recorded in the chart [72].

Studies conducted by using the IFT suggested that the incidence of wakefulness (connected consciousness) may approach up to 40% of patients [73]. More recently, Sanders et al. [74] showed, through the ConsCIOUS-1 study, an incidence of IFT responsiveness around the time of intubation of 4.6%. Interestingly, no participant referred explicit recall of intraoperative events when questioned after surgery. Despite the exact incidence of implicit recall is difficult to obtain, the occurrence of AWER (i.e., implicit memory) has been indicated to be significantly higher than the incidence of AAWR (i.e., explicit recall) [75].

In children, the incidence of AAWR after general anesthesia is slightly higher. According to the Davidson et al.'s studies, it may range from 0.2% to 1.2% [76, 77]. Approximately 70% of cases occur at induction or emergence from anesthesia [78]. Clinical characteristics are extremely variable and approximately 50% of experiences are distressing. However, it seems that compared to adults, there are fewer psychological effects and complications [79]. Nevertheless, the interpretation of the data must be done with great accuracy due to the potential bias related to developmental factors and efficacy of postoperative interviews.

Concerning incidence related to different anesthesia approaches, in a retrospective analysis, the authors showed that the incidence of AAWR was not statistically different in patients receiving general anesthesia compared with those underwent sedation [80]. However, data from the Anesthesia Awareness Registry [81] and the NAP5 indicated that the occurrence of awareness in patients receiving sedation can be associated with long-term psychological consequences [65].

4 Clinical Features

The patient's experiences reported after the intervention are very disparate. The majority of patients report hearing voices during the surgery without feeling anything else. Although reports are usually vague, certain patients are able to remember the detailed moment of surgery or the surroundings such as whole conversations in the theatre. In other cases, they describe tactile sensation such as surgical manipulation or endotracheal tube insertion, or feeling of pressure. Painful sensations are also described (e.g., sore throat due to the endotracheal tube or pain at the incision site) in about 10% of patients [82], although the larger part of patients who experience awareness generally does not feel pain. When recalled, pain is occasionally described as severe. The immobilization experiences during the intervention are much more complex; these experiences are frequently associated with a profound discomfort of the patient. Patients can report that they felt buried alive, unable to communicate and imprisoned in their bodies. Distressing experiences of air hunger and difficult breathing until a sense of suffocation, complete the clinical picture of the awareness with paralysis. Again, distress can be also associated with less complex experience (e.g., tactile or hearing sensations).

Approximately one-third of patients with awareness report discomfort. The occurrence of anxiety and distress during an episode of awareness has a paramount clinical impact. Lennmarken et al. [83], demonstrated that all the patients who were still severely disabled due to psychiatric/psychological manifestations after 2 years from the GAA experienced severe discomfort. These psychological manifestations can be expressed as anxiety, depression, intrusive thoughts, flashbacks, recurrent dreams or nightmares, avoidance of stimuli associated with the trauma, sleep disturbance (e.g., insomnia), exaggerated startle response, hypervigilance, as well as a permanent aversion to surgery and anesthesia. Symptoms can be isolated or grouped into syndromes such as the PTSD (*see* Vulser's Chapter 5).

Concerning the duration of reports, sometimes patients are quite precise about the duration of the phenomenon. Although most experiences recalled were less than 5 min in duration, recalls lasting the whole surgical procedure have been also described [84].

5   Diagnosis and Classification

5.1   Diagnosis of Awareness

The diagnosis of GAA may often represent a clinical challenge. During anesthesia, it should be difficult to recognize a potential episode of GAA, especially when NMBAs are used, and in absence of any depth of anesthesia (DoA) monitoring system. To this latter regard, however, it is important to emphasize that, to date, there is no system helpful to assess with extreme precision the exact level of DoA [85]. Again, signs of cardiovascular activation (e.g., tachycardia), respiratory problems (e.g., bronchospasm, decreased compliance), and increased sympathetic tone as well as lacrimation, and spontaneous movement may be indicative for the occurrence of intraoperative awakening but not for awareness occurrence. Moreover, all these clinical manifestations can be masked by the use of drugs such as β-blockers and NMBAs. Yet, antimuscarinic drugs and opioids, as well as eye tape/ointment may mask pupillary dilatation and reactivity to light.

Thus, the clinical signs must act as an alarm bell and induce the search (and correction) of the cause of the inadequate anesthesia. Early recognition of IA is likely to play a fundamental role in avoiding consolidation. Furthermore, combined with the correction of its cause, theoretically, it may be possible to interfere with the memory processing [86]. Interestingly, Timić et al. [87], showed, through in vivo experiments, the possibility of using BDZs to interfere with retrograde memory. Previously, Semba et al. [88], reported that midazolam (with propofol) may induce retrograde amnesia by increasing serotonergic transmission. On these bases, the American Society of Anesthesiologists (ASA) recommended the intraoperative administration of BDZs to patients who may have become conscious [89]. This approach is also helpful for avoiding further anterograde acquisitions.

After the end of the surgery, the diagnosis is made on the patient's report. In some circumstances it may result difficult to establish whether the patient's story can be interpreted as a true episode of AAWR. Vague reports or the description of situations probably occurred in the immediate preoperative or postoperative period (i.e., people talking, application of dressing) should be carefully evaluated as these experiences do not represent AAWR events [69].

Furthermore, awareness must not be confused with intraoperative dreaming which represents a paraphysiological event in the course of anesthesia and sedation [90], and probably influenced by preoperative suggestions [90]. The occurrence of dreaming during anesthesia is a well-known phenomenon; it has been reported in up to 50% [91] in case of general anesthesia, and in approximately 25% of individuals underwent sedation [92]. However, the exact correlation between dreaming and awareness is an unsolved rebus, so far. The incidence of AAWR, for instance, is reported more commonly among patients who experienced a dream during surgery [93].

A milestone for the diagnosis of awareness is the tool developed by Brice et al. [94], and commonly termed as the "Brice's questionnaire." The tool was developed in the 1970 and its modified versions are commonly used for research and clinical purposes. In its original version, the interview encompassed five questions: "*What was the last thing you remembered happening before you went to sleep?*" "*What is the first thing you remember happening on waking?*" "*Did you dream or have any other experiences whilst you were asleep?*" "*What was the worst thing about your operation?*" "*What was the next worst?*"

A recall may not be expressed at the emergence but reported at distance from the end of the intervention. As an extreme case, for instance, it has been described a patient who recalled a specific detail 5 years after the operation [95]. Thus, regardless of the test used, several evaluations should be performed: at the emergence, after 24–48 h, 7–8 days and 1 month after surgery. As a consequence, because suspicion of awareness occurrence can also derive after the patient is discharged from the hospital, surgeon and primary care physician should be vigilant in identifying the complication as well as potential early signs of a psychological discomfort.

5.2 Awareness Classification

About classification, based on the spectrum of clinical manifestations of the phenomenon, the Michigan instrument for awareness detection is a useful tool that conjugates clinical features and severity of the event. The instrument considers 6 classes of events. In particular, Class 0 is no awareness; Class 1 refers to isolated auditory perceptions; Class 2 tactile perceptions; Class 3 are reports of pain; Class 4 are episodes of paralysis such as inability to move, speak, or breathe while conscious. Finally, Class 5 are severe episodes featuring the combination of paralysis and pain. An additional designation of "D" for distress was also included for patient reports of fear, anxiety, suffocation, sense of doom, sense of impending death, or other explicit descriptions [96].

Later on, Wang et al. [97] proposed a classification of intraoperative cognitive states of consciousness and correlated them to postoperative immediate or late (more than 1 month) complications. This approach included five intraoperative conscious states, from a wakefulness with obliterated explicit and implicit memory, to a wakefulness state with subsequent implicit memory, to a state with implicit emotional memory, and finally, two states of consciousness with explicit recall featured, or not, distress and/or pain. Although the Wang's approach is an interesting attempt to correlate states of intraoperative consciousness, awareness type (explicit/implicit), and postoperative psychological sequelae, the correlation between consciousness, intraoperative memorization and emotional responses in not linear and, in turn, depends on a huge number of predictable and unpredictable factors.

6 Risk Factors

The risk of developing GAA is likely to be influenced by several risk factors (Table 2). These risk factors can be grouped into patient-related risk factors and non-patient–related risk factors.

6.1 Patient-Related Risk Factors

Patient-related risk factors can be further divided into modifiable and nonmodifiable risk factors. This nosographic approach can be helpful for addressing the issue of GAA prevention.

6.1.1 Nonmodifiable Patient-Related Risk Factors

Nonmodifiable conditions reported as consistent risk factors include patient age, sex, and resistance or tolerance to general anesthetics.

Concerning age, the NAP5 investigations identified a higher incidence of awareness in patients aged between 25 and 45 years [65]. This datum is universally accepted, as it is believed that young subjects are more predisposed due to their better cognitive performances. For instance, the Linassi et al. [18] meta-analysis showed

Table 2
Risk factors for general anesthesia awareness

Patient-related risk factors

Modifiable (when possible)
    ASA physical status > III
    Limited hemodynamic response
    MAC increment conditions[a]

Nonmodifiable
    Patient age (younger adults)
    Female gender
    Resistance (genetics)
    Tolerance to general anesthetics[b]
    Previous awareness

Non-patient–related risk factors
    Anesthetic techniques (e.g., TIVA)
    Junior anesthetists
    Use of muscle relaxant drugs
    Misuse of neuromuscular monitoring[c]
    Type of surgical procedure
    Organizational factors (out-of-hours operating)

MAC minimal alveolar concentration, *ASA* American Society of Anesthesiologists, *TIVA* total intravenous anesthesia
[a]Modifiable risk factors: Hypercapnia, hyperthyroidism, hypertension. Nonmodifiable: young age, pregnancy
[b]Due to drug induction mechanisms (e.g., opioids, benzodiazepines, cocaine)
[c]Especially for the occurrence of awareness at the emergence. Obstetric (especially Caesarean section), cardiac surgery, thoracic surgery, emergencies

that IFT-positive patients (connected consciousness) decreased significantly with increasing age.

About potential gender differences, a retrospective analysis based on reviewing clinical charts of over 60,000 surgical patients, indicated that female patients were more likely to experience GAA than males, although the difference was not statistically significant [98]. These findings confirmed other previously reported data. In particular, investigations reported that the awareness incidence can be three times higher in women than in men [99]. On the other side, an American study ($n = 177,468$) demonstrated that compared to female patients the complication was two times higher in male patients. Moreover, patients expected awareness were in older age. Of note, the authors adopted a modified Brice's tool and reported a very low incidence (0.0068%) of awareness [100]. Despite results from this latter study, epidemiological data suggest that GAA may occur more frequently in female patients. However, due to the heterogeneous methodology used in the studies (e.g., self-reporting or interview), it is difficult to obtain definite data on the role of sex, and the age for GAA susceptibility.

It seems that body weight and level of education are not risk factors for the complication, as they were not found to be significantly related to the awareness complication. However, obesity may be associated with a higher risk for several reasons, such as the possibility of prolonged time for endotracheal intubation, and error in drug dosage [101].

6.1.2 Modifiable Patient-Related Risk Factors

Modifiable patient-related risk factors include clinical conditions that may influence the patient's status. For instance, because patients presenting with high ASA physical status (>III) and those admitted to intensive care unit (ICU) for postoperative care are at significantly higher risk of awareness, improving clinical conditions, when possible, can certainly reduce the risk of GAA, in addition to the overall improvement in outcomes. A limited hemodynamic reserve is another patient-related risk factor, although it is not always possible to correct it, especially in the emergency setting.

Many factors may influence the response to the anesthetics. For instance, minimal alveolar concentration (MAC) varies among and even within individuals, and is dependent on age, genetics, and temperature. Several factors are associated with an increment in MAC including hypercapnia, hyperthyroidism, hypertension, as well as the nonmodifiable factors young age and pregnancy.

A special issue concerns the tolerance and the resistance to anesthetics which can explain the finding that in approximately 15% of cases the GAA occurrence has no easily recognizable cause [102]. To simplify the topic, in this paper, the "tolerance" to anesthetics is referred to acquired mechanisms inducing less drug effect. On the other side, the "resistance" to anesthetics indicates a congenital less efficacy due to genetic conditions.

Patients at increased risk for awareness include those with a history of substance use, or abuse (e.g., opioids, BDZs, amphetamines, centrally acting drugs, and cocaine), and chronic pain patients treated with high doses of opioids. Tobacco smoking, heavy alcohol consumption, and centrally acting drugs (e.g., monoamine oxidase inhibitors, tricyclic antidepressants) are other examples of acquired tolerance to anesthetic agents. These substances act by inducing the metabolism of inhaled anesthetics (but also BDZs and opioids) through cytochrome P450 2E1 activation. Moreover, in terms of awareness pathogenesis, the "tolerance" of anesthetics includes several other acquired mechanisms other than those more strictly implicated in the metabolic pathways. For instance, long-term exposure to alcohol, or persistent seizures, may induce alteration in hippocampal GABA subtype receptors (i.e., $GABA_A$ slow), implicated in memory blocking processes and responsive to the anesthetic (e.g., etomidate) action. The impairment of this inhibitory pathway may trigger memory consolidation under anesthesia [103]. Finally, clinical conditions such as pyrexia and hyperthyroidism may induce tolerance through metabolic activation.

Furthermore, a "physiological resistance" to anesthetic agents has been also described. This innate, or genetic resistance, probably acts through a pharmaco-induction mechanism and represents a fascinating and poorly understood pharmacodynamic phenomenon. Altered gene expression or function of target receptors of anesthetics are mechanisms implicated in the postulated congenital resistance to anesthetics. For instance, preclinical experimental data, obtained from mutant analysis in Drosophila, showed that numerous genes (e.g., encoding for second messengers, memory formation substrates, ion channels, synaptic proteins) and related isoforms are implicated in the normal response to anesthetics [104]. Single or multiple mutations may probably underlie the mechanism of resistance, even in humans.

Although Aranake et al. [105] demonstrated a fivefold increased risk of intraoperative awareness in patients with a history of awareness, a previous event of GAA is not always a risk factor for a new episode. Obviously, if the episode was linked to a human error, or subsequent to an equipment problem, there is no increase in risk. On the other hand, the risk may considerably increase in the case of awareness due to resistance (or tolerance) to anesthetics.

Among patient-related risk factors, there are also the use of β-blocker agents, and clinical conditions (e.g., diabetes) that may mask responses to stimuli and clinical signs, indicative for light anesthesia and unexpected awakening.

6.2 Non-Patient–Related Risk Factors

These risk factors are related to anesthetic techniques and drugs used, and type of surgical procedures.

6.2.1 Anesthetic Techniques and Medications

Anesthesia with propofol-based and use of NMBAs during the operation was associated with a significantly higher incidence of GAA compared to volatile anesthetics and those received no muscle relaxation, respectively. Rapid sequence induction, and misuse of neuromuscular monitoring are other important anesthesia-related risk factors.

The role of different methods of GA on awareness occurrence has been the object of controversies. For instance, Morimoto et al. [106] published results from a survey of 172 anesthesiologists who had performed approximately 85,000 anesthetics in Japan, and indicated that total intravenous anesthesia (TIVA) modality was used in 88% of the patients who reported awareness complication. However, the majority of the patients included in the survey received inhaled agents. Despite these disputes, it seems that TIVA is probably associated with a higher risk for GAA compared with techniques based on the administration of inhaled general anesthetics. For instance, the effects of TIVA on GAA incidence were also investigated by Yu et al. [107]. They compared TIVA to the anesthetic combination of intravenous and inhaled anesthesia (CIIA) and found that the occurrence of awareness was significantly higher in the TIVA group than that in the CIIA group ($p = 0.29$). The matter of the correlation between anesthesia approach and the GAA risk is a complex issue. It concerns the readable monitoring, through biochemical parameters of efficacy, of the anesthetic working into the brain. In other words, it concerns the linkage between pharmacokinetics and pharmacodynamics of anesthetics. Thus, while the exhaled concentration of the anesthetic agents allows real-time dose adjustments, blood concentrations of intravenous agents are not easily obtained during TIVA. Perl et al. [108] published, in 2009, an interesting paper on the determination of serum propofol concentrations by breath analysis using ion mobility spectrometry. Although it could be a very useful strategy for biochemical monitoring of intravenous anesthesia, further research for a practical application is needed.

The most important risk factor is the use of NMBAs as muscle paralysis removes one of the physiologic signs of patient awareness which is the purposeful movement. The Ghoneim's meta-analysis demonstrated that up to 85% of the patients suffering from GAA received muscle relaxants [109]. In a prospective case study conducted on 11,780 patients, Sandin et al. [69] showed that the GAA incidence was almost double in patients treated with NMBAs, compared to those who did not receive muscle relaxants (0.1%). Moreover, the occurrence of awareness in patients treated with NMBAs was associated with anxiety and psychological sequelae.

Indeed, because complete paralysis worsens the psychological trauma due to the awareness experience, the potential for long-term psychological complications may be increased. The recent NAP5 study confirmed a higher incidence of AAWR when NMBA was administered (approximately 1 in 8200) [67]. While this research investigated on the awareness with recall, the use of NMBAs is also associated with a higher intraoperative implicit memory (i.e., AWER) formation [110].

The use of NMBAs may also interfere with the readability of brain monitoring devices. For instance, Nunes et al. [111] found significant reductions in several indices of DoA, such as the entropies (i.e., Response Entropy and State Entropy), and the Cerebral State Index, in volunteers after the use of succinylcholine [112]. Again, investigations conducted in awake volunteers who did not receive any anesthetic demonstrated that low Bispectral Index (BIS) values (suggestive for a surgical anesthesia status) after administration of a dose of muscle relaxant [113]. Certainly, avoiding NMBAs is an effective strategy to reduce awareness incidence. To this regard, a recent study demonstrated a low incidence of awareness in patients ventilated with a laryngeal mask under spontaneous ventilation [98].

The misuse of neuromuscular monitoring is another important risk factor for GAA, especially during emergence from anesthesia [114, 115], especially in the case of butyrylcholinesterase deficiency when succinylcholine or mivacurium are used [116]. Finally, particular approaches, such as the procedure of rapid sequence induction, have been associated with a high risk of awareness [65].

6.2.2 Types of Surgery and Awareness Risk

The type of surgery is a significant non-patient–related risk factor. Compared to other surgical branches, obstetric (especially Caesarean section), bariatric, cardiac, and thoracic surgeries are at a higher risk of awareness. Furthermore, major traumas characterized by hypovolemia and hypotension are at higher risk for awareness. Probably, in these settings it is more often possible to fall in circumstances which it is judged unsafe to administer sufficient anesthesia. Furthermore, in cardiac and thoracic surgeries muscle relaxants are essential and many patients have other risks due to their general health. Interventions that are carried out at night are also indicated as at high risk for awareness [117].

7 Causes

The main cause of awareness is the anesthetic underdosing relative to the patient's specific requirements (Table 3). The insufficient administration can be the consequence of several conditions such as a dosage error (e.g., dosage miscalculation because of lack of

Table 3
Causes of general anesthesia awareness

Anesthetic underdosing
Primitive dosage error
Dosage errors in specific circumstances[a]
Underappreciated higher patient's needs (acquired tolerance to anesthetics)
Misunderstood higher patient's needs (acquired resistance)[b]
Failures in the delivery of anesthesia due to equipment malfunctioning
Intravenous agents
Unseen intravenous disconnection
Extravasation
Mistakes with syringe drivers
Inhaled agents
Gas alarm malfunctioning
Mistakes with volatile delivery (e.g., failure to fill vaporizers)
Other causes
Distractions caused by another member of the staff

[a]For example, difficult airway management and intubation; for example, opioids, benzodiazepines, cocaine, centrally acting drugs, or metabolic clinical conditions (e.g., pyrexia)
[b]Genetic factors

understanding of offset times of medications) or technical problems (e.g., failure to fill vaporizers). The delivery of volatile anesthetics at a MAC indexed for ages value less than 0.8 can expose patients to an increased risk of awareness occurrence, especially in young subjects. However, it is not possible to refer to precise dosages for general anesthetics that could be used to alert anesthesiologists for potentially insufficient anesthesia [118], whereas several cofactors are often associated in the pathogenetic chain. Sometimes, a specific patient's needs are underappreciated. These conditions include a congenital resistance to general anesthetics or a misunderstood acquired tolerance to anesthetic agents.

A special issue and a frequent cause of awareness concern mistakes or failures in the delivery of anesthesia [72]. It should not be surprising as a recent incidents report found a rate of drug administration error of approximately 7 procedures every 1000 GA (0.7%) [119]. Equipment malfunctioning falls in this group and may regard the TIVA technique (e.g., infiltration of the intravenous catheter), or inhaled anesthesia (e.g., gas alarm malfunctioning).

A difficult intubation is a well-recognized cause of awareness. An increased occurrence of GAA has been described during prolonged intubation attempts, and as a complication of airway management failure [120]. Patients whose airways are difficult to intubate (or manage) may develop awareness probably because insufficient attention is done to ensuring adequate anesthesia.

8 Awareness Management

The report of an awareness episode must necessarily trigger a precise management program that requires a multidisciplinary approach through the involvement of several professionals (Fig. 2). The first step is to gather as much information as possible about the event. This phase is of considerable importance because it allows excluding episodes that can confuse diagnosis such as dreaming or the memory of events immediately preceding or following anesthesia. Nevertheless, the patient's story must be accepted as a genuine experience. At the same time, a root cause analysis involving anesthetists, nurses, surgeons, and other personnel in the theater, must be performed. The details of recalled events (e.g., hearing conversations, feeling the incision) should be noted in the

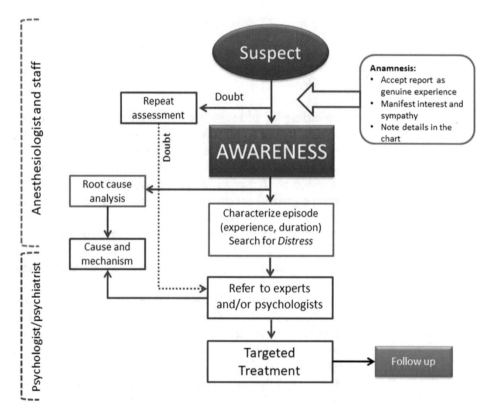

Fig. 2 Algorithm for awareness management. Managing awareness requires a multidisciplinary approach which involves several professionals. At the beginning (red pathway), the issue can be addressed by the anesthesiologist and the staff who perform the assessment. In case of doubt, the evaluation must be repeated (at 1 or 2 days, or after a week). If the patient's story is suggestive for an episode of awareness, its precise characterization (experience and duration) is necessary. The search for distress is fundamental. After this phase, the help of other professionals (psychologists, psychiatrists) is needed for targeted treatments and follow up (blue pathway). The collaboration from specialists is necessary even when the doubt for diagnosis persists

patient's records. This information may be of great importance should medicolegal issues arise.

Once the diagnosis has been established with certainty, the detailed anamnesis allows characterizing the episode according to one of the proposed classifications (e.g., the Michigan tool). The patient's experience must be taken seriously, and sympathy should be manifested. Rejecting the veracity of the patient's report could get a worse psychological outcome. This doctor-patient approach requires particular communication skills, often beyond the anesthesiologist's skills. Thus, it would be also advisable to refer the patient to reference centers, or at least to anesthesiologists and/or psychologists who are experts in the matter. Experts can help to identify the presence of distress associated with the event, as a report with contents of distress calls for early treatment. These cases, indeed, require careful evaluation to avoid the development of serious psychological sequelae.

The plan of action recommended by the NAP5 for dealing with a report of GAA encompasses investigation on flashbacks, nightmares, new anxiety states, and depression [65]. Even for this purpose, the collaboration of a specialist psychologist or a psychiatrist is suggested.

Each ascertained episode of awareness requires active follow-up and must be treated through professional intervention and psychological support. Nevertheless, follow-up counseling must be also offered in suspicious cases of awareness or when the diagnosis is doubtful. Finally, changes in performances or behaviors, reported by the patients themselves or through the surgeon, the general practitioner, or the family, should be carefully addressed as they may represent an expression of AWER occurrence.

9   Prevention

Prophylactic interventions can be divided into *preoperative* and *intraoperative* strategies (Table 4).

9.1   Preoperative Strategies

The milestone of awareness prevention is a careful attention to the administration of anesthetic, avoiding underdosing during the intraoperative course. Preoperative clinical assessment is fundamental and mandatory. In particular, it allows identifying high-risk patients, including those with a history of drug abuse, or those treated with central acting medications. This step also must include a careful evaluation of other potential risk factors, including predictable difficulty in airway management. The ASA task force on intraoperative awareness recommended informing high-risk patients of the possibility of awareness as this strategy should not increase the actual risk of the complication [89].

Table 4
Prophylactic interventions

Preoperative strategies
    Check of devices and instruments
    Careful patient assessment for awareness risk
    Information of high-risk patients

Intraoperative strategies
    Check of devices and instruments
    Set threshold alarms[a]
    BDZ premedication in high-risk patients
    Avoid anesthetics underdosing
    MAC 0.8–1.3 for inhaled agents
    Use curare with caution
    Neuromuscular monitoring
    Avoid nitrous oxide alone
    Processed EEG devices for:
        TIVA
        High-risk patients
    Maintain professionalism in the theatre
    Use BDZ in case of awareness suspicious

Postoperative strategies[b]   Prophylactic interventions for avoiding sequelae

BDZs Benzodiazepines, *MAC* minimal alveolar concentration
[a]According to device manufacturer's specifications
[b]After awareness diagnosis

Concerning premedication, according to the ASA guidelines, the decision to administer BDZs should be made on a case-by-case basis. Moreover, these drugs could be helpful in case of unexpected intraoperative awakening [89]. On the other hand, attention must be paid to the potential BDZs-induced triggering of postoperative cognitive complications, such as postoperative delirium [121]. The preoperative strategies must also include the check of all devices (e.g., respirators, vaporizers, respiratory system, and infusion pumps) used in the anesthetic procedure.

9.2   Intraoperative Strategies

Anesthesia devices and instruments must be checked during the surgical procedure. Furthermore, several suggestions regard the anesthetic conduct in terms of anesthetics dosing, and monitoring.

About medications, it must be avoided anesthetic underdosing, as it represents the main cause for awareness. In clinical circumstances featuring hemodynamic instability (e.g., hypotension) a safe anesthetic conduct should be focused on the correction of the issue (e.g., fluids or vasopressors) while maintaining the surgical DoA status.

Concerning NMBAs, when used these medications must be administered upon the guide of monitoring of the neuromuscular function, maintaining T1 above 5% [122].

Inhalation anesthetics must be administered via gas monitoring (end-tidal anesthetic concentration, ETAC). In the course of inhalation anesthesia, ETAC expresses concentrations of anesthetics which estimate their arterial blood partial pressure with acceptable precision [123]. It shows the concentration of an inhalation agent closed to the MAC value. For each inhaled agent administered, a MAC 1 value describes the concentration required, at 1 atm ambient pressure, to prevent 50% of subjects moving in response to a stimulus. Although this definition does not embrace awareness, investigations proved that the MAC requirement for movement suppression is usually higher than MAC for suppression of consciousness (MAC-awake), or memory (MAC-amnesia) [124]. In turn, inhaled agents should be administrated obtaining a MAC never inferior to 0.8, which represents a postulated value of safety margin. In his regard, a recent large-sized clinical trial (the MEETS-PANDA study) [125] confirmed the results of previous studies [126] and showed that the ETAC-guided anesthesia with a MAC ranging at 0.7–1.3 was able to reduce the incidence of awareness. Of note, NO must be never used as the only anesthetic to perform anesthesia. On the other side, the administration of intravenous agents must be performed, when possible, upon brain monitoring guidance [65].

Description of functioning and features of processed EEG-based devices falls out of the scope of this chapter (*see* Chapter 2). Concerning their efficacy, DoA monitors did not represent a definitive solution for awareness prevention [127]. Furthermore, a recent Cochrane analysis concluded that the use of DoA monitors did not reduce the risk of the complication, compared to standard clinical and instrumental monitoring [128]. These results support the recommendations from the ASA [65], the Australian and New Zealand College of Anaesthetists (ANZCA) [129], and the Association of Anaesthetists of Great Britain and Ireland [130] which underlined that the decision to use DoA devices must be made on a case-by-case basis by the individual practitioner for selected patient such as those at higher risk and in case of total intravenous anesthesia, and NMBAs use. This is an important recommendation. For example, a Cochrane analysis demonstrated that BIS may reduce the incidence of awareness by 75% in patients at high risk [131]. DoA devices can be also helpful in particular clinical settings, for instance in the case of patients undergoing deep sedation [65, 90].

The standard clinical and instrumental monitoring is certainly useful, but several factors limit (e.g., NMBAs or β-blocker agents) their power to prevent awareness. Among the preventive strategies, although the IFT is a proof of consciousness during anesthesia, there is no evidence that these "experiences" correlate to a positive memory processing leading, in turn, consolidation [18]. However, the approach is certainly helpful, as it is the only method to ascertain consciousness during surgery.

In summary, the best strategy to prevent awareness seems to be a dynamic — adaptable to the different phases of the intervention—combination of elements. This optimal approach thus must combine preoperative pathways for identifying patients at risk, with intraoperative approaches based on careful administration of drugs and clinical instrumental monitoring, including (in selected cases) EEG-based brain monitoring. Finally, regardless of the type of monitoring used, threshold alarms must be set following the device manufacturer's specifications.

References

1. Kotsovolis G, Komninos G (2009) Awareness during anesthesia: how sure can we be that the patient is sleeping indeed? Hippokratia 13 (2):83–89

2. Winterbottom EH (1950) Insufficient anaesthesia. Br Med J 1(4647):247–248

3. Meyer BC, Blacher RS (1961) A traumatic neurotic reaction induced by succinylcholine chloride. N Y State J Med 61:1255–1261

4. American Psychiatric Association (ed) (2013) Diagnostic and statistical manual of mental disorders, 5th edn. American Psychiatric Association, Arlington, VA

5. Domino KB, Posner KL, Caplan RA, Cheney FW (1999) Awareness during anesthesia: a closed claims analysis. Anesthesiology 90 (4):1053–1061

6. Cascella M (2014) What about memory, consciousness, recall, and awareness in Anesthesia? Iran J Med Sci 39(3):311–312

7. Mashour GA (2004) The cognitive binding problem: from Kant to quantum neurodynamics. Neuro Quantology 1:29–38

8. Edelman GM (1987) Neural Darwinism. The theory of neuronal group selection. Basic Book, New York

9. Baars B (1988) A cognitive theory of consciousness. Cambridge University Press, New York

10. Crick F, Koch C (1990) Towards a neurobiological theory of consciousness. Semin Neurosci 2:263–275

11. Thagard P, Stewart TC (2014) Two theories of consciousness: semantic pointer competition vs. information integration. Conscious Cogn 30:73–90

12. Tononi G (2004) An information integration theory of consciousness. BMC Neurosci 5:42

13. Graziano MSA, Kastner S (2011) Human consciousness and its relationship to social neuroscience: a novel hypothesis. Cogn Neurosci 2:98–113

14. Rauss K, Pourtois G (2013) What is bottom-up and what is top-down in predictive coding? Front Psychol 4:276

15. Cascella M, Schiavone V, Muzio MR, Cuomo A (2016) Consciousness fluctuation during general anesthesia: a theoretical approach to anesthesia awareness and memory modulation. Curr Med Res Opin 32(8):1351–1359

16. Mashour GA, Orser BA, Avidan MS (2011) Intraoperative awareness: from neurobiology to clinical practice. Anesthesiology 114 (5):1218–1233

17. Cascella M, Bifulco F, Viscardi D, Tracey MC, Carbone D, Cuomo A (2016) Limitation in monitoring depth of anesthesia: a case report. J Anesth 30(2):345–348

18. Linassi F, Zanatta P, Tellaroli P, Ori C, Carron M (2018) Isolated forearm technique: a meta-analysis of connected consciousness during different general anaesthesia regimens. Br J Anaesth 121(1):198–209

19. Pandit JJ (2014) Acceptably aware during general anaesthesia: 'dysanesthesia'--the uncoupling of perception from sensory inputs. Conscious Cogn 27:194–212

20. Chung HS (2014) Awareness and recall during general anesthesia. Korean J Anesthesiol 66:339–345

21. Ullman MT (2004) Contributions of memory circuits to language: the declarative/procedural model. Cognition 92:231–270

22. Rauch SL, Shin LM, Phelps EA (2006) Neurocircuitry models of posttraumatic stress disorder and extinction: human neuroimaging research--past, present, and future. Biol Psychiatry 60(4):376–382

23. Veselis RA, Reinsel RA, Feshchenko VA, Johnson R Jr (2004) Information loss over

time defines the memory defect of propofol: a comparative response with thiopental and dexmedetomidine. Anesthesiology 101 (4):831–841

24. Dutton RC, Smith WD, Smith NT (1995) Brief wakeful response to command indicates wakefulness with suppression of memory formation during surgical anesthesia. J Clin Monit 11(1):41–46

25. Dutton RC, Smith WD, Smith NT (1995) Wakeful response to command indicates memory potential during emergence from general anesthesia. J Clin Monit 11(1):35–40

26. Takamiya S, Yuki S, Hirokawa J, Manabe H, Sakurai Y (2019) Dynamics of memory engrams. Neurosci Res. pii: S0168-0102(19)30089-6. https://doi.org/10.1016/j.neures.2019.03.005

27. Newell BR, Shanks DR (2014) Unconscious influences on decision making: a critical review. Behav Brain Sci 37(1):1–19

28. Veselis RA (2015) Memory formation during anaesthesia: plausibility of a neurophysiological basis. Br J Anaesth 115(Suppl 1):i13–i19

29. Labossière DI, Leboe-McGowan JP (2018) Specific and non-specific match effects in negative priming. cta Psychol (Amst) 182:138–153

30. Hagood EW, Gruenewald TL (2018) Positive versus negative priming of older adults' generative value: do negative messages impair memory? Aging Ment Health 22:257–260

31. Sanders RD, Raz A, Banks MI, Boly M, Tononi G (2016) Is consciousness fragile? Br J Anaesth 116:1–3

32. Deeprose C, Andrade J (2006) Is priming during anesthesia unconscious? Conscious Cogn 15(1):1–23

33. Tunstall ME, Sheikh A (1983) Comparison of 1.5% enflurane with 1.25% isoflurane in oxygen for caesarean section: avoidance of awareness without nitrous oxide. Br J Anaesth 62:138–143

34. King H, Ashley S, Brathwaite D et al (1993) Adequacy of general anesthesia for cesarean section. Anesth Analg 77:84–88

35. Sanders RD, Tononi G, Laureys S et al (2012) Unresponsiveness? Unconsciousness. Anesthesiology 116:946–959

36. Andrade J, Deeprose C (2007) Unconscious memory formation during anaesthesia. Best Pract Res Clin Anaesthesiol 21:385–401

37. Bailey AR, Jones JG (1997) Patients' memories of events during general anaesthesia. Anaesthesia 52(5):460–476

38. Andrade J (1995) Learning during anaesthesia: a review. Br J Psychol 86:479–506

39. Jelicic M, De Roode A, Bovill JG, Bonke B (1992) Unconscious learning during anaesthesia. Anaesthesia 47(10):835–837

40. Rosendahl J, Koranyi S, Jacob D, Zech N, Hansen E (2016) Efficacy of therapeutic suggestions under general anesthesia: a systematic review and meta-analysis of randomized controlled trials. BMC Anesthesiol 16 (1):125. https://doi.org/10.1186/s12871-016-0292-0

41. Tulving E (2001) Episodic memory and common sense: how far apart? Philos Trans R Soc Lond Ser B Biol Sci 356(1413):1505–1515

42. Veselis RA, Pryor KO, Reinsel RA, Li Y, Mehta M, Johnson R Jr (2009) Propofol and midazolam inhibit conscious memory processes very soon after encoding: an event-related potential study of familiarity and recollection in volunteers. Anesthesiology 110 (2):295–312

43. Habib R, Nyberg L (2008) Neural correlates of availability and accessibility in memory. Cereb Cortexl 18(7):1720–1726

44. Pryor KO, Root JC, Mehta M, Stern E, Pan H et al (2015) Effect of propofol on the medial temporal lobe: a functional magnetic resonance imaging study in human subjects. Br J Anaesth 115:i104–i113

45. Deeprose C, Andrade J, Varma S, Edwards N (2004) Unconscious learning during surgery with propofol anaesthesia. Br J Anaesth 92:171–177

46. Veselis RA, Reinsel RA, Feshchenko VA, Wroński M (1997) The comparative amnestic effects of midazolam, propofol, thiopental, and fentanyl at equisedative concentrations. Anesthesiology 87:749–764

47. Johns FR, Sandler NA, Buckley MJ, Herlich A (1998) Comparison of propofol and methohexital continuous infusion techniques for conscious sedation. J Oral Maxillofac Surg 56:1124–1127

48. Zarnowska ED, Rodgers FC, Oh I et al (2015) Etomidate blocks LTP and impairs learning but does not enhance tonic inhibition in mice carrying the N265M point mutation in the beta3 subunit of the GABA (A) receptor. Neuropharmacology 93:171–178

49. Cheng VY, Martin LJ, Elliott EM, Kim JH, Mount HT et al (2006) Alpha5GABAA receptors mediate the amnestic but not sedative-hypnotic effects of the general anesthetic etomidate. J Neurosci 26:3713–3720

50. Friederich P, Dybek A, Urban BW (2000) Stereospecific interaction of ketamine with nicotinic acetylcholine receptors in human sympathetic ganglion-like SH-SY5Y cells. Anesthesiology 93:818–824

51. Forman SA, Chin VA (2008) General Anesthetics and molecular mechanisms of unconsciousness. Int Anesthesiol Clin 46:43–53

52. Nedaei SE, Pourmotabbed A, Mehrabi Nasab E, Touhidi A (2009) Ketamine induces anterograde and retrograde amnesia in rats. Physiol Pharmacol 12(4):328–335

53. Arrázola MS, Varela-Nallar L, Colombres M et al (2009) Calcium/calmodulin-dependent protein kinase type IV is a target gene of the wnt/beta-catenin signaling pathway. J Cell Physiol 221:658–667

54. Liu H, Xu GH, Wang K et al (2014) Involvement of GSK3β/β-catenin signaling in the impairment effect of ketamine on spatial memory consolidation in rats. Neurobiol Learn Mem 111:26–34

55. Chortkoff BS, Bennett HL, Eger EI 2nd (1993) Subanesthetic concentrations of isoflurane suppress learning as defined by the category-example task. Anesthesiology 79(1):16–22

56. Nishikawa K, MacIver MB (2001) Agent-selective effects of volatile anesthetics on GABAA receptor-mediated synaptic inhibition in hippocampal interneurons. Anesthesiology 94(2):340–347

57. Masquelier T, Hugues E, Deco G, Thorpe SJ (2009) Oscillations, phase-of-firing coding, and spike timing-dependent plasticity: an efficient learning scheme. J Neurosci 29(43):13484–13493

58. Perouansky M, Rau V, Ford T et al (2010) Slowing of the hippocampal θ rhythm correlates with anesthetic-induced amnesia. Anesthesiology 113:1299–1309

59. Jordan BD, Wright EL (2010) Xenon as an anesthetic agent. AANA J 78:387–392

60. Haseneder R, Kratzer S, Kochs E et al (2008) Xenon reduces N-methylD-aspartate and alpha-amino-3-hydroxy-5-methyl-4-isoxazolepropionic acid receptor-mediated synaptic transmission in the amygdala. Anesthesiology 109:998–1006

61. Lister RG (1985) The amnesic action of benzodiazepines in man. Neurosci Biobehav Rev 9:87–94

62. Lister RG, Weingartner H, Eckardt MJ, Linnoila M (1988) Clinical relevance of effects of benzodiazepines on learning and memory. PsychopharmacolSer 6:117–127

63. Mashour GA, Kent C, Picton P et al (2013) Assessment of intraoperative awareness with explicit recall: a comparison of 2 methods. Anesth Analg 116:889–891

64. Nordström O, Engström AM, Persson S, Sandin R (1997) Incidence of awareness in total i.v. anaesthesia based on propofol, alfentanil and neuromuscular blockade. Acta Anaesthesiol Scand 41(8):978–984

65. Pandit JJ, Andrade J, Bogod DG et al (2014) 5th National Audit Project (NAP5) on accidental awareness during general anaesthesia: summary of main findings and risk factors. Br J Anaesth 113(4):549–559

66. Bischoff P, Rundshagen I, Schneider G (2015) Undesired awareness phenomena during general anesthesia: evidence-based state of knowledge, current discussions and strategies for prevention and management. Anaesthesist 64(10):732–739

67. Tasbihgou SR, Vogels MF, Absalom AR (2018) Accidental awareness during general anaesthesia - a narrative review. Anaesthesia 73(1):112–122

68. Walker EM, Bell M, Cook TM, Grocott MP, Moonesinghe SR, SNAP-1 investigator group (2016) Patient reported outcome of adult perioperative anaesthesia in the United Kingdom: a cross-sectional observational study. Br J Anaesth 117:758–766

69. Sandin RH, Enlund G, Samuelsson P, Lennmarken C (2000) Awareness during anaesthesia: a prospective case study. Lancet 355:707–711

70. Sebel PS, Bowdle TA, Ghoneim MM et al (2004) The incidence of awareness during anesthesia: a multicenter United States study. Anesth Analg 99:833–839

71. Mashour GA, Shanks A, Tremper KK et al (2012) Prevention of intraoperative awareness with explicit recall in an unselected surgical population: a randomized comparative effectiveness trial. Anesthesiology 117:717–725

72. Cascella M, Viscardi D, Schiavone V et al (2016) A 7-year retrospective multisource analysis on the incidence of Anesthesia awareness with recall in Cancer patients: a chance of collaboration between Anesthesiologists and psycho-oncologists for awareness detection. Medicine (Baltimore) 95(5):e2757. https://doi.org/10.1097/MD.0000000000002757

73. Schneider G, Wagner K, Reeker W et al (2002) Bispectral index (BIS) may not predict awareness reaction to intubation in surgical patients. J Neurosurgical Anesthesiology 14:7–11

74. Sanders RD, Gaskell A, Raz A et al (2017) Incidence of connected consciousness after tracheal intubation: a prospective, international, multicenter cohort study of the isolated forearm technique. Anesthesiology 126:214–222

75. Mashour GA, Avidan MS (2015) Intraoperative awareness: controversies and non-controversies. Br J Anaesth 115: Si20–SS26

76. Davidson AJ, Smith KR, Blussé van Oud-Alblas HJ et al (2011) Awareness in children: a secondary analysis of five cohort studies. Anaesthesia 66:446

77. Davidson AJ, Huang GH, Czarnecki C et al (2005) Awareness during anesthesia in children: a prospective cohort study. Anesth Analg 100:653

78. Sury MR (2016) Accidental awareness during anesthesia in children. Paediatr Anaesth 26 (5):468–474

79. Noël E, Van Der Linden P (2016) Le réveil peropératoire en anesthésie pédiatrique : revue de la literature. Rev Med Brux 37 (6):476–482

80. Mashour GA, Wang LY, Turner CR, Vandervest JC, Shanks A, Tremper KK (2009) A retrospective study of intraoperative awareness with methodological implications. Anesth Analg 108:521–536

81. Kent CD, Mashour GA, Metzger NA, Posner KL, Domino KB (2013) Psychological impact of unexpected explicit recall of events occurring during surgery performed under sedation, regional anaesthesia, and general anaesthesia: data from the Anesthesia awareness registry. Br J Anaesth 110:381–387

82. Jones JG (1994) Perception and memory during general anaesthesia. Br J Anaesth 73 (1):31–37

83. Lennmarken C, Bildfors K, Enlund G, Samuelsson P, Sandin R (2002) Victims of awareness. Acta Anaesthesiol Scand 46 (3):229–231

84. Kent CD, Posner KL, Mashour GA et al (2015) Patient perspectives on intraoperative awareness with explicit recall: report from a north American anaesthesia awareness registry. Br J Anaesth 115(Suppl 1):i114–i121

85. Cascella M (2016) Mechanisms underlying brain monitoring during anesthesia: limitations, possible improvements, and perspectives. Korean J Anesthesiol 69(2):113–120

86. Cascella M (2015) Anesthesia awareness. Can midazolam attenuate or prevent memory consolidation on intraoperative awakening during general anesthesia without increasing the risk of postoperative delirium? Korean J Anesthesiol 68(2):200–202

87. Timić T, Joksimović S, Milić M, Divljaković J, Batinić B, Savić MM (2013) Midazolam impairs acquisition and retrieval, but not consolidation of reference memory in the Morris water maze. Behav Brain Res 241:198–205

88. Semba K, Adachi N, Arai T (2005) Facilitation of serotonergic activity and amnesia in rats caused by intravenous anesthetics. Anesthesiology 102(3):616–623

89. American Society of Anesthesiologists Task Force on Intraoperative Awareness (2006) Practice advisory for intraoperative awareness and brain function monitoring: a report by the american society of anesthesiologists task force on intraoperative awareness. Anesthesiology 104(4):847–864

90. Cascella M, Fusco R, Caliendo D et al (2017) Anesthetic dreaming, anesthesia awareness and patient satisfaction after deep sedation with propofol target controlled infusion: a prospective cohort study of patients undergoing day case breast surgery. Oncotarget 8 (45):79248–79256

91. Gyulaházi J, Redl P, Karányi Z, Varga K, Fülesdi B (2016) Dreaming under anesthesia: is it a real possibility? Investigation of the effect of preoperative imagination on the quality of postoperative dream recalls. BMC Anesthesiol 16(1):53

92. Samuelsson P, Brudin L, Sandin RH (2008) Intraoperative dreams reported after general anaesthesia are not early interpretations of delayed awareness. Acta Anaesthesiol Scand 52:805–809

93. Stait ML, Leslie K, Bailey R (2008) Dreaming and recall during sedation for colonoscopy. Anaesth Intensive Care 36(5):685–690

94. Brice DD, Hetherington RR, Utting JE (1970) A simple study of awareness and dreaming during anaesthesia. Br J Anaesth 42(6):535–542

95. Osterman JE, van der Kolk BA (1998) Awareness during anesthesia and posttraumatic stress disorder. Gen Hosp Psychiatry 20 (5):274–281

96. Mashour GA, Esaki RK, Tremper KK, Glick DB, O'Connor M, Avidan MS (2010) A novel classification instrument for intraoperative awareness events. Anesth Analg 110:813–815

97. Wang M, Messina AG, Russell IF (2012) The topography of awareness: a classification of intra-operative cognitive states. Anaesthesia 67:1197–1201

98. Kuo PJ, Lee CL, Wang JH, Hsieh SY, Huang SC, Lam CF (2017) Inhalation of volatile

anesthetics via a laryngeal mask is associated with lower incidence of intraoperative awareness in non-critically ill patients. PLoS One 12 (10):e0186337

99. Ghoneim MM (2007) Incidence and risk factors for awareness during anaesthesia. Best Prac Res Clin Anaesthesiol 21:327–343

100. Pollard RJ, Coyle JP, Gilbert RL, Beck JE (2007) Intraoperative awareness in a regional medical system. Anesthesiology 106:269–274

101. Wennervirta J, Ranta SO, Hynynen M (2002) Awareness and recall in outpatient anesthesia. Anesth Analg 95:72–77

102. Bergman IJ, Kluger MT, Short TG (2002) Awareness during general anaesthesia: a review of 81 cases from the anaesthetic incident monitoring study. Anaesthesia 57 (6):549–556

103. Dai S, Perouansky M, Pearce RA (2009) Amnestic concentrations of etomidate modulate GABAA, slow synaptic inhibition in hippocampus. Anesthesiology 111(4):766–773

104. Al-Hasan YM, Krishnan HR, Ghezzi A et al (2011) Tolerance to anesthesia depends on synaptic proteins. Behav Genet 41 (5):734–745

105. Aranake A, Gradwohl S, Ben-Abdallah A et al (2013) Increased risk of intraoperative awareness in patients with a history of awareness. Anesthesiology 119:1275

106. Morimoto Y, Nogami Y, Harada K et al (2011) Awareness during anesthesia: the results of a questionnaire survey in Japan. J Anesth 25:72

107. Yu H, Wu D (2017) Effects of different methods of general anesthesia on intraoperative awareness in surgical patients. Medicine (Baltimore) 96(42):e6428

108. Perl T, Carstens E, Hirn A et al (2009) Determination of serum propofol concentrations by breath analysis using ion mobility spectrometry. Br J Anaesth 103:822

109. Ghoneim MM, Block RI, Haffarnan M et al (2009) Awareness during anesthesia: risk factors, causes and sequelae: a review of reported cases in the literature. Anesth Analg 108 (2):527–535

110. Nunes RR, Cavalcante SL, Lobo RF (2007) Memórias explícita e implícita em anestesias com bloqueio neuromuscular e BIS. São Paulo Med J 125(Suppl):129

111. Nunes RR, Cavalcante SL (2007) Influência do bloqueio neuromuscular despolarizante nas entropias. São Paulo Med J 125 (Suppl):126

112. Nunes RR, Porto VC, Miranda VT, de Andrade NQ, Carneiro LM (2012) Risk factor for intraoperative awareness. Rev Bras Anestesiol 62(3):365–374

113. Schuller PJ, Newell S, Strickland PA, Barry JJ (2015) Response of bispectral index to neuromuscular block in awake volunteers. Br J Anaesth 115(Suppl 1):i95

114. Cascella M (2018) Emergence from anesthesia: a winding way back. Anaesthesiol Intensive Ther 50(2):168–169

115. Cascella M, Bimonte S, Muzio MR (2018) Towards a better understanding of anesthesia emergence mechanisms: research and clinical implications. World J Methodol 8(2):9–16

116. Thomsen JL, Nielsen CV, Eskildsen KZ et al (2015) Awareness during emergence from anaesthesia: significance of neuromuscular monitoring in patients with butyrylcholinesterase deficiency. Br J Anaesth 115(Suppl 1): i78–i88

117. Radovanovic D, Radovanovic Z (2011) Awareness during general anaesthesia--implications of explicit intraoperative recall. Eur Rev Med Pharmacol Sci 15(9):1085–1089

118. Shanks AM, Avidan MS, Kheterpal S et al (2015) Alerting thresholds for the prevention of intraoperative awareness with explicit recall: a secondary analysis of the Michigan awareness control study. Eur J Anaesthesiol 32(5):346–353

119. Zhang Y, Dong YJ, Webster CS et al (2013) The frequency and nature of drug administration error during anaesthesia in a Chinese hospital. Acta Anaesthesiol Scand 57 (2):158–164

120. Cook TM, MacDougall-Davis SR (2012) Complications and failure of airway management. Br J Anaesth 109(1):i68–i85

121. Cascella M, Fiore M, Leone S, Carbone D, Di Napoli R (2019) Current controversies and future perspectives on treatment of intensive care unit delirium in adults. World J Crit Care Med 8(3):18–27

122. Reshef ER, Schiff ND, Brown EN (2009) A neurologic examination for Anesthesiologists: assessing arousal level during induction, maintenance, and emergence. Anesthesiology 130:462

123. Behne M, Wilke HJ, Harder S (1999) Clinical pharmacokinetics of sevoflurane. Clin Pharmacokinet 36:13–26

124. Antognini JF, Schwartz K (1993) Exaggerated anesthetic requirements in the preferentially anesthetized brain. Anesthesiology 79:1244

125. Wang J, Zhang L, Huang Q et al (2017) Monitoring the end-tidal concentration of sevoflurane for preventing awareness during anesthesia (MEETS--PANDA): a prospective clinical trial. Int J Surg 41:44–49

126. Avidan MS, Jacobsohn E, Glick D et al (2011) BAG-RECALL research group. Prevention of intraoperative awareness in a high-risk surgical population. N Engl J Med 365 (7):591–600

127. Avidan MS, Zhang L, Burnside BA et al (2008) Anesthesia awareness and the Bispectral index. N Engl J Med 358:1097–1108

128. Messina AG, Wang M, Ward MJ et al (2016) Anaesthetic interventions for prevention of awareness during surgery. Cochrane Database Syst Rev 10(10):CD007272. https://doi.org/10.1002/14651858.CD007272

129. Australian and New Zealand College of Anaesthetists (ANZCA) Guidelines on monitoring during anaesthesia. Available at: http://www.anzca.edu.au/documents/ps18-2015-guidelineson-monitoring-during-anaesthe.pdf. Updated December 2015. Accessed 8 June 2019

130. Checketts MR, Alladi R, Ferguson K et al (2016) Association of Anaesthetists of Great Britain and Ireland. Recommendations for standards of monitoring during anaesthesia and recovery 2015: Association of Anaesthetists of Great Britain and Ireland. Anaesthesia 71:85–93

131. Punjasawadwong Y, Phongchiewboon A, Bunchungmongkol N (2014) Bispectral index for improving anaesthetic delivery and postoperative recovery. Cochrane Database Syst Rev (6):CD003843

Chapter 2

Impact of Anesthetics on Brain Electrical Activity and Principles of pEEG-Based Monitoring During General Anesthesia

Marco Cascella

Abstract

Starting from the first half of the nineteenth century, investigations proved that anesthetics are able to induce significant modifications in the brain's electrical activity and, in turn, in the recorded electroencephalogram (EEG). The anesthesia-related electrical activity involves different EEG correlates. Stage of anesthesia and type of anesthetics used influence EEG expression. Nevertheless, the complex waves of unprocessed EEG may not be easily interpreted. Moreover, several issues obstacle the utilization of standard EEG in anesthesia. These limitations stimulated the search for easy and solid techniques and tools. The EEG raw signal has been sectioned to extract its core element and to simplify the interpretation of the huge amount of data that it contains. Technically, the increased flexibility, speed, and economy of digital circuits, as well as progress in computer hardware and signal-processing algorithms, induced radical changes in the field of signal processing. This evolutionary process involved the application of mathematical models such as the Fourier analysis and its improvement by the bispectral (BIS) analysis. Technical advances and algorithms allowed to process the EEG raw (processing EEG, pEEG) and extrapolate values (indices) which express the depth of anesthesia (DoA) status and other features (e.g., response to noxious stimuli). Apart from the pEEG-based brain monitoring devices, other instruments work on acoustic evoked potentials. Because many mathematical models are proprietary algorithms, it is usually problematic to precisely interpret mechanisms underlying of DoA monitors.

After its commercialization, in 1994, the BIS analysis-based monitor was intended as the best instrument to follow the cerebral activity during general anesthesia. The availability of this technology was enthusiastically considered as the discovery of the "Holy Grail" of anesthesia monitoring and the definitive solution of one among the major concerns of anesthesia: the awareness phenomenon. After two decades of debates and investigations in the field, the scientific community has understood that this is not completely true.

The aim of this chapter is to dissect the technology and functionality behind these monitors. Explanation of functionality assumes the recognition of the functional mechanisms of anesthetics, specifically their mechanisms of action, from the molecular level to neural correlates, responsible for the anesthetic-induced unconsciousness, maintenance, and recovery of consciousness. Furthermore, concepts of neurophysiology and anesthesia-related electrical activity are offered because they are the functional basis of the brain monitoring devices. The chapter also addresses the limitations of DoA devices and perspectives in brain monitoring.

Key words Electroencephalography, Intraoperative awareness, General anesthesia awareness, Intraoperative monitoring, Bispectral analysis, Depth of anesthesia, Brain monitoring

Marco Cascella (ed.), *General Anesthesia Research*, Neuromethods, vol. 150, https://doi.org/10.1007/978-1-4939-9891-3_2,
© Springer Science+Business Media, LLC, part of Springer Nature 2020

Abbreviations

AAGA	Accidental awareness during general anesthesia
AE	Anesthesia emergence
AEPs	Acoustic evoked potentials
BAEPs	Brainstem AEPs
BIS	Bispectral
BS	Burst suppression
CNS	Central nervous system
CSI	Cerebral state index
DoA	Depth of anesthesia
DSA	Density Spectral Array
ECoG	Electrocorticogram
EEG	Electroencephalogram
EMG	Electromyography
ETAC	End-tidal anesthetic concentration
FFT	Fast Fourier transform
fMRI	Functional magnetic resonance imaging
fNIRS	Functional near-infrared spectroscopy
FT	Fourier transform
GA	General anesthesia
GABA	γ-Aminobutyric acid
hd-EEG	High-density electroencephalography
LLAEPs	Long-latency AEPs
LoC	Loss of consciousness
MAC	Minimal alveolar concentration
MLAEPs	Middle-latency AEPs
NMBAs	Neuromuscular blocking agents
NMDA	N-methyl-D-aspartate
NO	Nitrous oxide
pEEG	Processed electroencephalogram.
PID	Proportional–integral–derivative
POCD	Postoperative cognitive dysfunction
POD	Postoperative delirium
PSI	Patient State Index
RE	Response entropy
RoC	Recovery of consciousness
SE	State entropy
S-EEG	Stereo-electroencephalography
SQI	Signal quality index
SR	Suppression ratio

1   Introduction

General anesthesia (GA) is a reversible, medically induced status encompassing loss of consciousness (LoC) with hypnosis, amnesia, analgesia, and mitigation of reflexes in the autonomic nervous system. Pharmacological reversible neuromuscular blockade with muscle relaxation is another feature of GA, although not always indispensable. Despite important signs of progress in the knowledge of the mechanisms of anesthetic functioning and dynamics of anesthesia, many aspects of the matter must necessarily still be elucidated [1]. Uncertainty still exists regarding the definition and modality of evaluation of the right anesthesia status, which is called "the surgical status." For this aim, standard clinical monitoring is usually mixed with strategies adopted for assessing the depth of GA, or "depth of anesthesia" (DoA), which represents a continuum of central nervous system (CNS) depression and decreased responsiveness to stimuli, from the awake conditions to a deep anesthesia status. In this context, the term "DoA monitoring" is widely used for indicating instrumental modalities for assessing unconsciousness in the course of anesthetics administration.

Apart from clinical evaluation, there are several modalities useful for the DoA instrumental evaluation. Among these is the monitoring of the end-tidal anesthetic concentration (ETAC) of inhalation agents performed through algorithms elaborated by the anesthetic machine. The ETAC expresses concentrations of anesthetics, which estimate their arterial blood partial pressure with acceptable precision, and indicates the concentration of an inhalation agent close to the minimal alveolar concentration (MAC) value. For each inhaled agent administered, a MAC 1 value describes the concentration required, at 1 atm ambient pressure, to prevent 50% of subjects moving in response to a stimulus. MAC requirement for movement suppression is usually higher than MAC for suppression of consciousness (MAC-awake) or memory (MAC-amnesia).

In 1937, Gibbs et al. published a pioneering study entitled "*Effects on the Electroencephalogram of Certain Drugs Which Influence Nervous Activity*" [2] and postulated that the potential role of electroencephalogram (EEG) as a measure of the DoA. After almost a century of research, nowadays technologies based on processed electroencephalogram (pEEG) represent the main approach adopted for DoA monitoring. Several commercial monitors for evaluation of brain functioning—in terms of consciousness/unconsciousness states—have been developed, and to date, more than 10 devices are available for clinical use. They should guide the administration of general anesthetics for avoiding complications due to an inadequate anesthetic depth such as accidental awareness during general anesthesia (AAGA), or at the other extreme, issues related to an excessive anesthetic depth such as delayed emergence, and postoperative delirium (POD) [3, 4]. Furthermore, the association between postoperative cognitive dysfunction in elderly and DoA should be better investigated [5]. For example, it has been demonstrated that patients who received Bispectral (BIS)-guided anesthesia showed less postoperative cognitive impairment (i.e., postoperative cognitive dysfunction, POCD), compared to those who received no–BIS-guided GA [6].

The aim of this chapter is to try to dissect the technology and functionality behind these monitors. Because general anesthetics induce EEG modifications, the functioning of DoA devices is strictly correlated to the functional mechanisms of anesthetics, namely the mechanisms of action of anesthetics, from the molecular level to neural correlates of the anesthetic-induced unconsciousness. Furthermore, concepts of neurophysiology (e.g., principles of EEG, and evoked potentials) and anesthesia-related electrical activity are offered for elucidating the dynamics of anesthesia but, above all, because they are the functional bases of the brain monitoring devices. Limitations of DoA devices and perspectives in brain monitoring are also discussed.

2   Functional Mechanisms of Anesthetics

Anesthesia-induced LoC and anesthesia emergence (AE) until recovery of consciousness (RoC) are quite different processes with, in part, distinct neurobiology. Both processes represent a neurobiological example of "hysteresis" because a process (i.e., emergence) develops following a path that does not necessarily coincide (in the opposite direction) with the path of the process (i.e., induction) that generated it [7]. In other words, AE culminates in the awakening state by following different patterns from those of the induction. Thus, AE's neurobiological events should not be easily considered as reverse processes of those occurring in the induction stage of GA [8].

2.1   Induction and Maintenance of Anesthesia

Induction and maintenance of GA are achieved through the administration of two categories of general anesthetics including intravenous (or injected) agents and inhaled (or volatile) medications. Among the former group, there are propofol, etomidate, midazolam, thiopental, and ketamine. The group of inhaled drugs includes different halogenated agents such as halothane, desflurane, isoflurane, and sevoflurane, and the gasses nitrous oxide (NO), and xenon. As a general scheme, because general anesthetic medications are small and uncharged molecules, they are able to cross the blood-brain barrier. At the molecular level, they interact with specific ion channels involved in synaptic transmission and membrane potentials in several target CNS areas. Different ligand-gated ion channels such as the Cys-loop nicotinic acetylcholine receptor, the γ-aminobutyric acid class A ($GABA_A$) receptor, N-methyl-D-aspartate (NMDA) subtype of glutamate receptor, and voltage-gated channels including different voltage-gated sodium and potassium channels, the tandem pore potassium channels, and nonspecific cation channels (e.g., the thalamic "Ih" channel) are implicated.

The binding of an anesthetic to one or more targets—depending on channel types and conformations—induces alterations in neuronal activities through neuronal hyperpolarization. The effect is an expanded neural inhibition or decreased synaptic excitation [9].

This complex matter of the operating mechanism of anesthetics can be explained by accepting that these agents are able to obstruct with cortical and subcortical signals. This interaction induces, in turn, changes in the functional/effective connectivity across brain regions. In this regard, pieces of evidence suggested that during GA alterations in functional and connectivity from different brain areas (e.g., from frontal to parietal regions) occur [10]. Volatile agents, for example, may impair frontal–posterior interactions by interfering with the gamma (20–60 Hz) oscillations. As proved in preclinical in vivo investigations, this pattern has a key role in arousal and maintenance of consciousness [11]. Furthermore, studies based on high-density electroencephalography (hd-EEG)—a dense array scalp EEG recording system widely used for research purposes—demonstrated that propofol-induced LoC provokes an increase in frontal delta power due to a cortical propagation of processes arising from subcortical regions such as lateral sulci and cingulate gyrus [12]. These slow-delta oscillations raise asynchronously forward and across the cortex, inducing a functional disruption of the connectivity between distinct cortical areas.

By summarizing, connectivity changes within separate brain regions impact the DoA. In particular, loss of feedbacks between the thalamus and the cortex could be the starting point for LoC. Subsequently, alterations in the corticocortical functional connectivity, and in the functioning of several brainstem nuclei (e.g., the ventrolateral preoptic nucleus which operates as an on/off switch on other brainstem nuclei and regulates natural sleep), as well as in the connectivity between these nuclei and cerebral cortex, finalize the process of anesthesia induction and stabilize the DoA.

2.2 Anesthesia Emergence and Recovery of Consciousness Mechanisms

The ending stage of anesthesia, indicated as AE, is characterized by the transition from unconsciousness to complete wakefulness and RoC. During this stage, processes responsible for LoC and anesthesia maintenance gradually reverse. Besides, specific awakening processes switch on. These mechanisms include distinct ascending arousal brain pathways, mainly involving the thalamus [13] through the activation of voltage-gated potassium channels [14]. Other structures such as the substantia nigra, ventral tegmental area, dorsal raphe, locus coeruleus, laterodorsal tegmental areas, and hypothalamus (e.g., orexinergic neurons) are also implicated in this active arousal process [4].

3   Anesthesia-Related Electrical Activity

The electrical brain's response to anesthetics which is recorded with
scalp EEG is the cortical synaptic activity of both excitatory and
inhibitory postsynaptic potentials from cortical or thalamic neurons
[15]. EEG is a noninvasive modality commonly used for clinical
purposes and research. Other neurophysiological methods such as
the electrocorticogram (ECoG) which is the EEG measured
directly from the cortical surface, the stereo-EEG (S-EEG), namely
the EEG performed via depth probes, are applied in specific clinical
settings (e.g., S-EEG in epilepsy) or for preclinical in vivo research.
Furthermore, neurophysiological changes in the brain under GA
may be also investigated through a combination of EEG (including
hd-EEG methods) with brain activity measures such as functional
near-infrared spectroscopy (fNIRS) [16], and neuroimaging modalities (e.g., functional magnetic resonance imaging, fMRI) [17], or
by combining EEG with electrodiagnostic methods, including elec-
tromyography (EMG) and evoked potentials [18].

3.1 EEG Findings During Induction and Maintenance of Anesthesia

Anesthesia-related electrical activity encompasses a huge number of
EEG patterns, mainly depending on the anesthesia stage (induction, maintenance, and emergence), and type of anesthetics. Before
induction, in the awake patient (with eyes closed) a prominent
alpha activity (8–12 Hz) which is maximal over parieto-occipital
scalp locations, is recorded. After inducing anesthesia, EEG pattern
features an increase in beta activity (13–30 Hz), until the LoC is
achieved [19]. During the maintenance phase of anesthesia, several
EEG patterns can be detected due to different anesthesia levels.
During light anesthesia, for instance, a decrease in EEG beta band
and an increase in both EEG alpha and delta activities (0–4 Hz) can
be recorded. Subsequently, as the anesthesia becomes deeper,
the beta activity decreases, and there is an increment in the delta
and alpha oscillations. Because the latter rhythm progressively localized anteriorly, this phenomenon is called "alpha anteriorization"
[20]. A further deepening of the anesthesia is electrically expressed
as an EEG pattern which encompasses flat periods electrical (suppression) alternated with high amplitude alpha and beta electrical
activity (bursts). This characteristic EEG finding is known as burst
suppression (BS) and can show several patterns, depending on the
duration of suppression/burst phases. It can be also recognized in
deeper coma status due to various conditions including cerebral
anoxia, cancers, drug intoxications, encephalopathies, or hypothermia [21]. The anesthesia-induced BS is probably related to a state
of cortical hyperexcitability induced by lowered inhibition [22]. As
the anesthesia status further deepens, EEG shows a progressive
stretching between the alpha activities. The amplitudes of the
alpha and beta activities progressively decrease, and in turn, the

EEG assumes isoelectric form. In this context, the deepest anesthesia status has been reached [23].

This general scheme is particularly suitable for explaining the dynamics of halogenated volatile anesthetics and the intravenous propofol. On the contrary, opioids, and ketamine usually induce less marked EEG modifications. Again, other injected agents show a completely different trend. Early studies, indeed, proved that etomidate and barbiturates induce a rapid shift toward the high voltage delta and theta (4–7 Hz) rhythms [24].

3.2 EEG Findings During Emergence

Classically, during AE and RoC a loss in the delta activity combined with a progressive decrease in frontal alpha power and increased higher frequencies can be detected [25]. Some electrical findings can be similar in LoC and AE. Indeed, from ECoG-based investigations emerged that slow oscillation in large-scale functional networks is maintained during both LoC and RoC stages [26].

Of note, the EEG pattern recorded during emergence can manifest itself in different modalities. In this regard, Chander et al. [27] described different AE sequences. At the beginning of AE, they found in the majority of patients (95%) a pattern of high power of alpha and beta bands and termed it as "Slow-Wave Anesthesia." In the remaining 5% of patients, the authors recorded very low spindles and delta power and called this EEG pattern "Non-Slow-Wave Anesthesia." Furthermore, they also showed that EEG patterns vary between the start of emergence and RoC. On this subject, four trajectories between the beginning of AE and the RoC were indicated. More recently, Liang et al. [28] classified EEG patterns during AE in sevoflurane anesthesia. Using an integrated approach, achieved through a multivariate statistical model, they characterized four modalities of emergence EEG patterns. It must be emphasized that because some of these emergence modalities are strictly age-related, these findings offer an interesting neurophysiological perspective for studying the correlation between AE features and postoperative alterations of the neurocognitive trajectory such as POD, and POCD. Probably, the occurrence of different EEG features at the emergence is the effect of different degrees of influence which the brainstem system exerts on cortical functioning through the thalamus mediation [13].

4 Operating Principles of DoA Devices

Since the Gibbs's studies in the first half of the nineteenth century, clinicians have been aware that anesthetics can induce significant modifications in the brain's electrical activity and, in turn, in the EEG. The overall dynamics of GA, indeed, inhibit higher cortical functions, decreasing, in turn, neuronal firing, slowing the waveforms and enhancing its synchrony.

In spite of that finding, the complex waves of raw EEG (unprocessed EEG) may not be easily interpreted. Thus, several issues have limited the application of standard EEG monitoring to anesthesia. Among these issues, EEG does not modify in a linear or monotonic manner with changing anesthetic status. Yet, as previously discussed, not all anesthetic agents produce similar EEG patterns. Thus, the interpretation of raw EEG requires skills that anesthesiologists cannot have [29].

These limitations stimulated the search for simple and reliable techniques and tools useful for brain monitoring in the theatre. The complex EEG signal has been progressively dissected for extracting its core component, and important advances have been achieved concerning the interpretation of the huge amount of information that the raw signal contains. Technically, the increased speed, flexibility, and economy of digital circuits, as well as signs of progress in computer hardware and signal processing combined with algorithms designing induced revolutionary changes in the field of EEG processing. Technologies allowed to compress, simplify, and display different "processed" summaries of EEG data (pEEG). Consequently, many of these algorithms have been applied in commercially available DoA devices. However, because many mathematical models are proprietary algorithms, it is often difficult to precisely understand the mechanisms underlying of DoA monitors. The basic functioning of DoA devices consists of the recording of limited EEG activity from forehead electrodes and subsequent real-time processing of raw EEG signals through complex algorithms.

4.1 The Technical Evolution of the Raw EEG Analysis: Fourier Analysis and Indices

A paramount step forward the development of DoA devices was the signal digitization of the unprocessed EEG. The digital signal, indeed, can represent discrete points in the time at which values are derived by a fixed, rather than continuous, resolution. In other words, digital signals are quantized in time, whereas analogic signals vary smoothly from moment to moment. On the other side, the process of analog-to-digital translation may involve a loss of fidelity in the resulting digital signal. Thus, for accurate data processing, it is essential to transform the whole EEG spectrum. One of the mathematical models used for this purpose is the Fourier analysis (or Fourier transform, FT) [30]. This process operates the so-called power spectrum analysis, as it decomposes the EEG time series into a voltage-by-frequency spectral graph commonly called "power spectrum," with "power" being the square of the EEG magnitude, and magnitude being the integral average of the amplitude of the EEG signal, measured from $(+)$ peak to $(-)$ peak, across the time or epoch sampled. The epoch length determines the frequency resolution of the Fourier, with a 1-s epoch providing a 1 Hz resolution (± 0.5 Hz resolution), and a 4-s epoch providing a ¼ Hz, or ± 0.125 Hz, resolution.

In mathematical terms, the Fourier analysis can transform a set of signal values $x(t_i)$ sampled at time moments to within a signal sample into a set of an equal number of complex values $X(f_i)$ corresponding to a set of frequencies f_i:

$$X\left(f_i\right) = \sum_{t_i} x(t_i) e^{-i2\pi f_i t_i}$$

This formula allows for generating frequency spectrum $X(f_i)$, which is simply a histogram of amplitudes or phase angles as a function of the frequency of the signal $x(t_i)$.

For explaining this complex transformation, a practical example is commonly used. The concept, indeed, may be explained by referring to the effect of a white light that crosses through a glass prism, thus creating a rainbow (i.e., the spectrum). Each color of light corresponds to a single frequency photon, and the relative brightness of the colors is a measure of the energy amplitude at each frequency. Each measured signal, such as the spectral components $X(f_i)$, transformed by the FT into the frequency domain will have both an amplitude and a phase component for each harmonic frequency.

Because this basilar approach to the elaboration of the data (e.g., the raw signal) is laborious, processing of the Fourier series is commonly performed through specific mathematical strategies, such as the Fast FT (FFT) algorithms (e.g., the Cooley-Tukey algorithm) which decompose a sequence of values (e.g., space and time) into components of different frequencies [31, 32].

As a subsequent evolutionary step, the FT was improved by the bispectrum measures. This process is termed as BIS analysis. While the algorithms of FT performs the processing of the phase of component frequencies relative to the start of the epoch, BIS analyses allow for phase correlation between different frequency components. Furthermore, the BIS analysis has several additional features that may be advantageous for processing EEG signals. For example, it enhances the signal by limiting sources of noise.

Technically, the bispectrum quantifies the relationships among the underlying sinusoidal components of the EEG and examines the relationship between the sinusoids at two primary frequencies, f_1 and f_2, and a modulation component at the frequency $f_1 + f_2$. This set of three frequency components (f_1, f_2, and $f_1 + f_2$) configures the triplet. For each triplet of the bispectrum, the analysis can precisely calculate the amount incorporating both phase and power information. Thus, the bispectrum can be also intended as a statistic approach used to obtain evidence of nonlinear interactions in irregular signals.

Based on these assumptions, all commercial DoA devises work analyzing and quantifying spontaneous (or elicited signals in case of evoked potential based devices) raw EEG data using mathematical methods. Specific algorithms allow obtaining several indices helpful for easier reading of the anesthesia status. These dimensionless indices are constructed, abstract quantities not directly linked to any physiological parameter. Among their other features, they are characterized by inherent time delays, consistent with the processing time, and the calculation and removal, through specific algorithms, of certain artifacts.

5 Depth of Anesthesia Monitors

5.1 Bispectral Index™ (BIS™) Monitor

In 1994, Sigl and Chamoun [33] firstly described the use of the BIS technology for the purpose of brain monitoring. Their BIS system (Bispectral Index™ (BIS™), Medtronic, UK) was a real revolution in anesthesia. It was a very easy-to-use device that provided a noninvasive adhesive sensor, a patient interface cable, a digital signal converter, and a monitor with the microprocessor. This microprocessor elaborates the amount of EEG data from a single EEG channel signal. EEG is obtained from the patient's forehead according to an algorithm that combines select EEG features for producing a single dimensionless number, which is the "BIS Index." This value may range from 0 (equivalent to EEG silence) to 100. An index value between 40 and 60 indicates an appropriate level for GA with a low probability of consciousness. The technology was also available as an external module adaptable to other manufacturers' monitoring systems.

Researchers have wondered what was behind this technology. Although the BIS algorithm is not publicly available, it has been shared that this value is a statistically based (and empirically derived) complex parameter equivalent to the weighted sum of several EEG subparameters, including a time domain, a frequency domain, and several high-order spectral subparameters [34].

In addition to the BIS Index (value), the device displays other data, such as the BIS trend, which is a graphic representation of BIS Index values over time, the signal quality index (SQI) bar, and the suppression ratio (SR). While SQI measures the reliability of the signal (a higher SQI number indicates more reliable BIS values), the SR number is the percentage of time over the last 63-s during which the EEG signal was suppressed (flat-lined). The monitor also displays a single-channel EEG waveform and an EMG bar graph, which, similar to the SQI, is used to help determine whether the BIS index values are reliable.

The BIS monitor was the first EEG-based DoA monitor and remains the most widely used system for assessing the brain monitoring of GA. The technology has been progressively improved. For instance, compared to older models, recent BIS monitors indicate lower BIS values, for the same level of hypnosis. This difference is due to the reduced level of noise, interference, and EMG activity, resulting in lower BIS values. To date, several BIS devices are commercialized, such as the BIS™ Complete 2-Channel Monitor; the BIS™ Complete 4-Channel Monitor which may detect hemispheric differences in the brain. Moreover, the BISx™ OEM module can be integrated into other medical devices.

Nevertheless, during the last 15–20 years, other EEG-based technologies have been studied and have become commercially available as noninvasive devices. The difference among these monitors concerns the parts of the EEG data selected (frequency, phase, and amplitude), processed and analyzed, and how the output measures are constructed and displayed into indices, or by graphical solutions.

5.2 Entropy™

The Entropy device (M-Entropy® Module, GE, Healthcare, Helsinki, Finland) works by a proprietary algorithm to elaborate the entropy of the EEG signal correlated to the behavioral responses of the patients. It was designed in Finland, in 1999, by Viertiö-Oja and colleagues [35]. The theoretical assumption is that during GA irregularity in the EEG signal tends to decrease. This irregularity can then be inferred by the entropy (the measure of the disorder) and used to estimate the DoA. The mathematical basis of spectral entropy originates from a measure called "Shannon entropy" [36]. This mathematical model, elaborated in 1948, has been successfully applied to the power EEG spectrum signal. The spectral entropy has been computed for each temporal variable (epoch) of the signal within a particular frequency range (f_1, f_2). From the Fourier's elaboration (i.e., FT) $X(f_i)$ of the signal $x(t_i)$, the power spectrum $P(f_i)$ is calculated by squaring the amplitudes of each element $X(f_i)$ of the FT:

$$P(f_i) = X(f_i) * X^{(f_i)}$$

In this formula, $X^{(f_i)}$ represents the complex conjugate of the Fourier component $X(f_i)$. The algorithm has been designed through a mathematical normalization of the overall frequency range of values between 1 (maximum irregularity) and 0 (complete regularity). However, this transformation operates in parallel with the variable time, producing time-frequency-balanced spectral entropy in which the signal values $x(t_i)$ are sampled within a finite time window (epoch) of a selected length with a particular sampling frequency. That is, to elaborate the signal rapidly, the algorithm allows for the study of each frequency in the context of an optimal time window.

Through this algorithm, data of the degree of disorder of EEG and frontal EMG signals, are converted into two values (indices) that indicate the DoA status. The first index (i.e., response entropy, RE) provides an indication on the patient's responsiveness to external stimuli and is useful for indicating an early awakening. It is computed from EEG and EMG dominant part of the spectrum of forehead muscles with frequency ranges from 0.8 to 47 Hz. The second value (i.e., state entropy, SE) is obtained after processing in a range from 0.8 to 32 Hz which includes the EEG dominating part of the spectrum and, thus, is indicative of the patient's cortical state. As a consequence, SE is a more stable parameter that can be used to assess the hypnotic effects of anesthetic agents.

More ordered signals, with less variation in the wavelength and amplitude over time, produce higher entropy values and may indicate that the patient is awake. Regular signals, with a constant wavelength and amplitude over time, produce low or zero entropy values, indicating a low probability of AAGA and suppression of brain electrical activity. The RE scale ranges from 0 to 100 (patient fully awake), and the SE scale ranges from 0 (cortical suppression/ deep anesthesia) to 91 (patient fully awake). The clinically relevant target for RE is 40–60, whereas SE values near 40 indicate a low probability of consciousness.

5.3 Narcotrend®

The Narcotrend® technology, developed at the University Medical School of Hannover, in Germany, was introduced in the year 2000. It analyses raw EEG data through spectral analysis and produces several parameters. Multivariate statistical methods, which use proprietary pattern recognition algorithms, are applied to provide an automatically classified EEG signal on the basis of a visual assessment of the EEG, as related to Loomis's sleep classification system, elaborated in 1937 [37]. The EEG visual classification scale ranges from stage A (awake) to stage F (very deep hypnosis). In particular, stage E indicates the appropriate DoA for surgery. Stages from A to F were further subdivided into three substages; thus, the Narcotrend device can automatically distinguish between 14 and 15 stages (depending on the software version used). Also, the Narcotrend software includes a dimensionless Narcotrend index, ranging from 100 (awake) to 0 (electrical silence) [38]. When compared to the BIS, the Narcotrend technology seems to perform better during the AE phase as it regains its baseline value upon discontinuation of the drug effect [39].

5.4 Conox®

The Conox® device (Fresenius Kabi, Bad Homburg, Germany) measures the DoA status and the patient's reaction to external stimuli through two derived indices: qCON and qNOX. In particular, the qCON index is derived from spectral analysis from a single-channel EEG and BS rate, processed through an artificial neural network and fuzzy logic system. The BS analysis is of paramount importance as burst intervals can be interpreted as the patient's awakening.

The qCON index is a measure of hypnosis and correlates to the patient's level of consciousness. Because the second index (qNOX) expresses the probability of response to a noxious stimulus, it correlates to the intraoperative analgesia [40]. Both indexes range from 0 to 99 (qCON: 0–39 deep anesthesia; 40–60 adequate anesthesia; 61–79 sedation or light anesthesia; 80–99 awake status. pNOX: 0–39 very low probability for the patient to respond to noxious stimuli; 40–60 patients unlikely to respond to noxious stimuli; 61–99 patients likely to respond to noxious stimuli). The device also displays EMG and BS data. The latter is expressed as a BS ratio (BSR) which is calculated as a ratio between the duration of suppression and the duration of bursting.

5.5 Patient State Index

A great limitation of different DoA indices lies in their neurophysiological basis. They assume, indeed, that anesthetics induce slowing of the EEG oscillations with increasing doses. Thus, slower oscillations should clearly correspond to a more profound state of GA. However, not all anesthetics show this EEG pattern. For example, ketamine shifts the alpha peaks of bicoherence induced by propofol to higher frequencies, perhaps through modulation of the nonlinear subcortical reverberating network [41]. Furthermore, NO suppresses low-frequency power, which can influence EEG monitoring parameters by increasing the value of the indices at clinically accepted doses [42].

From these premises, devices have been developed to try to overcome these gaps. Among these monitors, Physiometrix Inc. introduced, in 2002, the PSA4000. The device indicates the Patient State Index (PSI™), which is a dimensionless number suggestive for the level of unconsciousness. The value is calculated by a high-resolution four-channel EEG device that elaborates information on the frequency and phases of brain electric activity according to anterior-posterior relationships in the brain. The algorithm guides the analysis and processing of specific frequency components between the frontal cortex and the posterior lobes. Moreover, the algorithm analyses coherence between bilateral brain regions. Finally, BS data and plausibility analysis are applied for final index derivation [43].

The PSI™ delivers a continuous numeric value derived from systematic studies based on alterations in brain state observed in loss and return of consciousness, independently from the anesthetic drug used. As a consequence, the variables adopted for incorporation in the index show high heterogeneity of variance at different levels of sedation/hypnosis (sensitivity) but nonsignificant differences across anesthetic agents at a specified level (specificity) [44]. PSI™ scale consists of 0–100, with an optimal depth between 25 and 50. Devices (e.g., SedLine®; Masimo) offers also other details. The Density Spectral Array (DSA), for instance, represents EEG power and provides in a graphical way data on bihemispheric activity including asymmetry.

Comparative studies showed that the PSI offered a better correlation with the ETAC of desflurane. Furthermore, investigations found that PSI values were less affected by electrical interference, mostly those induced by electrocautery. In terms of anesthesia stages, although the PSI showed better performance during both induction and AE, in course of maintenance the PSI values were equivalent to those of the BIS [45, 46].

5.6 Wavelet Analysis

In the early years of the 2000s, Zikov et al. [47] studied a wavelet-based decomposition of the EEG signal, associated with a statistical function for information extraction. The result of their job was the NeuroSENSE Monitor (NeuroWave Systems Inc., Cleveland Heights, OH). The device provides the WAVcns index (ranging from 1 to 100) which is obtained via wavelet analysis of the EEG signals (from a single-channel frontal) in the gamma frequency band. The wavelet coefficients calculated from the EEG are pooled into a biostatistical analysis. The processing provides a further comparison of the value to two well-defined states: the awake state with normal EEG activity and the isoelectric state with maximal cortical depression.

5.7 The SNAP II™ Monitor

The SNAP II™ Monitor (Stryker, Inc., Kalamazoo, MI) is another device adopted for brain monitoring of GA. It displays an index that varies from 100 (fully awake state) to 0 (no brain activity). The algorithm of this instrument encompasses the processing of low frequencies (0 to 18 Hz) and ultrahigh EEG frequencies (80 to 420 Hz) (high-frequency index), which are involved in the formation of consciousness, and ignores contaminated frequencies from 40 to 80 Hz. The use of these frequencies can be helpful for ameliorating responsiveness during increased brain activity. Thus, the device should be useful in evaluating transition phases between the awake and anesthetized states, and vice versa (e.g., for optimizing emergence and RoC). Concerning its efficacy, Ruiz-Gimeno et al. [48] indicated that SNAP index ranging from 58 to 70 is probably equivalent to a targeted BIS index value ranging from 40 to 60. The SNAP-II index has been also used in mechanically ventilated pediatric patients for assessing the level of sedation [49].

5.8 Evoked Potentials

Apart from the pEEG-based brain monitoring devices, another group of instruments works on acoustic evoked potentials (AEPs). In contrast to spontaneous potentials detected by EEG, evoked potentials (also termed as an evoked response) are electrical potentials recorded following the presentation of a stimulus such as repetitive auditory inputs [50] or visual stimuli, and evaluate the integrity of the neural pathways that carry information from the periphery (cochlea) to the cortex. When the auditory pathway is testing (by auditory clicks) the AEPs are recorded and investigated. The auditory stimulation produces a waveform which is the summary of 11 EEG waves. Depending on the CNS site responsible for their production, these waves are divided into three groups:

brainstem AEPs (BAEPs), middle-latency AEPs (MLAEPs), and long-latency AEPs (LLAEPs). In particular, the BAEPs correspond to the way of transduction of the stimulus in the brainstem, from the acoustic nerve to the cochlear nucleus, and thus to the medial geniculate body via superior olivary complex, lateral lemniscus, and inferior colliculus. These potentials generate within 10 ms after the stimulus. The MLAEPs (also defined as early cortical potentials) generates from the medial geniculate body to the primary auditory cortex within 100 ms after stimulation. Finally, the LLAEPs are the expression of the neuronal activity of the frontal cortex and association areas. These components manifest a different degree of impairment during anesthesia and depending on the anesthetic agents used. For instance, BAEPs peaks may be modified in course of inhaled anesthesia, but not when injected anesthetics are used. On the contrary, because MLAEPs show more predictable variations with lesser intraindividual and interindividual changes, this subtype of AEPs is used for DoA measures.

The first commercial monitor based on AEP was developed by Danmeter, in 2001. Compared to the raw EEG, AEPs are less sensitive to artifacts that are random in time of occurrence. Functionally, the AEP response to increased anesthetic concentrations (a deeper DoA status) is characterized by is increased signal latency and decreased amplitude. In particular, the AEP monitoring technique isolates the neurophysiological signal generated during stimulation of the eighth cranial nerve using a repetitive auditory stimulus (e.g., with a bilateral click stimulus of 70 dB intensity and 2 ms duration delivered through headphones (in: Alaris AEP Monitor; Alaris Medical Systems Inc., San Diego, CA, USA). Repeated sampling allows the signal to be extrapolated from the background EEG noise.

The AEP Monitor/2™ (Danmeter A/S, Odense, Denmark) arranges a dimensionless number, the composite AEPs index (A-line autoregressive index (cAAI)). Newest AEP technologies elaborate AEPs from EEG signals. For instance, the AEP Index (aepEX) monitor (Medical Device Management Ltd., Essex, UK) that measures hypnotic depth during anesthesia by the means of direct EEG-derived values, obtained from both cortical EEG and AEP data [51]. This composite index represents a picture of the overall balance between anesthesia status, and control of noxious stimulation with analgesia. In contrast to the conceptualization of pEEG-based monitors, the transition from asleep to awake is characterized by a sudden increase in the AEP Index.

The AEP used for index construction has a latency of 144 ms and contains 256 samples. The formula is as follows:

$$\text{AEP Index} = k \cdot \sum_{i=1}^{N-1} \sqrt{|\, xi - x_{i+1} \,|}$$

There is also the derived A-line ARX Index with derives from an integrate correction for BAEP, LLAEP, and muscular response.

Other devices commercialized for brain monitoring during anesthesia are the Index of Consciousness (IoC) monitor (Morpheus Medical Company) which uses an algorithm designed through the symbol dynamics analysis (a mathematical approach used in data storage and transmission which divides events such as EEG into a finite number of elements and assigns a symbol to each partition), and the Cerebral State Monitor™ (CSM), introduced by the Danmeter Company, in 2004. The CSM's algorithm derives by the EEG-algorithm of its predecessor AEP-Monitor/2 (it also uses the same electrodes). CSM monitor (model CSM 2) also performs BS analysis and displays an index (Cerebral State Index, CSI), a signal quality information (SQI), and EMG activity. Interestingly, the CSI (0–100, where 0 indicates a flat EEG and 100 corresponds to the awake state; the surgical anesthesia is between 40 and 60) is obtained through a fuzzy analysis which elaborates EEG subparameters (α-ratio, β-ratio, and αβ differences) and BS interference.

6   Limitations and Perspectives

After its commercialization, the BIS device was intended as the best instrument for monitoring cerebral activity during GA. The availability of this technology was enthusiastically considered as the discovery of the "Holy Grail" of anesthesia monitoring and the definitive solution of one among the major concerns of anesthesia: the AAGA complication. As a consequence, in 2007, half of all American operating rooms had a BIS monitor. Unfortunately, doubts began to arise. In 2008, Avidan et al. [52] reported the same incidence of AAGA, independently from BIS monitoring. After two decades of investigations on the topic, the scientific community agrees that the use of the brain monitoring must be done with the "awareness" of its limits. In 2006, the task force on intraoperative awareness of the American Society of Anesthesiologists concluded that "(instrumental brain monitoring) *is not routinely indicated for patients undergoing general anesthesia, either to reduce the frequency of intraoperative awareness or to monitor the depth of anesthesia.*" Again, "*the decision to use a brain function monitor should be made on a case-by-case basis by the individual practitioner for the selected patient*" [53]. More recently, the Australian and New Zealand College of Anaesthetists (ANZCA) ["*Monitoring of anaesthetic effect on the brain—When clinically indicated, equipment to monitor the anesthetic effect on the brain should be available for use on patients, especially those at high risk of awareness, during general anesthesia*"] [54] have issued similar recommendations, whereas the Association of Anaesthetists of

Great Britain and Ireland recommended the use of DoA devices in case of total intravenous anesthesia and neuromuscular-blocking agents (NMBAs) administration [55]. Furthermore, the European Society of Anaesthesiology recommended brain monitoring in the elderly patient to prevent the risk of POD (Grade A) [56]. Finally, DoA devices can be helpful in particular clinical settings, for instance in the case of patients undergoing deep sedation [57, 58].

6.1 Limitations

Although these devices are equipped with algorithms for overcoming interferences such as eye movements and muscle activity, several patient-dependent variables such as age, neurological disorders (e.g., brain ischemia, seizures), medications (e.g., psychoactive drugs), and pathophysiologic states such as hypothermia, hypoglycemia, acid-base abnormalities, as well as external factors (e.g., poor skin contact, mains or power line interference, electrocautery) may still affect the validity of the instrumental brain monitoring. Among medications, the use of neuromuscular relaxants may also interfere with the readability of brain monitoring devices. For instance, Nunes et al. [59] demonstrated significant reductions in several indices of DoA, such as RE, SE, and CSI, in volunteers after the use of succinylcholine [60]. Again, investigations conducted in awake volunteers who did not receive any anesthetic, demonstrated lower BIS values (suggestive for a surgical anesthesia status) after administration of a single dose of muscle relaxant [61]. Thus, despite the refinement of algorithms for improving devices already in use, and the development of new technologies, right now it is impossible to abolish the influence of all confounding variables [62].

Furthermore, a paramount limitation of DoA monitors concerns their functional bases and, in turn, the impossibility of application to all anesthesia regimens. A significant issue, indeed, is that different anesthetics act at different molecular targets, impact different neural circuits, and, in turn, induce different brain states. Because the neurophysiological presupposition predicts that anesthetics induce slowing of the EEG oscillations with increasing doses, slower oscillations should clearly indicate a more profound state of GA. However, not all general anesthetics work by following this pattern. The potential combination of different general anesthetics, or through general anesthetics and other drugs (e.g., opioids) complicate the matter. For example, ketamine shifts the alpha peaks of bicoherence induced by propofol to higher frequencies, by interfering on the subcortical reverberating network [43]. Ketamine, moreover, has special features. This drug acts on several receptors (e.g., NMDA and the hyperpolarization-activated cyclic nucleotide-gated potassium channel 1, HCN1, receptor), influencing in a positive and negative manner different neuronal pathways (including cholinergic, aminergic, and opioid systems), has no effects on AEPs [63]. The gas NO suppresses low-frequency power, which can influence EEG monitoring parameters by increasing the value of the indices at clinically accepted doses of anesthetics [42]. On the contrary, dexmedetomidine can produce profound slow EEG oscillations and low index values despite the patient can be easily aroused from what is a state of sedation rather than in a surgical DoA status [64]. Inhaled and intravenous anesthetics produce EEG signatures featuring oscillations in the raw EEG and its spectrogram.

6.2 Perspectives

Probably, the main gap of DoA technologies lies in their impossibility of discriminating with precision between consciousness and unconsciousness. Indeed, their functioning is based on EEG signal analysis rather than considerate the corticocortical connectivity and communication processes [65]. This is a complex topic encompassing (1) consciousness/unconsciousness transition and mechanisms, and (2) the operating mechanism of general anesthetics. Neuronal and functional mechanisms that regulate the switch from consciousness to unconsciousness still require investigation. Moreover, although research showed that the anesthetic effect is obtained through ligand-receptor interactions in different brain areas, a major question concerns the physiological mechanisms by which these medications interfere with cortical and subcortical signals and consciousness mechanisms. Research in this field involves different study groups. For instance, the Boly's investigations on the spectral EEG changes after propofol administration are very important [66], whereas Patrick Purdon and the Emory Brown's team published their fascinating research on the dynamics of the unconscious brain under GA, explaining how EEG patterns (and its changing) may be indicated in real-time of the patient transition from consciousness to the anesthesia status [67–69].

An interesting perspective is the closed-loop control system where a DoA monitor may work to automatically control the number of general anesthetics to be administered to the patient [70]. For this purpose, a proportional-integral-derivative (PID) controller (also termed as a three-term controller), which is a control loop feedback mathematical approach used in industrial control systems and in other applications in which a continuously modulated control is required, is used. Basically, the PID controller continuously calculates an error value ($e(t)$) like the difference between the desired setpoint and a measured process variable. Subsequently, the system applies a correction by proportional (p), integral (i), and derivative (d) operations and coefficients (K). Thus, the controller attempts to minimize the error over time ($u(t)$).

$$u(t) = K_p \, e(t) + K_i \int_0^t e(t')\mathrm{d}t' + Kd\frac{\mathrm{d}e(t)}{\mathrm{d}t}$$

In anesthesia, this modality is performed through systems (algorithms) developed for infusion of intravenous anesthetics. Of note, Rugloop© is a Windows-based Target Controlled Infusion (TCI) program developed by Tom de Smet and Michel Struys. It is an experimental tool for research and teaching purposes and encompasses a complete TCI model database for most common intravenous anesthetic agents. A large amount of clinical research is reported in the literature on the application of anesthesia in closed-loop control systems. Studies were performed through different algorithms. For instance, Liu et al. [71] published the results of a randomized multicenter study conducted on a closed-loop modality based on BIS-guided coadministration of propofol and remifentanil with an interface between the DoA monitor and the anesthetic infusor. Ngai Liu together with Nicolas Choussat and Thierry Chazot conducted several studies by using the EasyTIVA system (Medsteer) which uses a proprietary algorithm adaptable with various brain monitor systems available on the market. Since 2006, several investigations have been conducted in different clinical settings such as pediatric and adolescent populations [72] and obese patients [73]. Apart from the BIS, other DoA monitors have been used for this purpose. West et al. [74] adopted the NeuroSENSE Monitor, whereas an investigation on the closed-loop coadministration of propofol and remifentanil guided by qCon and qNox indexes is ongoing (NCT03540875).

Although a recent meta-analysis of a randomized clinical trial found that BIS-guided anesthetic delivery (for intravenous anesthesia) reduced propofol requirements during induction, stabilized the right DoA status, and improved recovery time, this interesting approach still has to go through rigorous testing, and further validation is needed [75].

To date, we can only indirectly evaluate the state of anesthesia (and consciousness) through EEG, AEP, and other parameters such as EMG, but the data they provide are not exhaustive and cannot allow for establishing with certainty the multiple functional changes occurring in the brain during anesthesia.

References

1. Hutt A, Hudetz AG (2015) Editorial: general anesthesia: from theory to experiments. Front Syst Neurosci 9:105
2. Gibbs FA, Gibbs EL, Lennox WG (1937) Effect on the electroencephalogram of certain drugs which influence nervous activity. Arch Intern Med (Chic) 60:154–166
3. Cascella M, Fiore M, Leone S, Carbone D, Di Napoli R (2019) Current controversies and future perspectives on treatment of intensive care unit delirium in adults. World J Crit Care Med 8(3):18–27
4. Cascella M (2015) Anesthesia awareness. Can midazolam attenuate or prevent memory consolidation on intraoperative awakening during general anesthesia without increasing the risk of postoperative delirium? Korean J Anesthesiol 68(2):200–202
5. Cascella M, Muzio MR, Bimonte S, Cuomo A, Jakobsson JG (2018) Postoperative delirium and postoperative cognitive dysfunction: updates in pathophysiology, potential translational approaches to clinical practice and

further research perspectives. Minerva Anestesiol 84(2):246–260

6. Chan MT, Cheng BC, Lee TM, Gin T, CODA Trial Group (2013) BIS-guided anesthesia decreases postoperative delirium and cognitive decline. J Neurosurg Anesthesiol 25:33–42

7. Cascella M, Bimonte S, Muzio MR (2018) Towards a better understanding of anesthesia emergence mechanisms: research and clinical implications. World J Methodol 8(2):9–16

8. Cascella M (2018) Emergence from anesthesia: a winding way back. Anaesthesiol Intensive Ther 50(2):168–169

9. Forman SA, Chin VA (2008) General anesthetics and molecular mechanisms of unconsciousness. Int Anesthesiol Clin 46(3):43–53

10. Lee U, Ku S, Noh G, Baek S, Choi B, Mashour GA (2013) Disruption of frontal-parietal communication by ketamine, propofol, and sevoflurane. Anesthesiology 118(6):1264–1275

11. Imas OA, Ropella KM, Ward BD, Wood JD, Hudetz AG (2005) Volatile anesthetics disrupt frontal-posterior recurrent information transfer at gamma frequencies in rat. Neurosci Lett 387(3):145–150

12. Murphy M, Bruno MA, Riedner BA et al (2011) Propofol anesthesia and sleep: a high-density EEG study. Sleep 34(3):283–291A

13. Scheib CM (2017) Brainstem influence on Thalamocortical oscillations during anesthesia emergence. Front Syst Neurosci 11:66

14. Lioudyno MI, Birch AM, Tanaka BS et al (2013) Shaker-related potassium channels in the central medial nucleus of the thalamus are important molecular targets for arousal suppression by volatile general anesthetics. J Neurosci 33(41):16310–16322

15. Marchant N, Sanders R, Sleigh J et al (2014) How electroencephalography serves the anesthesiologist. Clin EEG Neurosci 45(1):22–32

16. Yeom SK, Won DO, Chi SI et al (2017) Spatiotemporal dynamics of multimodal EEG-fNIRS signals in the loss and recovery of consciousness under sedation using midazolam and propofol. PLoS One 12(11):e0187743

17. Baria AT, Centeno MV, Ghantous ME, Chang PC, Procissi D, Apkarian AV (2018) BOLD temporal variability differentiates wakefulness from anesthesia-induced unconsciousness. J Neurophysiol 119(3):834–848

18. Hight DF, Voss LJ, García PS, Sleigh JW (2017) Electromyographic activation reveals cortical and sub-cortical dissociation during emergence from general anesthesia. J Clin Monit Comput 31(4):813–823

19. McCarthy MM, Brown EN, Kopell N (2008) Potential network mechanisms mediating electroencephalographic beta rhythm changes during propofol-induced paradoxical excitation. J Neurosci 28(50):13488–13504

20. Hight D, Voss LJ, Garcia PS, Sleigh J (2017) Changes in alpha frequency and power of the electroencephalogram during volatile-based general anesthesia. Front Syst Neurosci 11:36. https://doi.org/10.3389/fnsys.2017.00036

21. Amzica F (2015) What does burst suppression really mean? Epilepsy Behav 49:234–237

22. Kroeger D, Amzica F (2007) Hypersensitivity of the anesthesia-induced comatose brain. J Neurosci 27(39):10597–10607

23. Brown EN, Lydic R, Schiff ND (2010) General anesthesia, sleep, and coma. N Engl J Med 363 (27):2638–2650

24. Sloan TB (1998) Anesthetic effects on electrophysiologic recordings. J Clin Neurophysiol 15 (3):217–226

25. Purdon PL, Pierce ET, Mukamel EA (2013) Electroencephalogram signatures of loss and recovery of consciousness from propofol. Proc Natl Acad Sci U S A 110(12):E1142–E1151

26. Breshears JD, Roland JL, Sharma M et al (2010) Stable and dynamic cortical electrophysiology of induction and emergence with propofol anesthesia. Proc Natl Acad Sci U S A 107(49):21170–21175

27. Chander D, García PS, MacColl JN, Illing S, Sleigh JW (2014) Electroencephalographic variation during end maintenance and emergence from surgical anesthesia. PLoS One 9 (9):e106291

28. Liang Z, Huang C, Li Y et al (2018) Emergence EEG pattern classification in sevoflurane anesthesia. Physiol Meas 39(4):045006

29. Isley MR, Edmonds HL Jr, Stecker M, American Society of Neurophysiological Monitoring (2009) Guidelines for intraoperative neuromonitoring using raw (analog or digital waveforms) and quantitative electroencephalography: a position statement by the American Society of Neurophysiological Monitoring. J Clin Monit Comput 23:369–390

30. Castellanos NP, Makarov VA (2006) Recovering EEG brain signals: artifact suppression with wavelet enhanced independent component analysis. J Neurosci Methods 158 (2):300–312

31. Sorensen HV, Jones DL, Heideman MT, Burrus CS (1987) Real-valued fast Fourier transform algorithms. IEEE Trans Signal Process 35:849–863

32. Cooley JW, Tukey JW (1965) An algorithm for machine calculation of complex fourier series. Math Computation 19:297–301

33. Sigl JC, Chamoun NG (1994) An introduction to bispectral analysis for the electroencephalogram. J Clin Monit 10(6):392–404

34. Rampil IJ (1998) A primer for EEG signal processing in anesthesia. Anesthesiology 89(4):980–1002

35. Viertiö-Oja H, Maja V, Särkelä M (2004) Description of the entropy algorithm as applied in the Datex-Ohmeda S/5 entropy module. Acta Anaesthesiol Scand 48(2):154–161

36. Shannon CE (1948) A mathematical theory of communication. Bell System Techn J 27:623–656

37. Loomis AL, Harvey EN, Hobart CA (1937) Cerebral states during sleep as studied by human brain potentials. J Exp Psychol 21:127–144

38. Kreuer S, Wilhelm W (2006) The Narcotrend monitor. Best Pract Res Clin Anaesthesiol 20:111–119

39. Schmidt G, Bischoff P, Standl T et al (2003) Narcotrend and bispectral index monitor are superior to classic electroencephalographic parameters for the assessment of anesthetic states during propofol-remifentanil anesthesia. Anesthesiology 99(5):1072–1077

40. Jensen EW, Valencia JF, López A et al (2014) Monitoring hypnotic effect and nociception with two EEG-derived indices, qCON and qNOX, during general anaesthesia. Acta Anaesthesiol Scand 58:933–941

41. Ching S, Cimenser A, Purdon PL, Brown EN, Kopell NJ (2010) Thalamocortical model for a propofol-induced alpha-rhythm associated with loss of consciousness. Proc Natl Acad Sci U S A 107(52):22665–22670

42. Mashour GA (2014) Top-down mechanisms of anesthetic-induced unconsciousness. Front Syst Neurosci 8:115

43. Drover D, Ortega HR (2006) Patient state index. Best Pract Res Clin Anaesthesiol 20:121–128

44. Prichep LS, Gugino LD, John ER et al (2004) The patient state index as an indicator of the level of hypnosis under general anaesthesia. Br J Anaesth 92(3):393–939

45. Chen X, Tang J, White PF, Wender RH, Ma H, Sloninsky A, Kariger R (2002) A comparison of patient state index and bispectral index values during the perioperative period. Anesth Analg 95(6):1669–1674. table of contents

46. Schneider G, Mappes A, Neissendorfer A et al (2004) EEG-based indices of anaesthesia: correlation between bispectral index and patient state index? Eur J Anaesthesiol 21(1):6–12

47. Zikov T, Bibian S, Dumont GA, Huzmezan M, Ries CR (2006) Quantifying cortical activity during general anesthesia using wavelet analysis. IEEE Trans Biomed Eng 53:617–632

48. Ruiz-Gimeno P, Soro M, Pérez-Solaz A et al (2005) Comparison of the EEG-based SNAP index and the Bispectral (BIS) index during sevoflurane-nitrous oxide anaesthesia. J Clin Monit Comput 19(6):383–389

49. Nievas IF, Spentzas T, Bogue CW (2014) SNAP II index: an alternative to the COMFORT scale in assessing the level of sedation in mechanically ventilated pediatric patients. J Intensive Care Med 29(4):225–228

50. Plourde G (2006) Auditory evoked potentials. Best Pract Res Clin Anaesthesiol 20(1):129–139

51. Weber F, Zimmermann M, Bein T (2005) The impact of acoustic stimulation on the AEP monitor/2 derived composite auditory evoked potential index under awake and anesthetized conditions. Anesth Analg 101(2):435–439

52. Avidan MS, Zhang L, Burnside BA et al (2008) Anesthesia awareness and the bispectral index. N Engl J Med 358:1097–1108

53. American Society of Anesthesiologists Task Force on Intraoperative Awareness (2006) Practice advisory for intraoperative awareness and brain function monitoring: a report by the American Society of Anesthesiologists Task Force on Intraoperative Awareness. Anesthesiology 104(4):847–864

54. Australian and New Zealand College of Anaesthetists (ANZCA) Guidelines on monitoring during anaesthesia. Available at: http://www.anzca.edu.au/documents/ps18-2015-guidelineson-monitoring-during-anaesthe.pdf. Updated December 2015. Accessed 29 May 2019

55. Checketts MR, Alladi R, Ferguson K et al (2016) Association of Anaesthetists of Great Britain and Ireland. Recommendations for standards of monitoring during anaesthesia and recovery 2015: Association of Anaesthetists of Great Britain and Ireland. Anaesthesia 71:85–93

56. Aldecoa C, Bettelli G, Bilotta F et al (2017) European Society of Anaesthesiology evidence-based and consensus-based guideline on postoperative delirium. Eur J Anaesthesiol 34:192–214

57. Wang ZH, Chen H, Yang YL et al (2017) Bispectral index can reliably detect deep sedation in mechanically ventilated patients: a prospective multicenter validation study. Anesth Analg 125(1):176–183

58. Cascella M, Fusco R, Caliendo D (2017) Anesthetic dreaming, anesthesia awareness and patient satisfaction after deep sedation with propofol target controlled infusion: a

prospective cohort study of patients undergoing day case breast surgery. Oncotarget 8 (45):79248–79256

59. Nunes RR, Cavalcante SL (2007) Influência do bloqueio neuromuscular despolarizante nas entropias. São Paulo Med J 125(Suppl):126

60. Nunes RR, Porto VC, Miranda VT, de Andrade NQ, Carneiro LM (2012) Risk factor for intraoperative awareness. Rev Bras Anestesiol 62(3):365–374

61. Schuller PJ, Newell S, Strickland PA, Barry JJ (2015) Response of bispectral index to neuromuscular block in awake volunteers. Br J Anaesth 115(Suppl 1):i95

62. Cascella M, Bifulco F, Viscardi D, Tracey MC, Carbone D, Cuomo A (2016) Limitation in monitoring depth of anesthesia: a case report. J Anesth 30(2):345–348

63. Vereecke HE, Struys MM, Mortier EP (2003) A comparison of bispectral index and ARX-derived auditory evoked potential index in measuring the clinical interaction between ketamine and propafol anaesthesia. Anaesthesia 58:957–961

64. Akeju O, Pavone KJ, Westover MB et al (2014) A comparison of propofol- and dexmedetomidine-induced electroencephalogram dynamics using spectral and coherence analysis. Anesthesiology 121:978–989

65. Mashour GA, Pryor KO (2015) Consciousness, memory, and anesthesia. In: Miller RD (ed) Miller's Anesthesia, 8th edn. Churchill Livingstone, Philadelphia, pp 287–289

66. Boly M, Moran R, Murphy M et al (2012) Connectivity changes underlying spectral EEG changes during propofol-induced loss of consciousness. J Neurosci 32(20):7082–7090

67. Flores FJ, Hartnack KE, Fath AB et al (2017) Thalamocortical synchronization during induction and emergence from propofol-induced unconsciousness. Proc Natl Acad Sci U S A 114(32):E6660–E6668

68. Liu F, Stephen EP, Prerau MJ, Purdon PL (2019) Sparse multi-task inverse covariance estimation for connectivity analysis in EEG source space. Int IEEE EMBS Conf Neural Eng 2019:299–302

69. Brown EN, Purdon PL, Akeju O, An J (2018) Using EEG markers to make inferences about anaesthetic-induced altered states of arousal. Br J Anaesth 121(1):325–327

70. Jensen EW (2018) New findings and trends for depth of anesthesia monitoring. Korean J Anesthesiol 71(5):343–344

71. Liu N, Chazot T, Hamada S et al (2011) Closed-loop coadministration of propofol and remifentanil guided by bispectral index: a randomized multicenter study. Anesth Analg 112 (3):546–557

72. Orliaguet GA, Benabbes Lambert F, Chazot T et al (2015) Feasibility of closed-loop titration of propofol and remifentanil guided by the bispectral monitor in pediatric and adolescent patients: a prospective randomized study. Anesthesiology 122(4):759–767

73. Liu N, Lory C, Assenzo V et al (2015) Feasibility of closed-loop co-administration of propofol and remifentanil guided by the bispectral index in obese patients: a prospective cohort comparison. Br J Anaesth 114(4):605–614

74. West N, van Heusden K, Görges M et al (2018) Design and evaluation of a closed-loop anesthesia system with robust control and safety system. Anesth Analg 127(4):883–894

75. Pasin L, Nardelli P, Pintaudi M et al (2018) Closed-loop delivery systems versus manually controlled administration of total IV anesthesia: a meta-analysis of randomized clinical trials. Anesth Analg 124(2):456–464

Chapter 3

Perioperative Monitoring of Autonomic Nervous Activity

Theodoros Aslanidis

Abstract

The role of autonomous nervous system (ANS) is becoming more evident as accumulating information about human physiology arises. Therefore, monitoring its activity during periods of abrupt changes, like those in perioperative settings, becomes essential for understanding, predicting, and managing their effects. There are several methods for measuring ANS, although none of them is yet broadly applied in perioperative period.

Electrodermal activity tracks autonomic nervous activity via skin electrical properties. *Heart rate variability* and *surgical stress index* integrate mainly cardiovascular monitoring, *digital pupillometry* records pupil's dynamics; other methods use laboratory measurements. No matter the focus of each method, careful consideration for its limitation should be made before its use. Summarizing, it seems that, instead of searching for the "gold-standard," a combination of methods should be used.

The present chapter presents selected tools used for perioperative autonomic nervous activity monitoring. Future research perspectives are also offered.

Key words Autonomic nervous activity, Perioperative monitoring, Pupillometry

1   Introduction

Autonomous nervous system (ANS) is integrated in almost every organ. Monitoring of its activity becomes essential during perioperative period, as a plethora of stimulus and preexisting conditions can provoke tremendous abrupt changes. Moreover, numerous reports about negative effects of perioperative stress (i.e., high sympathetic tone) exist [1–5].

Stating otherwise, ANS monitoring is evaluation of stress-induced physiological changes.

There are several methods for measuring ANS, although none of them is yet broadly applied in perioperative period. Moreover, none can be considered as a "gold standard." Methods for ANS monitoring are complementary one to other; thus, a combination of them is often used in practice. In addition, indirect information about ANS activity can be obtained nearly by any monitor. And to further complicate things, ANS activity can be influenced by a

Marco Cascella (ed.), *General Anesthesia Research*, Neuromethods, vol. 150, https://doi.org/10.1007/978-1-4939-9891-3_3,
© Springer Science+Business Media, LLC, part of Springer Nature 2020

variety of factors: age, gender, ethnicity, drugs, and preexisting health status. The present chapter will focus on selected tools for perioperative ANS activity monitoring including *electrodermal activity of the skin, surgical plethysmographic (or stress) index, salivary biomarkers, digital pupillometry, and heart rate variability.* The physiological concept, the monitoring technique, and notes of each method are presented.

2   Selected Methods for Autonomous Nervous System Monitoring

2.1   Electrodermal Activity of the Skin (EDA)

2.1.1   Physiological Basis

Sweating is mainly a thermoregulatory mechanism. The output sweat is an electrolyte water solution. It consists primarily of sodium and chloride, though potassium, calcium, urea, lactate, and amino acids can also be found [6–8].

Palm and plantar sweating however is known to be independent of the ambient temperature, and it is elicited by emotional (fear, pleasure, agitation), physiological (inspiratory gasp, tactile stimulation, movements), and stressful (mental exercises) stimuli [6–8]. The secretory part of palm glands is innervated solely via the sympathetic branch of the ANS, which reaches the dermal part. In addition, the postganglionic synapse is cholinergic, having acetylcholine as synaptic transmitter [7–9]. Findings concerning the central innervations of sweat glands activity point to several centers, located at different levels of the CNS, and partly independent of one another [6, 8, 10]. Thus, through measuring of the electrical properties of the output (i.e., sweat) of the palm sweat glands, electrodermal activity (EDA) provides a simple gauge of the level and the extent of sympathetic activity (SNS).

Although EDA reports in the past were coming mostly from the fields of psychophysiology and neurology, recently EDA monitoring is also applied in perioperative setting.

2.1.2   Terminology

Currently, the term EDA is considered a very general term. It includes all electrical properties such as conductance (SC), resistance (SR), potentials (SP), impedance (SZ), and admittance (SY) which can be traced back to the skin [8]. The most often used EDA method is SC, measured in μS (micro-Siemens) or μΩ (micro-Ohm). The SC changes have usually two components: one with longer (tonic) and the other with shorter (phasic) time course. The tonic component relates to the slower acting components and background characteristics of the signal, including the overall level, the slow climbing, and the slow declinations over time. The most common measure of this component is the skin conductance level (SCL) because changes in the SCL are thought to reflect general changes in autonomic arousal [11–13]. The faster component is the skin conductance response (SCR or more generally EDR) and can be either stimulus/event related (ER-SCR) or not specific

(NS-SCR). Furthermore, every SCR signal has its own characteristics: amplitude, latency period (i.e., time interval between stimulus and onset of SC change), number of fluctuation of SC (NFSC) for given time, rise rate and time, decrease rate and time (especially half-recovery time = interval from peak SC to 50% decrease) and selected areas under the curve (AUC) of the signal [8, 11–13]. Other terms include galvanic skin response (GSR) and psychogalvanic reflex (PSR)-used in the past for SCR, skin potential response (SPR) and skin potential level (SPL). The latter two are method-specific terms (*see* below), their units are in µV (microvolt) and they are further divided to uniphasic or biphasic SPR, and sudorific (originated from sweat glands) or nonsudorific (associated with dermal and epidermal tissues) SPL [11].

2.1.3 Techniques for EDA Monitoring

The methods for EDA monitoring are mainly two types: endosomatic and exosomatic methods. While the former is seldom used, details are also given.

Endosomatic Method

Endosomatic method detects EDA changes between two measuring electrodes (active/inactive) along palmar or plantar skin without the application of external current. Skin potentials are measured in µV. Low drift Ag/AgCl electrodes are recommended. Preferred recordings sites are the glabrous skin of palms and the soles (*see* first paragraph of Subheading "Endosomatic Method"). Pretreatment of the site of the inactive electrode—placed in the volar/inner surface of the forearm or over the ankles—is needed [11, 13, 14]. Slight abrasion of the stratum corneum beneath the inactive electrode to lessen the difference in potential between the site and the body core (skin resistance is reduced from about 1 MΩ to 100 kΩ) [6, 14] is performed. Skin drilling methods were used in the past for that purpose. The site was first cleaned with antiseptic marked and then a dental spherical burr, covered with industrial diamonds was held steady and applied over the skin 3–4 times, until a gentle depression was made on the epidermis. Criteria for the right depth were visual inspection (slight glistering and cell fluid in the depression under bright light), a stinging sensation after swabbing with acetone, and the absolute potential difference between a pair of electrodes usually <1 mV and seldom <3 mV [14].

No pretreatment is needed for the active electrodes. They are fixed either to the volar phalanges of the fingers or to thenar and hypothenar sites on the palms of the nondominant hand. Alternatively, one of two sites are fixed at the inner aspect of the foot, over the abductor hallucis muscle adjacent to the sole of the foot and midway between the proximal phalanx of the big toe and a point directly beneath the ankle [13].

An ordinary bio-amplifier (although with appropriate input impedance $R_{input} >> R_{skin}$ and gain factor) may be used to detect SP responses.

SPRs can be both negative and positive (e.g., at the palms an increased negative potential or a less negative potential respectively) monophasic or biphasic. Triphasic SPRs have also been described in a condition in which the recovery from the "positive" component goes more negative than the peak of the initial negative wave.

In general, endosomatic recording may be considered as the most "physiological" measure of electrodermal activity, since no external current is applied. The latter also contributes to the absence of electrode polarization. Finally, skin potential is unaffected, thus no artifacts may be caused by variation in contact area (electrode/skin surface).

As limitations the SPL is highly sensitive to skin hydration temperature differences and the complexity of the SPRs creates difficulties in both analysis and interpretation. However, a major limitation of endosomatic method is its invasivity. Consequently, the method is not popular and it is not used in perioperative setting [8, 10, 13].

In order to evaluate the suitability of the electrodes, the Bias Voltage Test should be performed before their placement. The test is carried by placing the two electrodes are placed in direct face-to-face gel-to-gel contact, without skin and without applied direct current (DC) voltage. The measured DC voltage from the pair, consisting in the bias voltage or difference between the half-cell potentials, should be negligibly small with values <5 mV, except for endosomatic measurements, where it should be <1 mV [13].

Exosomatic Method

This approach measures EDA via application of external current to the skin: either DC or alternating current (AC).

In DC measurements, if voltage if kept constant (known as quasi-constant voltage method), EDA is recorded directly in SC units (μS), whereas if current is kept constant (quasi-constant current method) then SR units (Ohms (Ω)). Accordingly, in AC measurements, if effective voltage is kept constant, EDA is recorded as SZ, while SY results are used when the effective current is kept constant [8, 10].

Concerning technique and recording, the sensor system in both types (DC or AC) has at least two electrodes with equal properties, although in most commercially available systems it has three electrodes. Bias Voltage test can control the latter. The three-electrode system consists of a measuring electrode (M), a counter-current electrode (C), and a reference voltage electrode (R), which ensured a constant applied voltage across the stratum corneum beneath the M electrode. Currently, ready for use, commercially available, disposable nonpolarizing Ag/AgCl electrodes are favored. Disposable electrodes have some important advantages. They can be hygienic, hypoallergenic and antiseptic. They can have a good fixation system, be stored many months in an unopened package, be produced in large production runs with uniform electrical characteristics, and have a snap-action connector or, even better, be prewired with the connector remote from the electrode. Pre-gelled models have the additional advantage that the metal–electrolyte interface is stabilized and ready for use. In order to be EDA compatible the gel is containing chloride salt (NaCl or KCl), with NaCl being most often used. Preferred concentrations are between 0.050 and 0.075 molar (0.3%–0.4% by weight) because they approximate the concentration of NaCl in sweat that reaches the surface [10, 13].

In the hand, the preferred active sites are the thenar and hypothenar eminences and the medial and distal phalanges of the index and middle fingers (Fig. 1). Electrodes are fixed in site usually with adhesive tape.

The electrical circuits used to record and detect exosomatic responses, are more complex as compared with those used for endosomatic measures. As compared to endosomatic recording, the measures are less affected by electrode artifacts such as bias potential or drift, need reduced amplifier gain to detect the SC or SR responses, and are less sensitive to hydration effects. The vast majority of the devices use differential amplifiers, whereas in the past operational amplifiers were used. Known resistors (with 1% precision) in place of the participant allow calibration of the equipment. Resistors of 1 MW and 200 kW provide calibration conductances of 1 mS and 5 mS, respectively, which is the range for SCRs. An additional resistor of 50 kW provides 20-mS calibration, the approximate range of SCL. Ideally, these tests are conducted and recorded as the session begins but before electrodes are attached and the results retained to document the proper functioning of the equipment in that session. This record of calibration can be invaluable if one should discover that the equipment has not been working properly and need to determine which records are valid

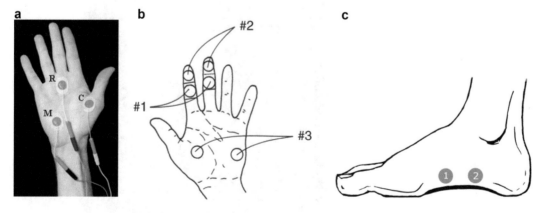

Fig. 1 Suggested measurement sites, (**a**) three-electrode system: thenar (current) and hypothenar (measurement) eminences and just below the middle finger (reference), (**b**) two-electrode systems: using either medial phalanges, either distal phalanges and either eminences (thenar and hypothenar), and (**c**) foot sites: inner side of the foot [6, 8, 13, 15]

[10, 13]. Possible presence of an external current may cause biased results. Hydration level of the skin can also affect recordings, yet in a lesser degree than in endosomatic recordings.

Responses coming from exosomatic measures always show a unidirectional (monophasic) course and therefore are easy to analyze. Apart from that, being the most studied methods, there is a larger knowledge database about their physiological correlates (*see* below).

Limitations and notes. Even though exosomatic technique is generally easy and not invasive, thorough knowledge of the device used for recording such as its type (e.g., DC or AC measurements), its capabilities, limitation of its use, is needed.

Most of the available data come from psychophysiology and neurology fields; thus, this can be both asset and drawback when using the method in intensive care units (ICU) or operation theatres (OR) (Fig. 2).

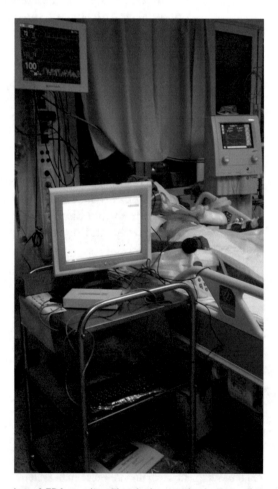

Fig. 2 Examples of EDA monitor (the device on the bottom of the image) use along with other ICU devices (personal archive)

Attention should be given to possible sources of bias in perioperative settings. For example, avoid placement of the electrodes in an edematous hand or in the same side where an arterial line is placed. Relatively strict patients' selection criteria should be used too. The latter is necessary in order to facilitate the analysis of the data. Thus, drug regiment (sedatives, inotropic agents, analgesics, etc.) should be carefully recorded, and health status (psychiatric or neurological diseases, neuropathy or myopathy, delirium, CNS or spinal cord injury, pregnancy) should be evaluated. Hemodynamic/respiratory instability and the presence of sensitive electrical life-sustainable devices such as cardiac pace, renal replacement therapy devices, intra-abdominal aortal counterpulsation pump, extracorporeal membrane oxygenation, and artificial liver are also important factors to consider [8].

Recording itself may be more demanding, since the environment in which it takes place (ICU, OR) is more dynamic.

Finally, while in psychophysiology, the existence of recommendations about analysis of the recording facilitates investigators [13], there are not enough data about, from perioperative settings [8].

2.2 Surgical Plethysmographic (or Stress) Index (SPI or SSI)

SPI is a multivariate index that is based on two cardiovascular parameters: the heart beat interval (HBI) and the pulse wave amplitude (PPGA). Information for these variables is delivered from the plethysmographic signal that is measured by a pulse oximeter. The raw time series data for heart rate and the plethysmographic amplitude are normalized and then a linear combination of the normalized values is computed as SPI according with the formula:

$$SPI = 100 - (0.7 * PPGAnorm + 0.3 * HBInorm)$$

where PPGAnorm is the normalized plethysmographic pulse wave amplitude and HBInorm the normalized heart beat interval [15–17].

First developed as intraoperative nociception index [16], SPI reflects a change of the autonomic nervous system balance in body. Indeed, the increase of the sympathetic activity increases SPI [17].

Concerning technique of recording, there is a commercially available SPI sensor, which serves also as SpO_2 monitor. The sensor is placed in the figure of the patient. There is additional preparation of the site, then that of the SpO_2 monitor (e.g., removal of any artificial nail colors). With the help of the appropriate embedded software in the main monitor, the observer gets a real time value of SPI and a trend over time [18]. The SPI is expressed in a scale 0–100, with values <30 to reflecting adequate and >60 inadequate analgesia respectively [15].

Yet SPI is to be used for unconscious and fully anesthetized adult (>18 years old) patients during general anaesthesia. In addition, SPI sensor is not available for all countries. For instance, it has not cleared for sales in USA [18].

In addition, several confounding factors for the SPI, including concomitant medication, volume status, pacemaker action, arrhythmias and posture changes, have been described [19].

Limitations and notes. Since SPI is solely based on hemodynamic measurement, its use in parallel with another method of ANS activity monitoring is recommended.

2.3 Salivary Biomarkers: Focus on a-Amylase (sAA)

With the development of point-of-care diagnostic technology, interest about saliva biomarkers had gained interest. For example, catecholamine can be identified in saliva ranging from 250 to 800 pg/mL^{-1}. However, it is unsuitable as an index for general sympathetic tone does, as it does not correlate well with serum levels [20]. Dihydroxyphenylglycol, a metabolite of catecholamine, shows better correlation with plasma level, yet its role is under research [21]. The same is valid also soluble fraction of receptor II of TNFα (sTNFαRII) and secretory IgA (sIgA) [22]. Lack of saliva/serum levels correlation has been reported for cortisol. In this case, the cause seems to be saliva 11 β-hydroxysteroid dehydrogenase which converts cortisol to cortisone [20]. Yet there are reports of saliva cortisol as potential diagnostic tools for detecting stress-induced cardiac diseases [23].

2.3.1 Physiological Basis of Salivary a-Amylase (sAA)

The enzyme sAA is secreted by the highly differentiated epithelial acinar cells of the exocrine salivary glands via activation, mainly, of beta-adrenergic receptors. Its release of salivary is regulated by autonomic innervations; thus many investigations viewed it as a measure of endogenous sympathetic activity [15]. To this regard, an association between sAA and plasma catecholamines was demonstrated [24, 25]. Generally, the physiological range of sAA in normal subjects is around 10 to 250 U mL^{-1} [26].

2.3.2 Material-Method of Recording

In general, saliva can be collected easily from human. However, the standardization of salivary collection has a great importance in saliva analysis because several factors may affect salivary flux and composition [27].

Though various strategies have been reported, the majority of them use stimulated saliva collection (i.e., need active cooperation of the patient) and may be unsuitable for perioperative setting. The most commonly used devices are sterile cotton dental rolls, as Salivette® (Sarsted, Newton, NC). The cotton roll is placed into the subject's mouth and is gently chewed for about 1–2 min, then placed into the vial. Saliva is obtained by expressing the saturated cotton using a needleless syringe or better by centrifugation [26, 27].

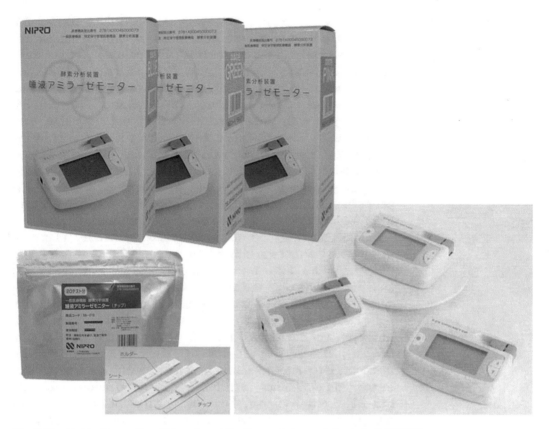

Fig. 3 Handheld sAA monitor. (Taken from http:/item.rakuten.co.jp/m-aoba/1437747)

More appropriate for perioperative setting are handheld monitors (e.g., COCORO meter salivary amylase monitor, NIPRO®, Osaka, Japan) which measure amylase activity in a batch type, using a disposable reagent test paper and an automated build in saliva transfer device (Fig. 3). The test paper is placed gentle under the tongue. Results are obtained usually within 60 s (displayed in built-in screen) and only 30 μL of saliva sample is needed [28]. With the latter method, the problem of sample storage is eliminated.

Limitations and notes. Several measurements are needed over time to create a reliable time trend. Thus, time of measurement should be strictly predefined in most cases, in regard to the time of the stimulus (e.g., surgical incision).

Apart from that, factors that affect sAA such as administered drugs, age, gender, and diurnal dynamics should also be noticed beforehand. Finally, special consideration should be applied in case of ear–nose–throat surgeries or in ICU patients where daily mouth care can affect the measurements [29–31].

Unfortunately, currently, there is no enough data available in literature about the effect range of the previous factors of sAA levels in perioperative setting. Future studies will eventually clarify the role of the method.

2.4   Digital Pupillometry

2.4.1   Physiological Basis

The size of the pupil of the eye is determined by the balance between the tone of two muscles, the constrictor (sphincter, the circular muscle) and the dilator (the radial muscle). Both of them are the cornerstone of the light reflex (PLR): the simultaneous and equal constriction of the pupil in response to illumination of one (direct) or the other eye (indirect or consensual). The reflex is multisynaptic with pulse speed 160 m/s and triggering time about 200 ms. The afferent pupillary pathway is controlled by the sympathetic ANS and originates in the retina [15, 32]. Hence, pupil's dynamics reflect the functional balance between sympathetic and parasympathetic ANS. In fact, in absence of injuries of the peripheral structures, pupillometry is an indirect estimation of functional status of ANS, especially at midbrain level [32].

Concerning method of recording, appropriate technique for perioperative settings is infrared digital pupillometry which is carried out by handheld portable pupillometers. There are several commercial devices of the kind and the basic concept is the same-automatic evaluation of pupil response to light. The latter is emitted when the device is held in front of the patient and centered in their eyes. The duration of emission is 100 ms, while the rest of the test parameters are manually selected. While with classical pupillometry the inter-rater variability is a problem, with infrared pupillometry intra-examiner variability can drop down to 1% [32–34]. Apart from that, the development of quantitative scale of pupil reaction (e.g., Neurological Pupil Index) allows a more rigorous and classification of pupil response. Others parameters can be recorded to create a more complete picture: the resting (maximum) and the minimum diameter, the percentage change (% Ch = maximum-minimum/maximum), the latency (time difference between of retinal light stimulation and onset of pupillary constriction), the average and maximum constriction velocity and the average dilatation velocity [32]. Other studies report more parameters, such as maximum redilatation velocity, the interval till the 75% of the redilatation ($t_{75\%}$), and the relative reflex amplitude [32].

These measurements allow not only the evaluation of pupil's diameter but also the neurophysiology of its reaction. For example, latency reflects the functionality of the nerve impulse transfer; the maximum constriction velocity reflects the parasympathetic ANS activation intense; the average dilation velocity reflects the sympathetic activation.

Limitations and notes. The technique of digital pupillometry is easy. Yet the investigator should have in mind that pupil size is affected by numerous of intrinsic and extrinsic factors. The intrinsic factors include stimulus variables such as light level, spectral composition, spatial configuration (field size, spatial structure of the field), accommodative state, monocular/binocular view and nonvisual stimuli (pain, noise). With prefigured automated devices

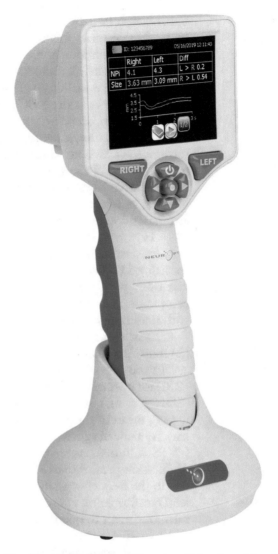

Fig. 4 Example of handheld digital pupillometer. (Taken from https:/neuroptics.com/npi-200-pupillometer/)

(Fig. 4) several of these problems are surpassed (light level, spectral composition, etc.), whereas others (e.g., nonvisual stimuli) remain. However, there are several reports in the literature about the application of digital pupillometry as perioperative analgesia index [35].

Extrinsic factors influencing the pupil size include individual differences, age, biomechanical cofactors like respirations and heart beats, day-to-day within observer variance and cognitive factors (arousal, attention, fear, workload, and hedonistic content) [32].

Thus, currently it is recommended that the method is to be used along with other monitoring methods [36].

2.5   Heart Rate Variability (HRV)

2.5.1   Physiological Concept

Variability in beat-by-beat heart period is an intrinsic characteristic of healthy cardiac functioning. ANS is the main determinant behind this phenomenon, which is further affected by several intrinsic and extrinsic factors.

Heart rate variability (HRV) is a term used for the interval between R waves in the ECG and is considered as complex reflection of the sympathetic-parasympathetic system balance activation (autonomic outflow), neuroendocrine influences and the ability of cardiovascular system to respond to the former factors (autonomic responsiveness) [15, 37].

It should always be kept in mind that, in reality, the R-R intervals variability (measured in msec) and HRV (i.e., a normalized estimation for given time measured in bpm) are not the same thing [38]. However, the terms are used interchangeably.

Concerning method of monitoring, recording is performed via specialized ECG monitors. Digitized ECG recordings of R-R intervals over given time are discarded from artifacts and premature ectopic beats and then analyzed in 2 ways: either with linear on with nonlinear methods.

Linear methods are mainly of two types: (1) time domain and (2) frequency analysis.

In the time domain, indexes of HRV are: mean R-R interval (RRi) for normal beats (in ms), standard deviation of all normal RRi (SDNN, in ms), standard deviation of the means of RRi (SDANN, in ms), mean of the 5-min standard deviations of RRi (SDNNi, in ms), root-mean square of differences between adjacent normal RRi (rMSSD, in ms), and percentage of adjacent RRi with a difference of duration greater than 50 ms (pNN50) [39].

In the frequency analysis, spectral power analysis via Fast Fourier Transformation is used to quantify heart rate oscillations (RRi) into four main frequency HRV components: The high-frequency (HF) (0.15–0.40 Hz), the low-frequency (LF) (0.04–0.15 Hz), the very low frequency (VLF) (0.0033–0.04 Hz), and the ultralow frequency (ULF) (<0.003 Hz) component. HF component usually reflect vagal activity, while LF component is considered to reflect both sympathetic (mainly) and parasympathetic activity. The ratio of high- and low-frequency oscillations LF/HF is assumed to reflect the absolute and relative changes between the sympathetic and parasympathetic components. The exact physiological mechanisms responsible for VLF and ULF are not yet established, but they may relate to the renin–angiotensin–aldosterone system, thermoregulation, and/or peripheral vasomotor tone [15, 37, 40].

Nonlinear HRV analysis methods (fractal dimension, entropy and complexity) have also been proposed, but literature about their use is still limited [15, 41].

Limitations and notes. HRV monitoring can be challenging in perioperative settings. As HRV deals with RR interval variations its measurement is limited to patients in sinus rhythm and to those with a low number of ectopic beats. In this sense, patients with frequent premature ventricular contractions, atrial arrhythmias (particularly atrial fibrillation) are excluded from any HRV analysis [37]. Besides that, intra-operative movement artifacts and electrical interference may complicate R- wave detection. No data also exist about application of the method during cardiothoracic operations (positioning of electrodes, interference, etc.).

Additional difficulties may be encountered also in data analysis. Time domain indexes can be used to assess autonomic nervous system activity; however, they cannot distinguish between changes in HRV due to increased sympathetic tone or withdrawal of vagal tone. Moreover, some indexes (SDNN, SDANN, and SDNNi indexes) require long-term recordings for accurate assessment, while others (rMSSD and pNN50) primarily quantify modulation of RRi driven by ventilation. On the other hand, spectral analysis can give important information about ANS cardiovascular control; yet care should be paid in the assumption that HF/LF ration reflect sympathovagal activity. Finally, when choosing nonlinear methods, fractal behavior of heart rate time series is assumed; though, it is difficult to assess the physiological meaning from obtained results and their relevance to cardiac autonomic control [38].

Despite the aforementioned difficulties, the method has been used (1) as intraoperative analgesia guide index [42–44], (2) as method of prediction of ANS dysfunction during anesthesia [15, 45], and (3) as prognostic index in critically ill trauma brain injury patients [46, 47].

2.6 Other Methods

Several other methods can also be considered as ANS monitors. Skin and muscle microneurography (SNA and MSNA), two methods used to visualize and record the normal traffic of nerve impulses in peripheral nerves; are used mainly as research tools of sympathetic nerve system [40].

Pulse rate variability (PRV) derived from continuous blood pressure measurements may provide a feasible alternative for HRV [48].

Few data are available on the perioperative applications of other methods such as quantitative sudomotor axon reflex test (QSART) [49, 50], resting sweating output (RSO) [51], or thermoregulatory sweat test (TST) [52].

3 Further Research Perspectives

Considering the aforementioned, the big challenges are ahead of us. No matter the method, currently, we are just recording measurements after stimuli. What is the best method for each setting (emergency department-ED, OR, ICU)? In which condition? In which patients? For a patient with ARDS, HRV may prove a better method in ED, while pupillometry should be used for the same patient when/if admitted to ICU. How then to compare the measurements before (HRV while in ED) and after (digital pupillometry in ICU). Is it better to use more than one method? When? Which combination? Why? We still do not know the answers. Apart from that data analysis of the recordings can be made by various ways? Which method is best in which condition? What information does every "way" give us? Moreover, for the time being (2019), we need to know/record the stimulus, in order to study its results on ANS activity. We are still in research of identifying stimulus-specific patterns, so that identifying the stimulus from the measurement. And certainly, we are far from connecting that measurement to the exact ANS activity: for example, for an EDA response, which part is due to the condition, which is due to emotional stress, which is due to pain, etc. In addition, what is the cutoff value for each method/condition/patient/setting for every factor (drugs, trauma) that can affect ANS activity? And we have no idea about the full predicting possibilities of those tools.

The list of questions is endless. Clinical implications: innumerable, given the fact that ANS is considered as the "life-sustaining" system. The former can be used not only for our better understanding of the pathophysiology of almost every disease (for example sepsis-related ANS dysfunction) but even as a tool to challenge fundamental definitions. The author of this chapter would like to focus on only one such definition: if ANS activity is used as part of the criteria for (brain) death, how will we define death after data from ANS activity monitors and, for example, functional neuroimaging (fMRI) are combined together? And what about sleep, anesthesia [53–55], locked-in syndrome?

4   Conclusion

Based on the variety of ANS monitoring methods and the aforementioned future possibilities, it becomes obvious that in order to assess a "globally integrated" system such as ANS, a multivariable monitor system is needed—a future device that tracks several parameters (e.g., skin conductance, HRV, SPI, and continuous sAA levels) at the same time.

On the other hand, further studies will clarify whether we need a "holistic" approach (and thus a complex multiparameter recording method) for perioperative ANS evaluation, or we should target to the organ/system of interest and spare the search for the golden standard perioperative monitor of the whole ANS activity.

References

1. Parker SD, Breslow MJ, Frank SM et al (1995) Catecholamine and cortisol responses to lower extremity revascularization: correlation with outcome variables. Perioperative Ischemia Randomized Anesthesia Trial Study Group. Crit Care Med 23:1954–1961

2. Myles PS, Hunt JO, Fletcher H et al (2002) Remifentanil, fentanyl and cardiac surgery: a double blinded, randomized, controlled trial of costs and outcomes. Anesth Analg 95:805–812

3. London MJ (2006) Beta-blockade in the perioperative period: where do we stand after all the trials? Semin Cardiothorac Vasc Anesth 10:17–23

4. Vanhorebeek I, Ingels C, Van den Berghe G (2006) Intensive insulin therapy in high-risk cardiac surgery patients: evidence from the Leuven randomized study. Semin Cardiothorac Vasc Anesth 18:309–316

5. Pud D, Amid A (2005) Anxiety as a predictor of pain magnitude following termination of first-trimester pregnancy. Pain Med 6:143–148

6. Bucsein W (2012) Electrodermal activity, 2nd edn. Springer Science + Business Media, New York, pp 2–31

7. Virtualmedicalcenter.com [Internet]: Virtual Medical Center; c2012: Sweating [Cited 2018 Dec 10]. Available from: http://www.virtualmedicalcentre.com/anatomy/sweating perspiration/75

8. Aslanidis T (2014) Electrodermal activity: applications in perioperative care. Int J Med Res Health Sci 3(3):687–695

9. Hall E, Lapworth R (2010) Use of sweat conductivity measurements. Ann Clin Biochem 47:390–391

10. Piacentini Rm (2001) Reaviling individual features in galvanic skin response signals. PhD. thesis. Reading University, Italy, pp 8–13

11. Christie M (1981) Electrodermal activity in the 80s: a review. J R Soc Med 41:616–622

12. Braithwaite J, Watson DG, Jones R et al (2013) Guide for analysing electrodermal activity (EDA) & skin conductance responses (SCRs) for psychological experiments. Technical Report: Selective Attention & Awareness Laboratory (SAAL) Behavioural Brain Sciences Centre, University of Birmingham, Birmingham. Available from: http://www.biopac.com/wp-content/uploads/EDA-SCR-Analysis.pdf. Accessed 20 Dec 2018

13. Boucsein W, Fowles DC, Grimnes S et al (2012) Society for Psychophysiological Research ad hoc Committee on Electrodermal measures. Publication recommendations for electrodermal measurements. Psychophysiology 49(8):1017–1034. https://doi.org/10.1111/j.1469-8986.2012.01384.x

14. Shackel B (1959) Skin-drilling: a method of diminishing galvanic skin-potentials. Am J Psychol 72(1):114–121

15. Aslanidis T (2015) Perspectives of autonomic nervous system perioperative monitoring–focus on selected tools. Int Arch Med 8:22):1–22):9

16. Huiku M, Uutela K, Van Gils M et al (2007) Assessment of surgical stress during general anaesthesia. Br J Anaesth 98(4):447–455

17. Haiku M, Kamppari L, Viertiö-Oja H (2016) Surgical pletysmographic index (SPI) in anesthesia practice. GE Healthcare White paper JB21042XX, 2014. Available from: http://www3.gehealthcare.co.uk/~/media/documents/us-global/products/clinical-consumables/brochures/general/spi%20whitepaperjb21042xx1nov191.pdf?Parent=%7B24E8DD90-59E0-4C23-B15C-1EFA3C7655A1%7D. Accessed 12 Dec 2018

18. Quick Guide Surgical Pleth Index. GE Healthcare (2014). Available from: http://www3.gehealthcare.co.uk/engb/products/categories/clinical_consumables/monitoring_solutions/spio2_sensors. Accessed 02 Jan 2017

19. Kang H (2015) Intraoperative nociception monitoring. Anesth Pain Med 10:227–234

20. Malon RSP, Sadir S, Balakrishnan M, Córcoles EP (2014) Saliva-based biosensors: noninvasive monitoring tool for clinical diagnostics. Biomed Res Int 2014:962903

21. Díaz Gómez MM, Bocanegra Jaramillo OL, Teixeira RR et al (2013) Salivary surrogates of plasma nitrite and Catecholamines during a 21-week training season in swimmers. PLoS One 8(5):e64043

22. Sobas EM, Reinoso R, Cuadrado-Asensio R et al (2016) Reliability of potential pain biomarkers in the saliva of healthy subjects: interindividual differences and intersession variability. PLoS One 11(12):e0166976

23. Cozma S, Dima-Cozma LC, Ghiciuc CM et al (2017) Salivary cortisol and α-amylase: subclinical indicators of stress as cardiometabolic risk. Braz J Med Biol Res 50(2):e5577

24. Nater UM, Rohleder N (2009) Salivary alpha-amylase as a non-invasive biomarker for the sympathetic nervous system: current state of research. Psychoneuroendocrinology 34(4):486–496

25. Thoma MV, Kirschbaum C, Wolf JM (2012) Acute stress responses in salivary alpha-amylase

predicts increases of plasma norepinephrine. Biol Psychol 91(3):342–348

26. Nater UM, Rohleder N, Gaab J et al (2005) Human salivary alpha-amylase reactivity in a psychosocial stress paradigm. Int J Psychophysiol 55(3):333–342

27. Chiappin S, Antonelli G, Gatti R (2007) Saliva specimen: a new laboratory tool for diagnostic and basic investigation. Clin Chim Acta 383 (1-2):30–40

28. Yamaguchi M (2010) Saliva sensors in point-of care testing. Sensor Materials 22(4):143–153

29. Bosch JA, Veerman EC, de Geus EJ (2011) α-Amylase as a reliable and convenient measure of sympathetic activity: don't start salivating just yet! Psychoneuroendocrinology 36 (4):449–453

30. Sahu GK, Upadhyay S, Panna SM (2014) Salivary alpha amylase activity in human beings of different age groups subjected to psychological stress. Reliable measure of SNS activity. Indian J Clin Biochem 29(4):485–490

31. Liu Y, Granger DA, Kim K (2017) Diurnal salivary alpha-amylase dynamics among dementia family caregivers. Health Psychol 36(2):160–168

32. Aslanidis T, Kontogounis G (2015) Perioperative digital pupillometry-the future? Greek E-Journal of Perioperative Medicine 13 (b):24–40

33. Rose D, Meeker M, Bacchetti P et al (2005) Evaluation of the portable infrared pupillometer. Neurosurgery 57:198–203

34. Taylor WR, Chen JW, Meltzer H et al (2003) Quantitative pupillometry, a new technology: normative data and preliminary observations in patients with acute head injury. J Neurosurgery 98:205–213

35. Abad-Torrent A, Sueiras-Gil A, Martínez-Vilalta M (2017) Monitoring of the intraoperative analgesia by pupillometry during laparoscopic splenectomy for splenic hydatid cyst. J Clin Anesth 36:94–97

36. Zafar SF, Suarez JI (2014) Automated pupillometer for monitoring the critically ill patient: a critical appraisal. J Crit Care 29(4):599–603

37. Deschamps Al DA (2007) Analysis of heart rate and blood pressure variability to assess autonomic reserves: its role in anesthesiology. Anesthesiology rounds 6(2):1–6.7

38. Draghici AD, Taylor JA (2016) The physiological basis and measurement of heart rate variability in humans. J Physiol Anthropol 235:22. https://doi.org/10.1186/s40101-016-0113-7

39. Kleiger RE, Stein PK, Bosner MS et al (1992) Time domain measurements of heart rate variability. Cardiol Clin 10:487–498

40. Montano N, Ruscone TG, Porta A et al (1994) Power spectrum analysis of heart rate variability to assess the changes in sympathovagal balance during graded orthostatic tilt. Circulation 90 (4):1826–1831

41. Chi X, Zhou J, Shi P et al (2016) Analysis methods of short-term non-linear heart rate variability and their application in clinical medicine. Sheng Wu Yi Xue Gong Cheng Xue Za Zhi 33(1):193–200

42. Nishiyama T (2016) Changes in heart rate variability during anaesthesia induction using sevoflurane or isoflurane with nitrous oxide. Anaesthesiol Intensive Ther 48(4):248–251

43. Boselli E, Logier R, Bouvet L et al (2016) Prediction of hemodynamic reactivity using dynamic variations of analgesia/nociception index (ΔANI). J Clin Monit Comput 30 (6):977–984

44. Ogawa Y, Kamijo YI, Ikegawa S et al (2017) Effects of postural change from supine to head-up tilt on the skin sympathetic nerve activity component synchronised with the cardiac cycle in warmed men. J Physiol 595 (4):1185–1200

45. Hanss R, Renner J, Ilies C et al (2008) Does heart rate variability predict hypotension and bradycardia after induction of general anaesthesia in high risk cardiovascular patients? Anaesthesia 63(2):129–135

46. Biswas AK, Scott WA, Sommerauer JF (2000) Heart rate variability after acute traumatic brain injury in children. Crit Care Med 28:3907–3912

47. Jurak P, Halamek V, Leinveber P et al (2012) Respiratory induced heart rate and blood pressure variability during mechanical ventilation in critically ill and brain death patients. Conf Proc IEEE Eng Med Biol Soc 2012:3821–3824

48. Iozzia L, Cerina L, Mainardi L (2016) Relationships between heart-rate variability and pulse-rate variability obtained from video-PPG signal using ZCA. Physiol Meas 37 (11):1934–1944

49. Siepmann T, Illigens BM, Reichmann H et al (2014) Axon-reflex based nerve fiber function assessment in the detection of autonomic neuropathy. Nervenarzt 85(10):1309–1314

50. Novak P (2011) Quantitative autonomic testing. J Vis Exp 19(53). pii 2502

51. Kleggetveit IP, Skulberg PK, Jørum E (2016) Complex regional pain syndrome following viper-bite. Scand J Pain 10:15–18

52. Biehle-Hulette SJ, Krailler JM, Elstun LT et al (2014) Thermoregulatory vs. event sweating--comparison of clinical methodologies, physiology and results. Int J Cosmet Sci 36(1):102–108

53. Cascella M, Bimonte S, Muzio MR (2018) Towards a better understanding of anesthesia emergence mechanisms: research and clinical implications. World J Methodol 8(2):9–16

54. Cascella M, Fusco R, Caliendo D et al (2017) Anesthetic dreaming, anesthesia awareness and patient satisfaction after deep sedation with propofol target controlled infusion: a prospective cohort study of patients undergoing day case breast surgery. Oncotarget 8(45):79248–79256

55. Cascella M (2016) Mechanisms underlying brain monitoring during anesthesia: limitations, possible improvements, and perspectives. Korean J Anesthesiol 69(2):113–120

Chapter 4

Monitoring Cerebral Oximetry by Near-Infrared Spectroscopy (NIRS) in Anesthesia and Critical Care: Progress and Perspectives

Antonio Pisano, Diana Di Fraja, and Concetta Palmieri

Abstract

Noninvasive estimation of cerebral regional oxygen saturation ($rScO_2$) by means of near-infrared spectroscopy (NIRS), first described more than 40 years ago, is currently commonly used as a cerebral and, more generally, hemodynamic monitoring tool in cardiovascular surgery and neonatal intensive care unit, and in recent years is spreading to other clinical settings in which brain injury and cognitive dysfunction are a major concern, such as interventional neuroradiology procedures, noncardiac surgery in the beach chair position or in high-risk (e.g., elderly) patients, cardiac arrest, and mechanical circulatory support. However, there is no agreement among clinicians about the usefulness and reliability of cerebral NIRS monitoring and, accordingly, its use in clinical practice varies widely worldwide. This is primarily due to the substantial lack of evidence showing improved outcomes with NIRS-guided management, combined with some limitations of the methodology such as the differences among the various commercially available devices and the lack of well-defined reference values and of clinically relevant thresholds for desaturations. In this chapter, we discuss the basic principles and the common clinical uses of cerebral oximetry and review the main evidence about the impact of NIRS-guided management on clinically relevant outcomes, in order to analyze the reasons that hinder a wider dissemination of a potentially useful monitoring tool and, accordingly, to outline the possible direction for future research.

Key words Near-infrared spectroscopy (NIRS), Cerebral oximetry, Cardiac surgery, Anesthesia, Cardiopulmonary bypass (CPB), Monitoring, intraoperative, Intraoperative neurophysiological monitoring, Postoperative cognitive dysfunction (POCD), Postoperative complications

1 Introduction

The use of near-infrared spectroscopy (NIRS) for noninvasive estimation of cerebral regional oxygen saturation ($rScO_2$) has been first described more than 40 years ago [1], and afterwards has had some spread in clinical practice as a cerebral and, more generally, hemodynamic monitoring tool, mainly in cardiovascular surgery [2–4] and in the neonatal intensive care setting [5–8]. By measuring "mixed" oxygen saturation in a small and shallow area of brain

Marco Cascella (ed.), *General Anesthesia Research*, Neuromethods, vol. 150, https://doi.org/10.1007/978-1-4939-9891-3_4,
© Springer Science+Business Media, LLC, part of Springer Nature 2020

(*see* below), cerebral oximetry has the potential to detect a mismatch between oxygen supply and consumption, which can be due to either "cerebral" causes (e.g., carotid artery occlusion, displacement of cardiopulmonary bypass cannulae, inadequate anesthesia, epilepsy) or "systemic" causes (e.g., low cardiac output or cardiopulmonary bypass flow, low perfusion pressure, hypoxia, anemia, hypocapnia) [3, 9, 10]. Accordingly, $rScO_2$ monitoring is particularly attractive in cardiac and vascular surgery, where the risk of postoperative neurological complications such as stroke, delirium, and postoperative cognitive dysfunction (POCD) is still high [11, 12], as well as in the neonatal intensive care unit (NICU), where neurocognitive outcome is also a key issue, the possibility of cerebral and hemodynamic monitoring is limited, and the brain of neonates is more easily probed by NIRS than in adults due to the thinner scalp and skull [2, 7, 8, 13]. In recent years, the use of cerebral oximetry is also spreading to other clinical settings in which brain injury and cognitive dysfunction may represent a major concern, such as interventional neuroradiology procedures [14], noncardiac surgery in the beach chair position (BCP) or in high-risk (e.g., elderly) patients [3], and adult intensive care unit (ICU), particularly during cardiac arrest [3, 15, 16].

However, there is no agreement among clinicians about the usefulness and reliability of cerebral NIRS monitoring and, accordingly, its use in clinical practice varies widely worldwide from centers routinely including cerebral oximetry among the monitoring tools on which to base therapeutic decisions to centers not using it at all [2–4, 17–21]. This is primarily due to the substantial lack of evidence showing improved outcomes with NIRS-guided management, combined with some technological and "pathophysiological" issues (e.g., the poor agreement among the different commercially available devices [22–24] and the lack of well-defined reference values and of clinically relevant thresholds for desaturations [2, 3, 18, 22, 23, 25]) and, maybe, with a not full awareness of the limitations of this technique and of the way it can best contribute to clinical management [21].

In this chapter, we discuss the basic principles and the common clinical uses of cerebral oximetry and review the main evidence about the impact of NIRS-guided management on clinically relevant outcomes (with particular emphasis on the randomized investigations and meta-analyses performed in the last decade), in order to analyze the reasons that hinder a wider dissemination of a potentially useful monitoring tool and, accordingly, to outline the possible direction for future research.

2   Basic Principles

A regional oximeter works, in part, like a pulse oximeter: both estimate oxygen saturation, that is, the ratio of the concentration of oxyhemoglobin (O_2Hb) to the sum of the concentrations of reduced hemoglobin (Hb) and O_2Hb (considering the presence of other forms of hemoglobin as negligible), by measuring the attenuation of light of appropriately chosen wavelengths as it passes through a certain body region [26, 27].

According to the so-called *Beer–Lambert law* [26], when a monochromatic (i.e., single-wavelength) light of intensity I_0 passes, by a length L, through a solution of a substance which absorbs light of that wavelength (*see* Fig. 1a), the intensity I of the transmitted light decreases according to the following equation:

$$I = I_0 e^{-\varepsilon CL}$$

where L is the length traveled through the solution (pathlength), ε is the *molar attenuation coefficient* of the substance for that wavelength, and C is the concentration of the solution. After defining the *absorbance A* of the solution as $\log I_0/I$, the above equation can be written, more simply, as

$$A = \varepsilon C L$$

that is, the higher the concentration (and the longer the distance traveled within the solution), the more the light absorbed.

Beer–Lambert law allows to measure the concentration of any substance provided that a wavelength which is absorbed only by that substance is used (and the value of ε for that wavelength is known). Unfortunately, both Hb and O_2Hb absorb light in a wide range of wavelengths. However, oxygen saturation can be estimated according to the ratio between the absorbances measured using two different wavelengths, appropriately chosen so that one is absorbed more by Hb than by O_2Hb, and the opposite occurs for the other one [26]. A common pulse oximeter uses two *light-emitting diodes (LEDs)* to alternately irradiate a finger with two wavelengths (red 660 nm and infrared 940 nm) and measure light intensity on the other side. The ratio between the absorbances for the two different lights is then matched with those observed in healthy volunteers for empiric calibration. To estimate arterial oxygen saturation, the cyclic change in light absorbance due to the systodiastolic variations in blood volume is analyzed (*photoplethysmography*) so as to consider only pulsatile signals [26, 28].

Cerebral oximetry is feasible thanks to the substantial transparence of brain to near-infrared (NIR) light (700–950 nm, that is, the portion of the infrared range closest to the visible spectrum), which is absorbed, with different absorption spectra, only by a few molecules contained therein (including hemoglobins) [2, 27]. However,

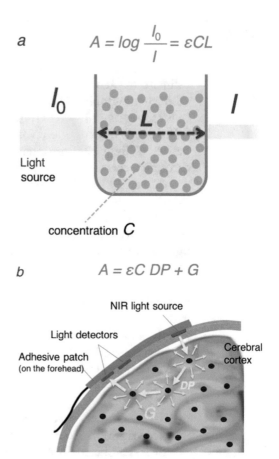

a

$$A = log \frac{I_0}{I} = \varepsilon C L$$

I_0

L

I

Light source

concentration C

b

$$A = \varepsilon C\, DP + G$$

NIR light source

Light detectors

Adhesive patch
(on the forehead)

Cerebral cortex

DP

G

Fig. 1 (**a**) Beer–Lambert law. *A*, absorbance; I_0 and *I*, light intensity before and after crossing through the solution, respectively; *ε*, molar attenuation coefficient; *L*, pathlength. (**b**) Modified Beer–Lambert law, on which regional oximeters rely for the estimation of light attenuation, and schematic representation of light scattering which accounts for the banana-shaped average route of light through brain tissue (*see* text). *DP* differential pathlength, *G* scattering loss

since positioning a light source and detector on the opposite sides of the head is not possible due to excessive pathlength, cerebral oximeters must rely on another physical phenomenon in addition to absorption: light *scattering*, namely the deviation (in any direction) of electromagnetic radiation when it encounters matter [27].

Most commercially available NIRS oximeters use self-adhesive patches containing LEDs or other types of light sources emitting two to five NIR wavelengths and a couple of light detectors located at different distances from the light source (*see* Table 1) [17, 29]. Hence, light which reaches detectors is that back-scattered after a certain (unknown) number of scattering events due to which it follows, on average, a "banana-shaped" route up to a depth of about 2–2.5 cm from skin (*see* Fig. 1b) [2, 3, 27].

Accordingly, as shown in Fig. 1b, light attenuation (as measured by the detectors) occurs not only due to absorption (mainly by O_2Hb and Hb), but also due to scattering, since most of the light emitted takes other directions and will never reach the detectors. Moreover, the actual pathlength is not equal to the geometrical distance L between light source and detectors (as in a pulse oximeter), but is certainly longer (and can only be estimated approximately). In practice, common regional oximeters rely on the following equation to estimate light attenuation (*modified Beer–Lambert law*) [2, 27]:

$$A = \varepsilon\,C\,L\,\text{DPF} + G$$

where A, ε, C, and L are, respectively, absorbance, molar attenuation coefficient, concentration of absorbing substance, and the geometrical distance between light source and detector, DPF (*differential pathlength factor*) is the coefficient of proportionality which, multiplied by L, gives the so-called *differential pathlength* (DP), that is, the actual pathlength traveled by light, and G accounts for the scattering loss, which depends on the geometry and the scattering properties of the tissue (but is assumed to remain constant once the sensor is positioned and, accordingly, can be eliminated using mathematical methods).

Estimation of DP is required in order to provide absolute values of oxygen saturation. Most devices use *spatially resolved* (e.g., *multidistance*) *spectroscopy* as a technique to estimate the DP: in practice, the attenuation of NIR radiation is measured by at least two detectors which are located at different distances from the emitting source, and the increase in attenuation along with the source-detector distance is measured and matched with the dependency of scattering on the wavelength. Other available NIRS oximeters use different technologies for this purpose (*see* Table 1), such as ultrashort pulsed laser (*time-resolved* spectroscopy) or radio-controlled intensity modulation (*frequency-resolved* or *frequency domain* spectroscopy) [2, 12, 27].

Finally, cerebral oximeters do not distinguish pulsatile from nonpulsatile signals and, accordingly, provide the mean oxygen saturation in the small volume of brain they illuminate (around 1 ml of prefrontal cortex when the adhesive sensors are appropriately positioned on the adult forehead) [3]. Although this small "sample" of brain, which is chosen rather arbitrarily, contains venous and arterial blood in a widely variable proportion [30], all commercially available oximeters assume a fixed ratio of venous-to-arterial content of 75:25 or 70:30 for calibration (*see* Table 1) [2, 23, 27]. As discussed below, this and the other assumptions which are made in the different devices' algorithms, together with a certain degree of technological complexity and with the lack of a reference value according to which to evaluate accuracy, are probably among the reasons of the relatively limited use of cerebral oximetry monitoring in clinical practice [25, 27].

Table 1
Main features of some NIRS oximeters available on the market

Device (manufacturer)	Wavelengths used (nm)	Venous to arterial ratio	Light source/detector distances (adult sensor)	Technology	Notes
INVOS™ 5100C/7100 (Medtronic, Dublin, Ireland[a])	730; 810	75:25	3 and 4 cm	Spatially resolved spectroscopy	INVOS series was the first to receive U.S. FDA registration and the most used in clinical studies until recently
EQUANOX™/SenSmart® (Nonin Medical Inc., Plymouth, MN)	730; 760[b]; 810; 880	70:30	2 and 4 cm	Spatially resolved spectroscopy	
FORE-SIGHT® ELITE (CAS Medical Systems, Branford, CT)	690; 730; 770; 810; 870	70:30	1.5 and 5 cm	Spatially resolved spectroscopy	Advertised as the most accurate regional oximeter[c]
NIRO 200 NX® (Hamamatsu Photonics, Hamamatsu City, Japan)	735; 810; 850	70:30	3.7 and 4.3 cm	Spatially resolved spectroscopy	
O3™ (Masimo, Irvine, CA)	N/A	70:30	3 and 4 cm	Spatially resolved spectroscopy	Very few clinical studies available to date
TRS-20 (Hamamatsu Photonics, Hamamatsu City, Japan)	760; 795; 830	–	3 cm	Time-resolved spectroscopy	Approved for clinical use only in Japan
OxiplexTS (ISS, Champaign, IL)	692; 834	–	2.5, 3, 3.5 and 4 cm	Frequency-resolved spectroscopy	Not for clinical use

FDA Food and Drug Administration, N/A Not available
[a]Formerly Covidien, Dublin, Ireland (while the first manufacturer of the INVOS™ series was Somanetics, Troy, MI)
[b]This wavelength is lacking in the three-wavelength version of EQUANOX™
[c]See text for comment

3 Clinical Use

In adults, cerebral oximetry is usually monitored using two disposable sensors positioned bilaterally on the forehead, approximately 1.5–2 cm above the eyebrows [3]. Although some investigations found an association between low preoperative values of $rScO_2$ and outcome [31–33], in patients undergoing surgical procedures intraoperative changes from baseline are regarded as more clinically relevant than any absolute value of $rScO_2$ (*see* below). Accordingly, baseline values should be always obtained, preferably before administration of oxygen and induction of anesthesia (since most clinical studies which identified/investigated the thresholds for clinically relevant cerebral desaturations defined baseline values in this way). Most commercially available regional oximeters display the percentage changes from baseline in addition to absolute values of $rScO_2$ once the baseline has been set on the device by pressing a button. A 20% reduction compared to baseline is generally considered as a clinically relevant cerebral desaturation [2, 3, 5, 34], that is, which may result in worse outcomes and, accordingly, should trigger some kind of intervention aimed at restoring $rScO_2$ values (*see* Table 2).

Below we briefly review the main clinical uses of NIRS cerebral monitoring in different perioperative and critical care settings.

3.1 Cardiac Surgery

As mentioned, the risk of neurological complications (and consequent increased mortality) after cardiac surgery is still very high, with an incidence of stroke of up to 3% (11.2% after surgery on aorta) and a rate of delirium and POCD (at discharge) of up to 45% and 80%, respectively [12]. Nevertheless, a recent survey of 796 members of the Society of Cardiovascular Anesthesiologists (SCA) showed that only about 35% of adult cardiac anesthesiologists and about 40% of pediatric cardiac anesthesiologists use cerebral oximetry routinely [21]. Moreover, respondents' answers on the best intervention to be implemented in different clinical scenarios involving cerebral desaturations were largely heterogeneous, thus confirming the lack of agreement on the usefulness and correct use of cerebral oximetry during cardiac surgery.

Since cerebral oxygen desaturation may reflect an imbalance between oxygen supply and consumption, the interventions to restore $rScO_2$ values should consider all the possible causes of increased oxygen consumption or inadequate oxygen delivery at both a local/hemispherical and global level (Table 2) [10, 35–38]. A protocolized approach to NIRS-guided management was proposed by Denault et al. [10] and its effectiveness in reversing cerebral desaturation has been recently shown in a multicenter randomized controlled trial (RCT) of 201 patients undergoing high-risk cardiac surgery [37]. A slightly modified treatment

Table 2
Possible causes of cerebral oxygen desaturation and treatment options during cardiac surgery

Possible causes of reduced rScO$_2$	Corrective options
Reduced cerebral oxygen supply	
Hypotension	Increase MAP (identify and correct the causes of hypotension and/or administer vasopressors)
Low arterial oxygen saturation	Check ventilation/oxygenator[a] Increase FiO$_2$ Treat causes of respiratory failure
Anemia	Consider RBC transfusion
Inadequate CO/CPB flow	Identify and treat causes of low CO Administer fluids/inotropes Increase pump flow[a] Consider switch from centrifugal to roller pump[a,b]
Cerebral vasoconstriction due to hypocapnia	Reduce ventilation Reduce gas flow to the oxygenator[a]
Reduced/obstructed cerebral venous return	Check head position[c] and venous cannulae Consider surgical causes of SVC obstruction (e.g., after bicaval anastomosis during heart transplantation)
Mechanically obstructed carotid flow	Check head position[c] Check for arterial cannulae malposition/rotation/migration[a] (e.g., endoclamp, aortic cannulae, SCP catheters) Consider air embolism (remove any source of air inlet; head-down tilt) Consider (iatrogenic) aortic dissection
Increased cerebral oxygen consumption	
Inadequate anesthesia depth[d]	Increase anesthetic depth
Seizures[d]	Administer antiepileptic drugs
Hyperthermia	Reduce temperature Treat causes of hyperthermia

[a]If on cardiopulmonary bypass (CPB)
[b]Laminar flow might reduce cerebral blood flow to a greater extent than pulsatile flow
[c]Excessive head axial rotation may obstruct cerebral venous return or cause carotid artery compression
[d]Associated use of bispectral index (BIS) monitoring can help in recognizing this cause of cerebral desaturation. rScO$_2$, cerebral regional oxygen saturation; *MAP* mean arterial pressure, *FiO$_2$* inspired oxygen fraction, *RBC* red blood cells, *CO* cardiac output, *SVC* superior vena cava, *SCP* selective cerebral perfusion

algorithm was proven to be effective in restoring rScO$_2$ levels during cardiopulmonary bypass (CPB) in a prospective multicenter observational study of 235 patients undergoing coronary artery bypass graft (CABG) or valve surgery [38].

One of the advantages of NIRS in cardiac surgery is its potential to provide useful information in all types of procedure and in all their stages, including CPB and deep hypothermic circulatory arrest (DHCA): in fact, unlike other monitoring tools, it does not need pulsatile blood flow to work (actually, it does not need flow at all) [10, 35].

In practice, cerebral oximetry can be used in cardiac surgery with two different "philosophies": in all patients, as an early warning of rare but potentially catastrophic events such as malposition of CPB or selective cerebral perfusion (SCP) cannulae, iatrogenic aortic dissection, obstructed cerebral venous return, air embolism, etc. (*see* Table 2) or, particularly in high-risk patients and complex procedures, to guide hemodynamic/CPB management (i.e., to maintain physiological variables and/or CPB settings at patient-tailored levels) [10, 19, 35, 36, 39]. In this regard, NIRS has been also proposed as a means to identify an impairment of cerebral autoregulation during CPB, which has been found to occur in about 20% of patients and to be associated with an increased risk for perioperative stroke [40]. In particular, the *lower limit of autoregulation* (*LLA*), namely, the value of mean arterial pressure (MAP) below which cerebral autoregulation is lost (i.e., cerebral blood flow becomes MAP-dependent), can be estimated according to a *cerebral oximetry index* (COx) which correlates MAP variations with cerebral oximetry changes and, hence, approaches 0 when autoregulation is intact and 1 when the LLA is exceeded [40, 41]. Using this approach, Joshi et al. [41] showed that the LLA during CPB is widely variable and poorly predictable according to factors such as age or previous cerebrovascular disease, while its estimation using NIRS may allow to individualize MAP targets during CPB. However, this analysis needs a sophisticated signal processing and it has never entered common clinical practice.

3.2 Noncardiac Surgery and Endovascular Procedures

Although the use of cerebral oximetry monitoring has been described and investigated in several adult surgical and endovascular treatment settings, such as orthopedic surgery in the beach chair position (BCP) [2, 3], thoracic surgery [3], major noncardiac surgery in elderly [3, 35], interventional neuroradiology procedures [14], and transcatheter aortic valve replacement (TAVR) [42], the most attractive and studied adult perioperative indication is *carotid endarterectomy* (CEA) [2, 3, 12, 43–46]. In fact, the risk of neurological complications such as stroke, cerebral hypoperfusion syndrome (CHS), and cognitive disorders after CEA is still a major concern (with a reported rate of up to 1.5–3% for stroke) [12, 45], and the reduction in ipsilateral $rScO_2$ over a certain threshold after carotid clamping has been suggested to be an early and sensitive alert of the need for shunt positioning or other interventions (such as raising MAP or inspired oxygen fraction) which may prevent brain injury or cognitive dysfunction [3, 43–46]. In a landmark (although relatively small and observational) study published about two decades ago, Samra et al. [43] used the first FDA-approved and marketed cerebral oximeter (INVOS 3100,

Somanetics, Troy, MI) in patients undergoing awake CEA under locoregional anesthesia and found that cerebral desaturations after clamping were associated with impairment of consciousness, with the best sensitivity and specificity (and a very high negative predictive value) for a cutoff of 20% from baseline. More recently, similar investigations using devices from the same developer (INVOS™ series) have yielded similar results [44, 45], while a small study using a FORE-SIGHT® oximeter (CAS Medical Systems, Branford, CT) reported high sensitivity and acceptable specificity for a cutoff of 9% from baseline [46]. However, the positive predictive value for cerebral ischemia of $rScO_2$ changes was generally low: this suggests that, while desaturations within the identified cutoffs represent a "safety zone," not all patients experiencing deeper desaturations necessarily require shunt positioning (which in turn carries an increased risk of stroke due to embolism) [2, 3, 43].

3.3 Critical Care, Cardiac Arrest, and Extracorporeal Life Support

Until recently, the interest of clinicians for cerebral oximetry monitoring in the adult ICU setting was very limited, and the literature on the subject is accordingly poor. A recent systematic review including a few small and heterogeneous studies found a weak signal of association between low $rScO_2$ and delirium in critically ill patients [47]. A prospective observational investigation including 33 comatose patients suggested that the aforementioned NIRS-derived COx could be a valid alternative to transcranial Doppler for the assessment of cerebral autoregulation in patients with acute brain injury [48], and a subsequent study from the same research group showed that COx values may predict both short- and long-term clinical outcomes (including mortality and severe disability) in this type of patients [49]. In an interesting pilot RCT, 102 neurocritically ill patients were randomized to receive red blood cell (RBC) transfusion according to Hb values alone or in combination with an $rScO_2$ threshold of 60% [50]. Patients in the cerebral oximetry group received significantly fewer RBC units, with no outcome differences between the two groups.

Apart from these few and very preliminary reports, in the critical care setting it is gaining increasing attention, in the last few years, the possible role of cerebral oximetry monitoring during *cardiac arrest* [3, 15, 16, 51–53] and veno-arterial (V-A) *extracorporeal membrane oxygenation* (ECMO) [54, 55], two clinical conditions in which the advantage of not requiring the presence of pulse or blood flow is particularly appreciated.

A meta-analysis including nine small and heterogeneously designed studies of patients suffering from either in-hospital or out-of-hospital cardiac arrest found a significant association between both initial and mean $rScO_2$ values during resuscitation and the rate of *return of spontaneous circulation* (ROSC) [51]. These findings were confirmed in two subsequent multicenter prospective investigations of 183 patients with in-hospital cardiac arrest [16] and 329 patients with out-of-hospital cardiac arrest [52], respectively, both using regional oximeters from the same manufacturer (EQUANOX™ 7600/SenSmart® X-100, Nonin Medical Inc., Plymouth, MN) (*see* Table 1). However, the clinical utility of cerebral oximetry in cardiac arrest is yet to be clarified before it can enter routine clinical practice [53].

Finally, two recent retrospective analyses found an association between cerebral desaturations during V-A ECMO (defined as a >25% reduction from baseline or an absolute $rScO_2$ value below 40% and as an absolute $rScO_2$ value below 60%, respectively) and the risk for acute cerebral complications in 18 patients monitored with a INVOS™ 5100c oximeter (Covidien, Dublin, Ireland) [54] and in 56 patients monitored with a FORE-SIGHT® ELITE device (CAS Medical Systems, Branford, CT) [55]. Further research is needed to evaluate the potential role of cerebral oximetry, and to define the best cutoff values of $rScO_2$, in the management of extracorporeal mechanical support.

3.4 Pediatric Anesthesia and Intensive Care Unit

As previously mentioned, cerebral oximetry monitoring is rather widely used in pediatric cardiac surgery. In this setting, in fact, potentially "catastrophic" events such as CPB cannulae malpositioning (e.g., malrotation of a j-tip aortic cannula leading to direct perfusion into brachiocephalic artery, which may be promptly signaled by a monolateral cerebral desaturation) occur more frequently than in adults, especially during correction of congenital heart defects [10, 35]. Moreover, perioperative $rScO_2$ values have been associated with brain injury and neurodevelopmental outcomes in infants [56].

In addition, ever since they became clinically available, NIRS oximeters are used in the NICU setting, in particular in newborns at risk for hypoxic brain injury and its consequent cognitive development abnormalities, for example, due to perinatal asphyxia or to the hemodynamic instability and the cerebral autoregulation impairment which are typical of preterm infants [7, 8, 57]. However, there is currently no clear evidence that NIRS-guided management in the NICU may favorably affect neurodevelopmental outcome in preterm newborns (*see* below) [7, 58, 59]. In the last few years, moreover, a few prospective observational studies investigated the use of cerebral oximetry as a perioperative monitoring tool in both infants and older children undergoing noncardiac (e.g., digestive) surgery, a setting in which the use of NIRS is not particularly widespread and skepticism dominated until recently [4]: as discussed in the next section, these investigations yielded promising [60, 61] but also less encouraging results [56].

4   Cerebral Oximetry and Outcome

Many clinical studies (mainly observational investigations) reported an association between either baseline $rScO_2$ values or cerebral desaturations compared to baseline and the risk of neurological complications or other adverse outcomes (e.g., acute kidney injury, myocardial dysfunction, and mortality) in various clinical settings, suggesting a prognostic role for cerebral oximetry but also the possibility that interventions aimed at restoring $rScO_2$ values could favorably affect outcome. However, several RCTs and meta-analyses of RCTs focused recently on the role of NIRS monitoring in guiding therapeutic management, especially during cardiac surgery, yielding mostly inconclusive results. Main evidences in this regard are discussed below.

4.1 Prognostic Role of Baseline Values

Low preoperative $rScO_2$ values have been associated with worse outcomes after cardiac surgery [31–33]. In a prospective observational study of 1178 patients undergoing cardiac procedures on CPB, preoperative $rScO_2$ values were measured using an INVOS™ 4100 or 5100 oximeter (Somanetics, Troy, MI) mostly in the cardiac surgery ward while patients were breathing ambient air and, afterwards, supplemental oxygen [31]. Lower $rScO_2$ values were associated with the severity of baseline cardiopulmonary dysfunction and with worse postoperative outcomes. In particular, preoperative $rScO_2$ values ≤50% (during oxygen supplementation) were found to be an independent risk factor for 30-day and 1-year mortality. In a substudy including 231 patients from the same cohort, low preoperative $rScO_2$ values were also shown to be associated with the risk of delirium, as assessed by the confusion assessment method for the ICU (CAM-ICU), during the first 3 days after surgery, with an $rScO_2$ of 59.5% as the best cutoff value [32]. More recently, a retrospective investigation of 210 patients undergoing *left ventricular assist device* (LVAD) implantation found that higher baseline $rScO_2$ values (measured by means of an INVOS™ 5100c oximeter, Somanetics, Troy, MI) were associated with significantly reduced 30-day mortality, but not with other outcomes such as major adverse cardiac events, need for renal replacement therapy (RRT), bleeding, and ICU length of stay (LOS) [33].

An association between low $rScO_2$ values at baseline and less favorable outcomes has been also described in other settings, such as vascular surgery [45] and cardiac arrest [51, 52]. In a recent prospective cohort study including 466 patients undergoing CEA (again monitored using an INVOS™ 5100 device), Kamenskaya et al. [45] found that preoperative $rScO_2$ values below 50% predicted a sixfold increase in the risk of perioperative and early postoperative stroke with a sensitivity of 90.7% and specificity of 66.7%.

Finally, as mentioned above, a meta-analysis including nine small studies in which NIRS oximeters from various manufacturers were used to monitor $rScO_2$ in patients with either in-hospital or out-of-hospital cardiac arrest [51], as well as a prospective investigation of NIRS monitoring (using an EQUANOX™ 7600 or a SenSmart® X-100 oximeter, Nonin Medical Inc., Plymouth, MN) during out-of-hospital cardiac arrest [52], found that patients with higher $rScO_2$ values at the first measurement had significantly increased chances of ROSC.

4.2 Impact of Cerebral Oximetry Changes on Outcome

There is a plenty of reports of an association between changes in $rScO_2$ values from baseline and various clinical outcomes in a wide variety of perioperative and critical care settings [16, 17, 40, 43–46, 49, 52, 54, 55, 60, 62–67].

In cardiac surgery, intraoperative cerebral desaturations have been found to be associated with an increased risk of POCD [17, 62–64], postoperative stroke [65] and various other complications [66], as well as with higher hospital LOS [62, 66]. In one of the first RCTs of cerebral oximetry monitoring in cardiac surgery, Slater et al. [62] monitored 265 CABG patients using an INVOS™ 5100 oximeter (Somanetics, Troy, MI) and randomized them to an intervention group, in which a prespecified protocol was used to treat cerebral desaturations below 20% from baseline, or to a control group in which $rScO_2$ values were recorded but not displayed. The authors calculated an *$rScO_2$ score* as the depth of desaturation below an absolute value of 50% multiplied by its duration (in seconds): as an example, an $rScO_2$ value of 40% for 3 min corresponded to a score of $(50\%–40\%) \times 180$ s $= 1800\%$s. They found that patients with a desaturation score $>3000\%$s had a significantly increased risk of early POCD (as assessed by a set of standardized neurocognitive tests) and an almost tripled risk of prolonged (>6 days) hospital LOS. However, there was no statistically significant difference in the rate of POCD between the intervention and the control group. An increased risk of POCD with more profound and prolonged intraoperative cerebral desaturations was also found in two prospective series of 61 patients undergoing CABG procedures [63] and 101 patients undergoing CABG and/or valve replacement procedures [64], respectively. In another relatively small series of 46 patients undergoing aortic arch surgery (with hypothermic circulatory arrest), cerebral desaturations between 14% and 24% from baseline during antegrade SCP showed a sensitivity up to 83% and a specificity up to 94% in predicting postoperative stroke [65]. In all three of the above investigations, $rScO_2$ was measured using an INVOS 4100 oximeter (Somanetics, Troy, MI).

Casati et al. [67] randomized 122 elderly patients undergoing major abdominal surgery under general anesthesia and continuous intraoperative monitoring with INVOS 4100 (Somanetics, Troy,

MI) to an intervention group, in which $rScO_2$ was maintained at values not lower than 75% of baseline, or a control group in which $rScO_2$ values were blinded. When considering only patients who experienced desaturations below 75% of baseline (about 21% of the study population), the Mini Mental State Examination (MMSE) at the seventh postoperative day was lower and the hospital LOS was higher in the control group as compared with the intervention group. However, there were no differences between the two groups in the MMSE or in the rate of complications, overall.

The impact of cerebral desaturations on neurological outcomes in patients undergoing vascular surgery (particularly CEA) [43–46] and in patients needing ECMO support [54, 55], as well as the association between the loss of cerebral autoregulation (as assessed by means of NIRS) and unfavorable outcomes after cardiac surgery [40] and in comatose critically ill patients [49], have been discussed above. Moreover, in one of the aforementioned investigations of cerebral oximetry monitoring during cardiac arrest, Genbrugge et al. [52] showed that an increase in $rScO_2$ by at least 15% during resuscitation as compared with the first measured values was significantly associated with a higher chance of ROSC (odd ratio (OR) 4.5; 95% confidence interval (CI) 2.75–7.41; $p < 0.001$).

Finally, two recent studies investigated the association between intraoperative cerebral desaturations and postoperative neurodevelopmental complications after neonatal/infant and pediatric surgery, respectively, with conflicting results [56, 60]. In fact, while Gómez-Pesquera et al. [60] found that intraoperative cerebral desaturations (within 20% from baseline) were associated with an increased risk of negative postoperative behavioral changes in 198 children aged 2 to 12 years undergoing noncardiac surgery, Olbrecht et al. [56] conducted a multicenter observational investigation of 453 infants under 6 months of age suggesting that severe intraoperative cerebral desaturations during noncardiac surgery are uncommon and not likely to be associated with postoperative neurocognitive abnormalities.

4.3 NIRS-Guided Therapeutic Management and Outcome

In the first RCT of cerebral oximetry monitoring ever performed, Murkin et al. [68] randomized 200 patients undergoing CABG surgery to either intraoperative $rScO_2$ monitoring (INVOS™ 5100; Somanetics, Troy, MI) in combination with a treatment protocol to maintain values ≥75% of baseline (intervention group) or blinded $rScO_2$ recording (control group). The most often effective interventions included raising CPB flow or MAP, deepening anesthesia, and increasing inspired oxygen fraction (FiO_2). Patients in the control group had significantly higher ICU LOS (1.87 ± 2.7 vs. 1.25 ± 0.8 days; $p = 0.029$) and greater incidence of major organ morbidity and mortality (MOMM), that is, at least one among death, stroke, myocardial infarction, kidney injury requiring RRT, mechanical ventilation >48 h, mediastinitis, reoperation (11 vs. 3%; $p = 0.048$), while no significant differences were found between groups in the rate of any of these adverse outcomes. However, this trial was clearly underpowered to detect differences in the incidence of events such as death or stroke.

Although the trial by Murkin et al. [68], as well as other subsequent investigations performed in cardiac surgery [37, 38, 69] and in the setting of preterm infant intensive care [58], demonstrated the effectiveness of NIRS-based treatment protocols in maintaining $rScO_2$ values within certain limits, other studies (some of which have already been discussed above [62, 67]) and, particularly, the RCTs performed in the last few years failed to show a favorable impact on outcomes of protocolized NIRS-guided management in various clinical settings, while the results of recent meta-analyses of RCTs, taken as a whole, are rather uncertain [34, 72, 73].

Two RCTs [70, 71] published in 2017 investigated the role of NIRS-guided management in adult cardiac surgery. Lei et al. [70] randomized 249 patients undergoing cardiac procedures on CPB to either a treatment protocol initiated for $rScO_2$ values below 75% of baseline or blinded $rScO_2$ recording. No difference in the rate of delirium was found between groups. In the PASPORT trial [71], 204 patients undergoing valve or combined valve and CABG surgery were randomized to a NIRS-based treatment algorithm (including a restrictive hematocrit threshold for RBC transfusion and aimed at maintaining $rScO_2$ at absolute values >50% or at >70% of baseline) or to standard care. The authors found no differences in cognitive function (assessed up to 3 months after surgery), RBC transfusion, biomarkers of brain, kidney, and myocardial injury, adverse events, and healthcare costs between the groups. An INVOS™ 5100 oximeter (Covidien, Dublin, Ireland) was used in both studies. A meta-analysis of 10 RCTs (1466 patients, overall), also published in 2017, confirmed the lack of any clinical benefit of NIRS-based treatment protocols in adult patients undergoing cardiac surgery, although the risk of bias was very high for most of the included trials and the level of evidence was low or very low for all the assessed outcomes [72].

Other two meta-analyses recently addressed the role of cerebral oximetry-based management in any adult perioperative setting, with inconclusive results [34, 73]. Zorrilla-Vaca et al. [34] analyzed 15 RCTs (2057 patients, overall) and found that NIRS-guided intraoperative management was significantly associated with a lower incidence of POCD (at one week after surgery) and a shorter ICU LOS, but not with hospital LOS and delirium. Moreover, subgroup analysis revealed that the effect on POCD rate was lost among patients undergoing noncardiac surgery. However, the authors found a high degree of heterogeneity among the included trials. Similarly, the results of the meta-analysis by Yu et al. [73], which included 15 RCTs with a total of 1822 patients undergoing abdominal, orthopedic, CEA, and cardiac surgery, leave great uncertainty about the impact of NIRS-guided management on delirium, stroke, and mortality.

Finally, the first (and the only so far) RCT of cerebral oximetry monitoring in preterm infants (SafeBoosC II trial) included 166 newborns from 8 NICUs across Europe [58]. Although this study showed that a dedicated treatment protocol was effective in maintaining $rScO_2$ values within a target range of 55–85%, a recently published follow-up found no benefits of NIRS monitoring on long-term neurodevelopmental outcome [59].

5   Limitations, Pitfalls, and Controversies

As mentioned, in addition to the lack of clear evidence that cerebral oximetry monitoring could favorably affect outcomes, some limitations of NIRS technology and criticisms of its reliability probably contributed to hinder a more widespread use in clinical practice in recent years [2–4, 17, 18, 20, 27]. These include, among others, the lack of reference values and of "accurate" absolute $rScO_2$ measurements [2, 25], the signal contamination by extracranial tissues [20, 23, 29, 74], the lack of information about brain areas away from the sensors [2, 75], the low agreement among the different commercially available devices [22–24] and, all considered, the lack of well-defined reference thresholds for both "normal values" and clinically relevant desaturations [3, 18, 22, 27].

Although the marketed regional oximeters are mostly based on the same technological principles and on similar basic assumptions (see above), devices from different manufacturers differ in many important aspects such as the algorithms used for signal acquisition and processing, the types of light sources, the wavelengths of light emitted, and the distances between light emitters and detectors (Table 1) [22, 29]. In an elegant study of 12 healthy volunteers, Davie and Grocott [29] showed that $rScO_2$ readings by three different NIRS oximeters (INVOS™ 5100C, Covidien, Dublin, Ireland; FORE-SIGHT®, CAS Medical Systems, Branford, CT; EQUANOX™ Classic 7600, Nonin Medical Inc., Plymouth, MN) were all significantly affected, although to a different extent, by extracranial tissue ischemia-hypoxia in the forehead region. It is even possible that regional oximetry values registered on the forehead do not reflect cerebral oxygen saturation at all due to excessive distance between skin and brain tissue, such as in case of cortical atrophy (e.g., elderly patients) [76].

Indeed, several investigations compared both the absolute $rScO_2$ values and their percentage changes from baseline as measured by different oximeters in various clinical settings (cardiac surgery, preterm infants) as well as in healthy volunteers and found a poor agreement among various devices [22–24, 77–79]. For

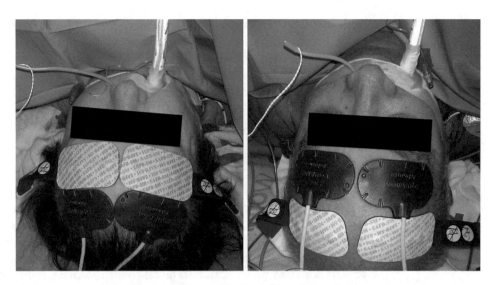

Fig. 2 Direct comparison between INVOS™ 5100C (Covidien, Dublin, Ireland) and EQUANOX™ 7600 (Nonin Medical Inc., Plymouth, MN) in patients undergoing cardiac surgery. (Adapted with permission from Pisano et al. [22])

example, Pisano et al. [22] collected $rScO_2$ values simultaneously from INVOS™ 5100C (Covidien, Dublin, Ireland) and EQUA-NOX™ 7600 (Nonin Medical Inc., Plymouth, MN) in 10 patients undergoing cardiac surgery (as shown in Fig. 2) at different pre-specified time points and whenever at least one of the two devices displayed a reduction in $rScO_2 \geq 20\%$ from baseline. The analysis of 140 coupled measurements suggested that the two oximeters are not interchangeable in measuring both absolute $rScO_2$ values and changes from baseline. The findings of these studies suggest caution when interpreting, in clinical practice, both absolute and trend $rScO_2$ values according to thresholds identified in studies which used devices from different manufacturers.

Most recently developed NIRS oximeters using four or more wavelengths such as FORE-SIGHT® (CAS Medical Systems, Branford, CT) or O3™ (Masimo, Irvine, CA) are commonly regarded (and claimed by manufacturers) as "modern" devices that measure absolute $rScO_2$ values more accurately [80, 81]. However, this claim is rather pretentious since "accuracy" is a feature which hardly fits to regional oximetry, or at least is very difficult to assess [3, 25, 27]. In fact, unlike pulse oximetry, which measures an existing physiological variable (i.e., oxygen saturation of arterial blood), $rScO_2$ represents (at best) the average oxygen saturation in a small arbitrarily chosen region of brain containing arteries and veins in a largely variable ratio [23, 30]. In a study of 20 children undergoing diagnostic or therapeutic cardiac catheterization, Watzman et al. [30] found that the arterial–venous ratio measured by NIRS ranged from 0:100 to 40:60 among different subjects, while as discussed above, commercially available oximeters assume a fixed ratio of 25:75 or 30:70 for calibration. Evidently, $rScO_2$ *does not exist unless measured* [25]: in other words, what is expected to be measured is "decided" only once the sensors have been placed on the patient's forehead [27]. Accordingly, evaluating the accuracy of NIRS oximeters by comparison with the average of arterial and venous saturation weighted according to the fixed artery-to-vein ratio they use for calibration, as it was done in some aforementioned investigations [23, 80], may be misleading.

Some clinicians advocate unreliability and uselessness of cerebral oximetry monitoring on the basis of the occurrence of postoperative neurological complications such as massive cerebral hemorrhage despite "normal" intraoperative $rScO_2$ values [75]. However, the lack of information about areas away from the sensors is certainly a limitation to be taken into account, but it does not imply that NIRS oximetry is not useful as an early warning of oxygen supply/demand imbalance at a global hemispherical level.

Finally, with regard to the lack of clear randomized evidence of a favorable impact of cerebral oximetry monitoring on outcomes, it was pointed out that, although the use of *pulse oximetry* has never shown any favorable impact on outcomes in RCTs, probably no anesthesiologist would be willing to give up it in the operating room [3, 36, 82].

6   Future Perspectives

Cerebral oximetry has the potential to become part of standard perioperative and ICU monitoring [82], but maybe in the not too near future [3]. In fact, although studies investigating its use in many surgical and critical care settings are increasingly accumulating and, for the first time, in 2019 the topic was covered in a national clinical guideline (Japanese Society of Cardiovascular Anesthesiologists, JSCVA) [12], many important issues still need to be clarified and, accordingly, should be addressed in future investigations in order to bring the positions of "supporters" and "skeptics" (sometimes real "haters") closer.

First, the high road of ever larger and well-designed RCTs aiming at showing a favorable impact of NIRS monitoring on outcomes should be always followed. However, the recent history of evidence-based medicine teaches us that this could be a very difficult task to accomplish. Several trials with this purpose are currently ongoing in different clinical settings [73]. For example, Lomivorotov et al. are investigating the risk of postoperative complications (myocardial infarction, stroke, delirium, POCD, wound infection, mediastinitis, prolonged mechanical ventilation, arrhythmias, reoperation for bleeding, acute kidney injury, need for RRT) and mortality in patients undergoing high-risk cardiac surgery randomized to either an intervention protocol for $rScO_2$ values <60% (measured using a FORE-SIGHT® oximeter, CAS Medical Systems, Branford, CT) or standard therapy (http://clinicaltrials.gov/show/NCT02155868). The estimated sample size of this study was 120 patients. A potentially larger RCT, currently recruiting, has planned to enroll 394 elderly patients undergoing major noncardiac surgery in order to investigate the impact of cerebral oximetry-guided management (using an O3™ regional oximeter, Masimo, Irvine, CA) on the incidence of MOMM (http://clinicaltrials.gov/show/NCT03861026). The results of these and other (smaller) ongoing trials will hopefully contribute to clarify the potential clinical benefits of NIRS monitoring, although much larger (multicenter) RCTs are probably needed.

Another important target of future research, maybe a little easier to achieve, is the attempt to identify thresholds for clinically relevant desaturations which could apply regardless of the model of oximeter used, or at least device-specific "reference" and cutoff values.

Finally, considered the above-discussed limitations of comparing different NIRS oximeters "in vivo," the performances of different commercially available devices could be better assessed and compared using experimental models and "phantoms" with homogeneous and known oxygen saturations [83].

References

1. Jobsis FF (1977) Noninvasive, infrared monitoring of cerebral and myocardial oxygen sufficiency and circulatory parameters. Science 198:1264–1267

2. Ghosh A, Elwell C, Smith M (2012) Cerebral near-infrared spectroscopy in adults: a work in progress. Anesth Analg 115(6):1373–1383

3. Green DW, Kunst G (2017) Cerebral oximetry and its role in adult cardiac, non-cardiac surgery and resuscitation from cardiac arrest. Anaesthesia 72(Suppl 1):48–57

4. Kasman N, Brady K (2011) Cerebral oximetry for pediatric anesthesia: why do intelligent clinicians disagree? Paediatr Anaesth 21 (5):473–478

5. Ghanayem NS, Hoffman GM (2016) Near infrared spectroscopy as a hemodynamic monitor in critical illness. Pediatr Crit Care Med 17 (8 Suppl 1):S201–S206

6. Van Bel F, Lemmers P, Naulaers G (2008) Monitoring neonatal regional cerebral oxygen saturation in clinical practice: value and pitfalls. Neonatology 94(4):237–244

7. Garvey AA, Dempsey EM (2018) Applications of near infrared spectroscopy in the neonate. Curr Opin Pediatr 30(2):209–215

8. Dix LM, van Bel F, Lemmers PM (2017) Monitoring cerebral oxygenation in neonates: an update. Front Pediatr 5:46

9. Frogel J, Kogan A, Augoustides JGT et al (2019) The value of cerebral oximetry monitoring in cardiac surgery: challenges and solutions in adult and pediatric practice. J Cardiothorac Vasc Anesth. https://doi.org/10.1053/j.jvca.2018.08.206. pii: S1053-0770(18)30860-7. [Epub ahead of print]

10. Denault A, Deschamps A, Murkin JM (2007) A proposed algorithm for the intraoperative use of cerebral near-infrared spectroscopy. Semin Cardiothorac Vasc Anesth 11:274–281

11. Newman M, Mathew J, Grocott H et al (2006) Central nervous system injury associated with cardiac surgery. Lancet 368:694–703

12. Yoshitani K, Kawaguchi M, Ishida K et al (2019) Guidelines for the use of cerebral oximetry by near-infrared spectroscopy in cardiovascular anesthesia: a report by the cerebrospinal Division of the Academic Committee of the Japanese Society of Cardiovascular Anesthesiologists (JSCVA). J Anesth. https://doi.org/10.1007/s00540-019-02610-y. [Epub ahead of print]

13. Greisen G, Leung T, Wolf M (2011) Has the time come to use near-infrared spectroscopy as a routine clinical tool in preterm infantsundergoing intensive care? Philos Trans A Math Phys Eng Sci 369(1955):4440–4451

14. Badenes R, García-Pérez ML, Bilotta F (2016) Intraoperative monitoring of cerebral oximetry and depth of anaesthesia during neuroanesthesia procedures. Curr Opin Anaesthesiol 29 (5):576–581

15. Wik L (2016) Near-infrared spectroscopy during cardiopulmonary resuscitation and after restoration of spontaneous circulation: a valid technology? Curr Opin Crit Care 22 (3):191–198

16. Parnia S, Yang J, Nguyen R et al (2016) Cerebral oximetry during cardiac arrest: a multicenter study of neurologic outcomes and survival. Crit Care Med 44(9):1663–1674

17. Zheng F, Sheinberg R, Yee MS et al (2013) Cerebral near-infrared spectroscopy monitoring and neurologic outcomes in adult cardiac surgery patients: a systematic review. Anesth Analg 116(3):663–676

18. Bickler P, Feiner J, Rollins M, Meng L (2017) Tissue oximetry and clinical outcomes. Anesth Analg 124(1):72–82

19. Vernick WJ, Gutsche JT (2013) Pro: cerebral oximetry should be a routine monitor during cardiac surgery. J Cardiothorac Vasc Anesth 27:385–389

20. Gregory A, Kohl BA (2013) Con: near-infrared spectroscopy has not proven its clinical utility as a standard monitor in cardiac surgery. J Cardiothorac Vasc Anesth 27:390–394

21. Zacharias DG, Lilly K, Shaw CL (2014) Survey of the clinical assessment and utility of near-infrared cerebral oximetry in cardiac surgery. J Cardiothorac Vasc Anesth 28(2):308–316

22. Pisano A, Galdieri N, Iovino TP et al (2014) Direct comparison between cerebral oximetry by INVOS(TM) and EQUANOX(TM) during cardiac surgery: a pilot study. Heart Lung Vessel 6(3):197–203

23. Bickler PE, Feiner JR, Rollins MD (2013) Factors affecting the performance of 5 cerebral oximeters during hypoxia in healthy volunteers. Anesth Analg 117:813–823

24. Tomlin KL, Neitenbach AM, Borg U (2017) Detection of critical cerebral desaturation thresholds by three regional oximeters during hypoxia: a pilot study in healthy volunteers. BMC Anesthesiol 17(1):6

25. Pisano A (2016) Can we claim accuracy from a regional near-infrared spectroscopy oximeter? Anesth Analg 122(3):920

26. Pisano A (2017) Light, air pollution, and pulse oximetry: the Beer-Lambert law. In: Pisano A. Physics for anesthesiologists. Springer, Cham, pp 117–127

27. Pisano A (2017) Scattering of electromagnetic waves: blue skies, cerebral oximetry, and some reassurance about X-rays. In: Pisano A. Physics for anesthesiologists. Springer, Cham, pp 129–141

28. Chan ED, Chan MM, Chan MM (2013) Pulse oximetry: understanding its basic principles facilitates appreciation of its limitations. Respir Med 107(6):789–799

29. Davie SN, Grocott HP (2012) Impact of extra-cranial contamination on regional cerebral oxygen saturation: a comparison of three cerebral oximetry technologies. Anesthesiology 116 (4):834–840

30. Watzman HM, Kurth CD, Montenegro LM et al (2000) Arterial and venous contributions to near-infrared cerebral oximetry. Anesthesiology 93(4):947–953

31. Heringlake M, Garbers C, Käbler JH et al (2011) Preoperative cerebral oxygen saturation and clinical outcomes in cardiac surgery. Anesthesiology 114(1):58–69

32. Schoen J, Meyerrose J, Paarmann H et al (2018) Preoperative regional cerebral oxygen saturation is a predictor of postoperative delirium in on-pump cardiac surgery patients: a prospective observational trial. Crit Care 15 (5):R218

33. Ghosal S, Trivedi J, Chen J et al (2018) Regional cerebral oxygen saturation level predicts 30-day mortality rate after left ventricular assist device surgery. J Cardiothorac Vasc Anesth 32(3):1185–1190

34. Zorrilla-Vaca A, Healy R, Grant MC et al (2018) Intraoperative cerebral oximetry-based management for optimizing perioperative outcomes: a meta-analysis of randomized controlled trials. Can J Anaesth 65(5):529–542

35. Fischer GW (2008) Recent advances in application of cerebral oximetry in adult cardiovascular surgery. Semin Cardiothorac Vasc Anesth 12(1):60–69

36. Edmonds HL Jr (2011) Central nervous system monitoring. In: Kaplan JA, Reich DL, Savino JS (eds) Kaplan's cardiac anesthesia, 6th edn. Elsevier Saunders, St. Louis, pp 485–490

37. Deschamps A, Hall R, Grocott H et al (2016) Cerebral oximetry monitoring to maintain normal cerebral oxygen saturation during high-risk cardiac surgery: a randomized controlled feasibility trial. Anesthesiology 124:826–836

38. Subramanian B, Nyman C, Fritock M et al (2016) A multicenter pilot study assessing regional cerebral oxygen desaturation frequency during cardiopulmonary bypass and

responsiveness to an intervention algorithm. Anesth Analg 122:1786–1793

39. Grocott HP (2019) Cerebral oximetry monitoring. To guide physiology, avert catastrophe or both? Eur J Anaesthesiol 36(1):82–83

40. Ono M, Joshi B, Brady K et al (2012) Risks for impaired cerebral autoregulation during cardiopulmonary bypass and postoperative stroke. Br J Anaesth 109(3):391–398

41. Joshi B, Ono M, Brown C et al (2012) Predicting the limits of cerebral autoregulation during cardiopulmonary bypass. Anesth Analg 114:503–510

42. Brodt J, Vladinov G, Castillo-Pedraza C et al (2016) Changes in cerebral oxygen saturation during transcatheter aortic valve replacement. J Clin Monit Comput 30(5):649–653

43. Samra SK, Dy EA, Welch K et al (2000) Evaluation of a cerebral oximeter as a monitor of cerebral ischemia during carotid endarterectomy. Anesthesiology 93:964–970

44. Ritter JC, Green D, Slim H et al (2011) The role of cerebral oximetry in combination with awake testing in patients undergoing carotid endarterectomy under local anaesthesia. Eur J Vasc Endovasc Surg 41:599–605

45. Kamenskaya OV, Loginova IY, Lomivorotov VV (2017) Brain oxygen supply parameters in the risk assessment of cerebral complications during carotid endarterectomy. J Cardiothorac Vasc Anesth 31(3):944–949

46. Jonsson M, Lindström D, Wanhainen A et al (2017) Near infrared spectroscopy as a predictor for shunt requirement during carotid endarterectomy. Eur J Vasc Endovasc Surg 53(6):783–791

47. Bendahan N, Neal O, Ross-White A et al (2018) Relationship between near-infrared spectroscopy derived cerebral oxygenation and delirium in critically ill patients: a systematic review. J Intensive Care Med 30:885066618807399. https://doi.org/10.1177/0885066618807399. [Epub ahead of print]

48. Rivera-Lara L, Geocadin R, Zorrilla-Vaca A et al (2017) Validation of near-infrared spectroscopy for monitoring cerebral autoregulation in comatose patients. Neurocrit Care 27(3):362–369

49. Rivera-Lara L, Geocadin R, Zorrilla-Vaca A et al (2019) Near-infrared spectroscopy-derived cerebral autoregulation indices independently predict clinical outcome in acutely ill comatose patients. J Neurosurg Anesthesiol. https://doi.org/10.1097/ANA.0000000000000589. [Epub ahead of print]

50. Leal-Noval SR, Arellano-Orden V, Muñoz-Gómez M et al (2017) Red blood cell transfusion guided by near infrared spectroscopy in neurocritically ill patients with moderate or severe Anemia: a randomized, controlled trial. J Neurotrauma 34(17):2553–2559

51. Sanfilippo F, Serena G, Corredor C et al (2015) Cerebral oximetry and return of spontaneous circulation after cardiac arrest: a systematic review and meta-analysis. Resuscitation 94:67–72

52. Genbrugge C, De Deyne C, Eertmans W et al (2018) Cerebral saturation in cardiac arrest patients measured with near-infrared technology during pre-hospital advanced life support. Results from Copernicus I cohort study. Resuscitation 129:107–113

53. Sandroni C, Parnia S, Nolan JP (2019) Cerebral oximetry in cardiac arrest: a potential role but with limitations. Intensive Care Med 45(6):904–906. https://doi.org/10.1007/s00134-019-05572-7. [Epub ahead of print]

54. Khan I, Rehan M, Parikh G et al (2018) Regional cerebral oximetry as an indicator of acute brain injury in adults undergoing veno-arterial extracorporeal membrane oxygenation—a prospective pilot study. Front Neurol 9:993

55. Pozzebon S, Blandino Ortiz A, Franchi F (2018) Cerebral near-infrared spectroscopy in adult patients undergoing veno-arterial extracorporeal membrane oxygenation. Neurocrit Care 29(1):94–104

56. Olbrecht VA, Skowno J, Marchesini V et al (2018) An international, multicenter, observational study of cerebral oxygenation during infant and neonatal anesthesia. Anesthesiology 128(1):85–96

57. Martini S, Corvaglia L (2018) Splanchnic NIRS monitoring in neonatal care: rationale, current applications and future perspectives. J Perinatol 38(5):431–443

58. Hyttel-Sorensen S, Pellicer A, Alderliesten T et al (2015) Cerebral near infrared spectroscopy oximetry in extremely preterm infants: phase II randomised clinical trial. BMJ 350:g7635

59. Plomgaard AM, Alderliesten T, van Bel F et al (2019) No neurodevelopmental benefit of cerebral oximetry in the first randomized trial (SafeBoosC II) in preterm infants during the first days of life. Acta Paediatr 108(2):275–281

60. Gómez-Pesquera E, Poves-Alvarez R, Martinez-Rafael B et al (2019) Cerebral oxygen saturation and negative postoperative behavioral changes in pediatric surgery: a prospective observational study. J Pediatr

208:207–213.e1. https://doi.org/10.1016/j.jpeds.2018.12.047. pii: S0022-3476(18)31821-3. [Epub ahead of print]

61. Beck J, Loron G, Masson C et al (2017) Monitoring cerebral and renal oxygenation status during neonatal digestive surgeries using near infrared spectroscopy. Front Pediatr 5:140

62. Slater JP, Guarino T, Stack J et al (2009) Cerebral oxygen desaturation predicts cognitive decline and longer hospital stay after cardiac surgery. Ann Thorac Surg 87(1):36–44

63. de Tournay-Jetté E, Dupuis G, Bherer L et al (2011) The relationship between cerebral oxygen saturation changes and postoperative cognitive dysfunction in elderly patients after coronary artery bypass graft surgery. J Cardiothorac Vasc Anesth 25:95–104

64. Yao FS, Tseng CC, Ho CY et al (2004) Cerebral oxygen desaturation is associated with early postoperative neuropsychological dysfunction in patients undergoing cardiac surgery. J Cardiothorac Vasc Anesth 18:552–558

65. Olsson C, Thelin S (2006) Regional cerebral saturation monitoring with near-infrared spectroscopy during selective antegrade cerebral perfusion: diagnostic performance and relationship to postoperative stroke. J Thorac Cardiovasc Surg 131(2):371–379

66. Fischer GW, Lin HM, Krol M et al (2011) Noninvasive cerebral oxygenation may predict outcome in patients undergoing aortic arch surgery. J Thorac Cardiovasc Surg 141:815–821

67. Casati A, Fanelli G, Pietropaoli P et al (2005) Continuous monitoring of cerebral oxygen saturation in elderly patients undergoing major abdominal surgery minimizes brain exposure to potential hypoxia. Anesth Analg 101:740–747

68. Murkin JM, Adams SJ, Novick RJ et al (2007) Monitoring brain oxygen saturation during coronary bypass surgery: a randomized, prospective study. Anesth Analg 104(1):51–58

69. Chan MJ, Chung T, Glassford NJ, Bellomo R (2017) Near-infrared spectroscopy in adult cardiac surgery patients: a systematic review and meta-analysis. J Cardiothorac Vasc Anesth 31(4):1155–1165

70. Lei L, Katznelson R, Fedorko L et al (2017) Cerebral oximetry and postoperative delirium after cardiac surgery: a randomised, controlled trial. Anaesthesia 72(12):1456–1466

71. Rogers CA, Stoica S, Ellis L et al (2017) Randomized trial of near-infrared spectroscopy for personalized optimization of cerebral tissue oxygenation during cardiac surgery. Br J Anaesth 119(3):384–393

72. Serraino GF, Murphy GJ (2017) Effects of cerebral near-infrared spectroscopy on the outcome of patients undergoing cardiac surgery: a systematic review of randomised trials. BMJ Open 7(9):e016613

73. Yu Y, Zhang K, Zhang L et al (2018) Cerebral near-infrared spectroscopy (NIRS) for perioperative monitoring of brain oxygenation in children and adults. Cochrane Database Syst Rev 1:CD010947

74. Sørensen H, Secher NH, Siebenmann C et al (2012) Cutaneous vasoconstriction affects near-infrared spectroscopy determined cerebral oxygen saturation during administration of norepinephrine. Anesthesiology 117(2):263–270

75. McAvoy J, Jaffe R, Brock-Utne J et al (2019) Cerebral oximetry fails as a monitor of brain perfusion in cardiac surgery: a case report. A A Pract 12(11):441–443. https://doi.org/10.1213/XAA.0000000000000963

76. Biedrzycka A, Lango R (2016) Tissue oximetry in anaesthesia and intensive care. Anaesthesiol Intensive Ther 48(1):41–48

77. Fellahi JL, Butin G, Fischer MO et al (2013) Dynamic evaluation of near-infrared peripheral oximetry in healthy volunteers: a comparison between INVOS and EQUANOX. J Crit Care 28:881

78. Hyttel-Sorensen S, Hessel TW, Greisen G (2014) Peripheral tissue oximetry: comparing three commercial near-infrared spectroscopy oximeters on the forearm. J Clin Monit Comput 28:149–155

79. Schneider A, Minnich B, Hofstätter E et al (2014) Comparison of four near-infrared spectroscopy devices shows that they are only suitable for monitoring cerebral oxygenation trends in preterm infants. Acta Paediatr 103(9):934–938

80. Redford D, Paidy S, Kashif F (2014) Absolute and trend accuracy of a new regional oximeter in healthy volunteers during controlled hypoxia. Anesth Analg 119:1315–1319

81. Eyeington CT, Ancona P, Osawa EA et al (2019) Modern technology–derived normative values for cerebral tissue oxygen saturation in adults. Anaesth Intensive Care 47(1):69–75

82. Moerman A, De Hert S (2015) Cerebral oximetry: the standard monitor of the future? Curr Opin Anesthesiol 28:703–709

83. Gunadi S, Leung TS, Elwell CE, Tachtsidis I (2014) Spatial sensitivity and penetration depth of three cerebral oxygenation monitors. Biomed Opt Express 5(9):2896–2912

Chapter 5

Post-Traumatic Stress Disorder Following Intraoperative Awareness

Helene Vulser and Gaele Lebeau

Abstract

Intraoperative awareness (IA) is an anesthesia complication defined as the unexpected and explicit recall by patients of events that occurred during anesthesia. This complication represents a potentially traumatic event as it may lead to psychological or psychiatric consequences, such as post-traumatic stress disorder (PTSD). PTSD is the most frequently reported psychiatric consequence of IA. However, the accurate proportion of PTSD following IA is not known. Due to the low incidence of IA, little is known about risk factors for developing psychiatric or psychological sequelae following IA. However, risk factors for the development of PTSD following trauma, in general, have been extensively explored and can be applied to IA. The common risk factors for PTSD are female gender, younger age, minority status, single status, low socioeconomic status, low level of education, low intelligence, psychiatric history, substance abuse or dependence, severity of the trauma, peritraumatic emotional responses, peritraumatic dissociation, lack of social support, low perceived support following the event, history of past traumatic event(s), life stress, childhood adversity, and family psychiatric history. PTSD symptoms may be experienced in the hours or days following IA but can also be delayed and appear months after the traumatic event. They are variable in duration, from a few weeks to several years. To make a diagnosis, they must be present for at least 1 month.

Although PTSD is the more frequently reported psychiatric consequence of IA, it is important to note that other syndromal or subsyndromal mental disorders such as the acute stress disorder and the subsyndromal PTSD may develop following IA. Early management strategies have to be provided in order to reduce the deleterious outcomes of IA, including taking complaints about IA seriously and offering early psychiatric care.

Key words Intraoperative awareness, Post-traumatic stress disorder, Acute stress disorder, Subsyndromal post-traumatic stress disorder, General anesthesia

---

1 Introduction

Intraoperative awareness (IA) with recall represents a potentially traumatic event. Frequently reported experiences such as paralysis, breathing difficulty or pain may be terrifying and make the patient think that he/she is going to die. This threatening situation may lead to psychological or psychiatric consequences, notably post-traumatic stress disorder (PTSD). We will review the

Marco Cascella (ed.), *General Anesthesia Research*, Neuromethods, vol. 150, https://doi.org/10.1007/978-1-4939-9891-3_5,
© Springer Science+Business Media, LLC, part of Springer Nature 2020

psychiatric/psychological symptoms reported after IA and their incidence. Then we will focus on risk factors and management of such situations.

2 Psychological and Psychiatric Symptoms

2.1 PTSD Symptoms Following Intraoperative Awareness

Although some patients who have experienced IA do not show any psychological symptoms and seem to be relatively indifferent to it, others develop severe mental health consequences. The most frequently described symptoms are acute and post-traumatic stress symptoms, such as anxiety, sleep disturbances, recurrent nightmares, flashbacks, or hyper-arousal [1–3]. These symptoms may constitute full-blown PTSD, a severe mental disorder that can develop following a situation of an exceptionally threatening nature. The experience of IA can undoubtedly be considered as an exceptionally distressing event. As this point is a necessary criterion to develop PTSD, the perception of a threat at the time of IA may be a key in the development of PTSD symptoms later on [4]. This perception may be explained by some cognitive appraisal, such as a catastrophic interpretation of the situation, which might be more frequent in patients with higher risk factors for PTSD (*see* Section 4).

A diagnosis of PTSD requires a number of criteria to be met among four categories of psychological symptoms: re-experience intrusion, avoidance, negative alterations in cognition and mood, and hyperarousal. Intrusion refers to persistent remembering or to re-experience of the traumatic situation by intrusive flashbacks, often leading to extreme distress. Avoidance behaviors are a consequence of the distress caused by secondary exposition to anything that reminds the patient of the event. In the case of IA, patients may avoid medical environment associated with the trauma, such as operating rooms, hospitals or doctors. Negative alteration in cognition and mood is a new criterion that has appeared in the fifth version of the Diagnostic and Statistical Manual of Mental Disorders (DSM). It includes negative beliefs about the incompetency of oneself and the dangerousness of the world that are commonly reported after trauma exposure [5]. Hyperarousal refers to heightened anxiety and altered arousal responses and includes symptoms such as sleep or concentration problems, hyper-vigilance or irritability. A fifth symptom, dissociation, may be present in PTSD patients but is not mandatory for the diagnosis. Dissociative symptoms may appear very early after IA. They include depersonalization, which may be described as the feeling of being a detached observer of oneself, and derealization which is an altered perception of the external world, which feels unreal. Dissociation may lead to an inability to remember an important aspect of the traumatic event (here IA). Early dissociative symptoms have been reported as an independent risk factor for subsequent PTSD [6] but are unfortunately rarely reported in studies. Some authors have underlined the need to better investigate post-traumatic dissociation following IA [6–8]. Indeed, dissociative symptoms may not only play a significant role in determining long-term psychological disability [7] but they might also explain why, in some cases, anxious symptoms are reported only after two or three interviews [8] or even a month after IA [7]. Aceto et al. [8] thus described that a dissociative state due to awareness episode may involuntarily mask hyperarousal symptoms of some patients.

PTSD symptoms may be experienced in the hours or days following IA but can also be delayed and appear months after the traumatic event. They are variable in duration, from a few weeks to several years. To make a diagnosis, they must be present for at least 1 month. In patients with PTSD, the symptoms seriously affect one's ability to function in at least one main domain such as social or occupational functioning (Table 1).

2.2 Other Psychological/ Psychiatric Symptoms Following Intraoperative Awareness

Although PTSD is the more frequently reported psychiatric consequence of IA, it is important to note that other syndromal or subsyndromal mental disorders may develop following IA.

2.2.1 Acute Stress Disorder

PTSD cannot be diagnosed in the month following IA, due to its duration criteria, but patients may experiment with acute stress disorder, another stress-related disorder. Acute stress disorder symptoms are close to those of PTSD but persist for at least 3 days and up to a month. This early disorder is an important risk factor for PTSD.

2.2.2 Subsyndromal PTSD

Although PTSD is regarded as dichotomous in terms of strict diagnostic criteria, PTSD symptoms, even at a subsyndromal level, may cause significant distress and seriously affect functioning and quality of life [6, 10]. For example, the symptoms may be sufficient to cause avoidance of medical settings [4] thus leading to a poorer quality of medical care. In the review of Ghoneim et al. [2], the most frequently reported symptoms were sleep disturbances (19%), nightmares (21%), daytime anxiety (17%) and fear about future anesthesia (20%). Unspecified "late psychological symptoms" were reported in 22% of cases in this review. However, previous studies had reported a higher incidence of late psychological symptoms, ranging from 33% to 84% [11]. The most frequently reported symptoms, in those studies too, were nightmares, flashbacks, and anxiety. Other symptoms, such as chronic fear, indifference, loneliness, and lack of confidence in a future life, have been reported less frequently after IA (11%, 7%, 7%, and 4%, respectively) [11].

Table 1
DSM 5 criteria for PTSD [9]

Exposure to trauma
Presence of one or more **intrusive symptoms** including: • Recurrent intrusive memories of the event; • Recurrent distressing dreams related to the event; • Dissociative reactions (e.g., flashbacks); • Psychological distress to reminders of the event; • Physiological reactions to reminders of the event
Persistent avoidance of—or efforts to avoid—stimuli associated with the traumatic event, either internal (memories, thoughts, feelings) or external (people, places, conversations, activities, objects, situations)
Presence of two or more **negative alterations in cognitions and mood** that are associated with the traumatic event: • Inability to remember an important part of the event; • Persistent negative beliefs about oneself, others, or the world; • Persistent distorted beliefs about the cause or consequences of the event; • Persistent negative emotional state; markedly diminished interest in activities; • Feelings of detachment from others; • Inability to experience positive emotions
Presence of two or more **arousal symptoms**: • Irritable or angry behavior; • Reckless or self-destructive behavior; • Hypervigilance; • Exaggerated startle response; • Concentration problems; • Sleep disturbance
The **duration** of these is more than 1 month
The disturbance causes **clinically significant distress or impairment** in social, occupational, or other important areas of functioning
Disturbance is not due to medication, substance use, or other illness
Specify if: with dissociative symptoms Presence of high levels of either of the following in reaction to trauma-related stimuli: • Depersonalization: Experience of being an outside observer of or detached from oneself (e.g., feeling as if "this is not happening to me" or one were in a dream) • Derealization: Experience of unreality, distance, or distortion (e.g., "things are not real")

2.2.3 Other Psychological or Psychiatric Symptoms

Finally, after IA, patients may feel a lack of trust in the medical staff. This may be due to both the IA episode per se and to the skepticism of doctors reported by some patients. For example, in the study of Samuelsonn et al. [11], 13 of the 39 patients reported that their awareness experiences had been greeted with skepticism. In several case reports, the doctor's response had an impact on the patient's psychological symptoms, with an association between doctors validating or supporting patients and a positive or less anxious patient response [3]. The lack of trust in medical staff may have many consequences on patients' lives, especially in those with chronic illness who require repeated surgeries and long-term medical care.

It is important to note that there may be a wide range of other psychological and psychiatric symptoms following IA, such as depressive symptoms, anxiety disorders, or suicidal thoughts. Those symptoms, notably anxiety, may emerge only after a certain period of time, for example when new general anesthesia is needed [4].

3   Incidence of PTSD Following Intraoperative Awareness

PTSD is the most frequently reported psychiatric consequence of IA. However, the accurate proportion of PTSD following IA is not known [4]. In their systematic review, Aceto et al. [8] reported a wide range of PTSD rate across the 7 studied publications of 0 to 71%. Longitudinal evaluations of patients originally recruited for prospective observational or interventional awareness studies have revealed a notable incidence of PTSD. The highest rate was reported in a small cohort of 7 IA patients in which 5 (71%) met the criteria for the diagnosis of PTSD [7]. In this study, 2 of the 5 patients who developed PTSD reported their experience at a 30-day interview, but not at the 2- to 4-h or the 24- to 36-h interview. In another cohort of patients who reported IA, interviews using qualitative approach were performed in 9 patients approximately 2 years after IA [12]. Four of them (44%) had PTSD that did not tend to improve with time, and two of them required medication. PTSD rate was also high in the study conducted by Osterman et al. [1] with 9 out of 16 patients (56.3%) meeting DSM-IV diagnostic criteria for PTSD years (mean 17.9 years), years after suffering awareness. In reports drawn from the Anesthesia Awareness Registry, Kent et al. [13] analyzed 56 patients' self-reported persistent psychological sequelae of IA and reported 42% of PTSD. However, medical-record confirmation of this diagnosis was not obtained in this study. In a prospective, multicentric, cohort study, PTSD symptoms of 35 patients who experienced IA were compared with those of 184 patients who did not experience IA [6]. Both the PTSD Checklist-Specific (PCL-S) and a modified Mini-International Neuropsychiatric Interview (MINI) telephone assessment were used to identify symptoms of PTSD. Using the PCL-S, 43% of participants with previous IA (versus 16% without) exceeded the screening cutoff score for PTSD. However, PTSD rates were reported in only 14% of those with IA (7.6% without) when the MINI was used.

Other studies have suggested that PTSD after IA is not a significant problem. For example, in the study of Schwender et al. [14], 3 patients (7%) among 45 who experienced IA, had developed PTSD resulting in the need for medical treatment. In a study of 4183 malpractice claims, entered into the ASA Closed Claims Project between 1961 and 1995, 10% of the 79 identified cases of IA had been assigned a diagnosis of PTSD [15]. In a cohort of 46 patients, 2 patients had persistent psychiatric symptoms and had needed psychiatric care. Only one of these patients was diagnosed with PTSD (2%) [11]. One important limitation of this study is the lack of a validated instrument for psychiatric assessment. It appears that the investigations which used not-standardized questionnaires were those with a lesser percentage of PTSD diagnosis [8]. However, using strict Diagnostic and Statistical Manual (DSM) criteria, neither PTSD nor other severe psychological sequelae were found in the 5 patients assessed by Ranta et al. [16], even though the interview was repeated at 2 and 6 months after surgery. Finally, a recent study used a battery of psychiatric diagnostic interviews and questionnaires to assess PTSD symptoms in 9 subjects, a median of 17.2 years after their documented IA episode and compared to 9 matched controls [17]. In this study, the authors did not report any case of PTSD following IA. Furthermore, no difference was found for the subsequent psychosocial outcome, other psychiatric morbidities, or quality of life. However, these findings may be due to the fact that the initial experiences of IA were not particularly traumatic [18].

Several explanations may explain the observed difference between the incidences of PTSD across studies. This difference may be due to study design (e.g., prospective or retrospective, presence or not of a control group), recruitment method, sample size, sociodemographic and clinical characteristics of the cohort, assessment of psychiatric symptoms (PTSD scales such as Clinician-Administered PTSD Scale or PCL-S, or nonspecific structured diagnosis interview such as the MINI or the Structured Clinical Interview for DSM (SCID), or even no psychiatric assessment tool), setting of the interview (by telephone, at home, in hospital, etc.) and type of interviewer (student, psychologist, psychiatrist, anesthesiologist, etc.). The diagnosis should be based on a structured interview conducted by an experienced professional [17]. The duration of the period between IA and the interview may explain a substantial part of the differences across studies, as the incidence of psychological reactive symptoms after a traumatic event is higher in the first weeks and then generally decreases. Finally, it is important to keep in mind that, in those studies, patients with PTSD following IA may have refused the interview to avoid reliving traumatic memories.

Considering subsyndromal symptoms of PTSD and other psychological symptoms (sleep disturbances, nightmares, anxiety, fear about future anesthesia, etc.), they have been estimated around 20–30% according to different studies [2, 11, 12]. However, these results should be taken with caution given the lack of utilization of standardized questionnaires and the great variability of studied symptoms among studies.

4   PTSD Risk Factors Following Intraoperative Awareness

Due to the low incidence of IA, little is known about risk factors for developing psychiatric or psychological sequelae following IA. However, risk factors for the development of PTSD following trauma, in general, have been extensively explored and can be applied to IA. The common risk factors for PTSD are: female gender, younger age, minority status, single status, low socioeconomic status, low level of education, low intelligence, psychiatric history, substance abuse or dependence, severity of the trauma, peritraumatic emotional responses, peritraumatic dissociation, lack of social support, low perceived support following the event, history of past traumatic event(s), life stress, childhood adversity, and family psychiatric history [19, 20].

It is important to consider that patients who undergo multiple surgeries are at higher risk of lifetime IA and are also more likely to report previous trauma and/or psychiatric diagnosis at the time of presentation [18]. For example, in one study of cardiac surgery patients, more than 40% of the population met criteria for a psychiatric disorder at their preoperative baseline, including 9% of PTSD [21].

As previously discussed, postoperative dissociation not only increases the risk for subsequent PTSD [6] but may also theoretically lead to an initial inability to remember IA and to delayed anxious symptoms by masking of the hyperarousal symptoms [6–8]. In a recent study on 303 patients followed-up 2 years postoperatively, both IA and dissociative symptoms were associated with a higher risk of subsyndromal and syndromal PTSD. Furthermore, perioperative dissociation was identified as a potential mediator for perioperative PTSD symptoms. The authors concluded that screening surgical patients with perioperative dissociation for postoperative PTSD symptoms could promote early referral, evaluation, and treatment [6].

Perceived life threat during the trauma and emotional distress (i.e., severity of the trauma) are major predictors of the development of PTSD [2, 20, 22]. In case of IA, it has thus been thought that some factors, such as pain or feeling of the inability to move, may increase the perceived distress during IA, leading to a higher risk of later PTSD. In the review of Ghoneim et al. [2], some factors, such as sensation of weakness or paralysis, inability to move, noises that were identified, hearing voices and above all "feeling of helplessness, anxiety, panic, impending death or catastrophe" were indeed significantly associated with later psychological symptoms, but pain was not. These results are consistent with PTSD criteria, namely, that PTSD occurs in patients who experienced a situation perceived as exceptionally threatening. In other words, patients who did not report a feeling of anxiety during IA are considerably less at risk of later PTSD. Long-term sequelae in patients who did not report distress during IA have indeed been estimated to be low, around 3%, versus 79% in those who reported distress [4]. Distress is more common in patients who reported pain, paralysis or breathing difficulty during IA, but is not associated with perceived duration of IA [4]. It is thought that a catastrophic interpretation of the situation, where the patient may think that he is going to die or be permanently paralyzed, for example, is thought to be central in the development of PTSD after IA [4]. Thus, neuromuscular paralysis, by preventing the patient from moving, may lead to the feeling of helplessness which may increase the risk of catastrophic interpretations of what is happening [4]. Patients with a psychiatric history, such as anxiety disorder, or patients with maladaptive coping strategies may have negative processes of cognitive appraisal during IA, also leading to catastrophic interpretation.

Finally, peritraumatic emotional responses and low perceived support following the event are factors of interest in the case of IA. Indeed, emotional responses to a traumatic IA may be affected by both postoperative course and doctor's response to IA. Certain postoperative factors, such as intensive care unit admission, mechanical ventilation, and in-hospital cardiac arrest have been described as risk factors for healthcare-associated PTSD [6] and may thus increase the risk of sequelae following IA. Considering the doctor's response, an explanation of the IA incident may decrease the risk of later psychiatric or psychological consequences [3]. In contrast, the feeling of not being believed by the medical staff, or the feeling that doctors did not consider this event as stressful as it was perceived, are extremely stressful for the patient. There is some evidence that doctors' skepticism with regard to the IA experience is linked with higher levels of anxiety in patients [3, 11].

Additional studies are needed to further explore these factors in the development of PTSD and other psychiatric or psychological sequelae following IA.

5 Management of Patients Following Intraoperative Awareness

First, if the medical staff suspects IA during the surgery, explanations should be provided about what is happening. Some studies have shown that paralysis or sensations of being unable to breathe were less distressing if they were understood [23, 24]. The anesthesiologist should thus focus on reassuring the patient instead of attempting to abolish memory retrospectively using drugs [4].

Postoperatively, when a patient reports an experience of IA, explanations should be provided by medical staff. All IA reports should be treated seriously, even when IA report is delayed or if a true IA is unlikely for the medical staff. Indeed, perioperative unpleasant experiences associated with important distress may have a severe psychological impact, sometimes reaching the same level as in "true IA" and require the same support.

If IA is reported to someone else, it is necessary to refer the case to the anesthesiologist responsible [4]. The anesthesiologist should meet the patient promptly. A second meeting is also recommended 2 weeks later [4]. During these consultations, it is important to listen carefully and empathically to the patient's account, without interruption or contradiction (even if there are inconsistencies). Express regret is also helpful for the patient. This is not an admission of error or medicolegal culpability [4]. Attention should also be paid to potential psychiatric symptoms presented by the patient (flashbacks, anxiety, etc.). The second consultation at 2 weeks will allow the anesthesiologist to highlight delayed symptoms that could have appeared since the first meeting. When present, early referral to an appropriate psychologist or psychiatrist should be offered. When not present, the possibility of later psychological symptoms should be explained with a general description of the symptoms of PTSD. It is also important to explain that early appropriate care improves the prognosis, whereas persistent symptoms run the risk of becoming chronic. However, referral to a psychiatrist should be offered with caution and should not put emphasis on putative patients' vulnerability rather than on the medical staff's responsibility.

Psychological interventions include trauma-focused cognitive behavioral therapy and eye movement desensitization and reprocessing (EMDR). Such interventions are commonly offered in patients with PTSD. However, a recent Cochrane review found limited evidence regarding their effects on PTSD in general [25]. In the case of PTSD following IA, cognitive behavioral therapy techniques may be useful, especially in patients who will need further medical care or repeated surgeries. They may comprise psycho-education, cognitive restructuring, and exposure (imaginal and in vivo) [3]. The psychotherapist may, for example, come with the patient in an empty operating room before another planned surgery. When medication is preferred by the patient, or in case of ineffectiveness of psychotherapy, selective serotonin reuptake inhibitor (SSRI) prescription should be considered. Benzodiazepines should not be offered to reduce symptoms in the first month after IA for several reasons. First, there is no evidence on the benefits of such a treatment on symptoms of traumatic stress after a recent traumatic event. Second, benzodiazepines may slow down the time to recover from potentially traumatic events. Third, there is a risk of dependence if symptoms and treatment are continued [26].

Finally, multidisciplinary coordination (anesthesiologists, surgeons, and psychiatrists) is necessary to avoid the subsequent poorer quality of medical care, as seen in other mental disorders [27]. Indeed, the fear of future anesthesia, hospital avoidance, or the altered trust in the medical staff may impede future medical care.

To conclude, IA may lead to severe psychological or psychiatric impact and alter both quality of life and future medical care. Early management strategies have to be provided in order to reduce the deleterious outcomes of IA, including taking complaints about IA seriously and offering early psychiatric care.

References

1. Osterman JE, van der Kolk BA (1998) Awareness during anesthesia and posttraumatic stress disorder. Gen Hosp Psychiatry 20(5):274–281

2. Ghoneim MM, Block RI, Haffarnan M et al (2009) Awareness during anesthesia: risk factors, causes and sequelae: a review of reported cases in the literature. Anesth Analg 108 (2):527–535

3. Bruchas RR, Kent CD, Wilson HD et al (2011) Anesthesia awareness: narrative review of psychological sequelae, treatment, and incidence. J Clin Psychol Med Settings 18(3):257–267

4. NAP5 Report - The National Institute of Academic Anaesthesia [Internet]. [cited 24th May 2017]. http://www.nationalauditprojects.org.uk/NAP5report

5. Zoellner LA, Bedard-Gilligan MA, Jun JJ et al (2013) The evolving construct of posttraumatic stress disorder (PTSD): DSM-5 criteria changes and legal implications. Psychol Inj Law 6(4):277–289

6. Whitlock EL, Rodebaugh TL, Hassett AL et al (2015) Psychological sequelae of surgery in a prospective cohort of patients from three intraoperative awareness prevention trials. Anesth Analg 120(1):87–95

7. Leslie K, Chan MTV, Myles PS et al (2010) Posttraumatic stress disorder in aware patients from the B-aware trial. Anesth Analg 110 (3):823–828

8. Aceto P, Perilli V, Lai C et al (2013) Update on post-traumatic stress syndrome after anesthesia. Eur Rev Med Pharmacol Sci 17 (13):1730–1737

9. American Psychiatric Association (2013) Diagnostic and statistical manual of mental disorders, 5th edn. American Psychiatric Association, Arlington, VA

10. Grubaugh AL, Magruder KM, Waldrop AE et al (2005) Subthreshold PTSD in primary care: prevalence, psychiatric disorders, healthcare use, and functional status. J Nerv Ment Dis 193(10):658–664

11. Samuelsson P, Brudin L, Sandin RH (2007) Late psychological symptoms after awareness among consecutively included surgical patients. Anesthesiology 106(1):26–32

12. Lennmarken C, Bildfors K, Enlund G et al (2002) Victims of awareness. Acta Anaesthesiol Scand 46(3):229–231

13. Kent CD, Mashour GA, Metzger NA et al (2013) Psychological impact of unexpected explicit recall of events occurring during surgery performed under sedation, regional anaesthesia, and general anaesthesia: data from the anesthesia awareness registry. Br J Anaesth 110 (3):381–387

14. Schwender D, Kunze-Kronawitter H, Dietrich P (1998) Conscious awareness during general anaesthesia: patients' perceptions, emotions, cognition and reactions. Br J Anaesth 80 (2):133–139

15. Domino KB, Posner KL, Caplan RA et al (1999) Awareness during anesthesia: a closed claims analysis. Anesthesiology 90 (4):1053–1061

16. Ranta SO-V, Herranen P, Hynynen M (2002) Patients' conscious recollections from cardiac anesthesia. J Cardiothorac Vasc Anesth 16 (4):426–430

17. Laukkala T, Ranta S, Wennervirta J et al (2014) Long-term psychosocial outcomes after intraoperative awareness with recall. Anesth Analg 119(1):86–92

18. Mashour GA, Esaki RK, Tremper KK et al (2010) A novel classification instrument for intraoperative awareness events. Anesth Analg 110(3):813–815

19. Brewin CR, Andrews B, Valentine JD (2000) Meta-analysis of risk factors for posttraumatic stress disorder in trauma-exposed adults. J Consult Clin Psychol 68(5):748–766

20. Ozer EJ, Best SR, Lipsey TL et al (2003) Predictors of posttraumatic stress disorder and symptoms in adults: a meta-analysis. Psychol Bull 129(1):52–73

21. Mashour GA, Wang LY-J, Turner CR et al (2009) A retrospective study of intraoperative awareness with methodological implications. Anesth Analg 108(2):521–526

22. Shalev AY, Tuval-Mashiach R, Hadar H (2004) Posttraumatic stress disorder as a result of mass trauma. J Clin Psychiatry 65(Suppl 1):4–10

23. Heier T, Feiner JR, Lin J et al (2001) Hemoglobin desaturation after succinylcholine-induced apnea: a study of the recovery of spontaneous ventilation in healthy volunteers. Anesthesiology 94(5):754–759

24. Topulos GP, Lansing RW, Banzett RB (1993) The experience of complete neuromuscular blockade in awake humans. J Clin Anesth 5 (5):369–374

25. Sin J, Spain D, Furuta M et al (2017) Psychological interventions for post-traumatic stress disorder (PTSD) in people with severe mental illness. Cochrane Database Syst Rev 1: CD011464

26. WHO. WHO releases guidance on mental health care after trauma [Internet]. [cited 24th May 2017]. http://www.who.int/mediacentre/news/releases/2013/trauma_mental_health_20130806/en/

27. Druss BG, von Esenwein SA (2006) Improving general medical care for persons with mental and addictive disorders: systematic review. Gen Hosp Psychiatry 28(2):145–153

Chapter 6

Mechanisms of Action of Inhaled Volatile General Anesthetics: Unconsciousness at the Molecular Level

Daniela Baldassarre, Giuliana Scarpati, and Ornella Piazza

Abstract

The mechanism by which general anesthetics prevent consciousness remains largely unknown because the mechanism by which brain physiology produces consciousness is yet unexplained. After its most evident goal, to allow surgery for million people in the world, the contribution of anesthesia to science is the unique and great opportunity to study consciousness.

General anesthetics drugs comprise inhaled volatile agents (e.g., isoflurane, sevoflurane, desflurane), and gases (nitrous oxide and xenon) as well as intravenous agents such as etomidate, propofol, thiopental, benzodiazepines, and ketamine. For the purposes of this chapter, because of the large number of compounds, we focus on the inhalational volatile anesthetics used in modern practice and on the fundamental question of how volatile halogenated anesthetics interact with their molecular target to produce unconsciousness.

Pieces of evidence suggest that volatile anesthetic agents do not shut down all brain activity, but they push the brain towards a distinct, highly specific and complex state. Interaction of general anesthetics with cytoskeletal microtubules, membrane and soluble protein are briefly summarized in this chapter. Nevertheless, notwithstanding considerable advances in our comprehension of the molecular properties of anesthetics, much remains to be studied about the deep and complex changes which occur at the level of neural structures during general anesthesia.

Key words Volatile anesthetic agents, Cytoskeletal microtubules, London dispersion force, GABA, Loss of consciousness

> *Existence means nerve existence, that is excitability*
>
> —*Gottfried Benn*

1 Introduction

General anesthetics are an assorted group of drugs which induces a behavioral state termed "anesthesia" that provides immobility, amnesia and loss of consciousness [1]. General anesthetics drugs comprise inhaled volatile agents (e.g., isoflurane, sevoflurane, desflurane, halothane), and gases (nitrous oxide and xenon) as well as intravenous agents such as etomidate, propofol, thiopental,

Marco Cascella (ed.), *General Anesthesia Research*, Neuromethods, vol. 150, https://doi.org/10.1007/978-1-4939-9891-3_6, benzodiazepines, and ketamine. For the purposes of this chapter, because of the large number of compounds, we will focus on the inhalational volatile anesthetics used in modern practice and to the fundamental question of how volatile halogenated anesthetics interact with their molecular target to produce unconsciousness.

Volatile anesthetic agents have low molecular weight and non-polar structure. All of these agents are relatively hydrophobic. The inhaled anesthetic reversibly produces all three of the therapeutic effects of anesthesia: amnesia, immobility and loss of consciousness, but the ability to generate these discrete therapeutic endpoints is not identical: inhaled general anesthetics easily affect explicit memory [2] while consciousness, evaluated by proper response to verbal commands, is preserved in the absence of memory during exposure to low doses of volatile anesthetics [3]. Sleep is produced by lower concentrations of general anesthetics than those required to stop movement in response to pain [4]. Anesthetics penetrate less on autonomic responses than on immobility [5].

Recently, reported sites of anesthetic action have been differentiated both anatomically and in terms of sensitivity to anesthetics. Sites for immobility during anesthesia are located in spinal cord [6] and those for amnesia largely in hippocampus [7] and amygdala [8]. Nevertheless, sites essential for consciousness appear to be widely positioned, particularly in thalamocortical projections but also in networks involving intracortical, corticocortical loops, prefrontal cortex, and other areas [9, 10].

There is the possibility to obtain the various component of anesthesia by different molecules used during anesthesia. Modern anesthesia is "balanced," with several intravenous and inhalational agents triggering or sustaining loss of consciousness, amnesia, muscle relaxation, analgesia, and anxiolysis. For example, anesthetists often use muscular relaxing drugs which selectively inhibit neuromuscular transmission to cut patient movement and allow surgery and other drugs to emphasize anxiolysis and anterograde amnesia or provide analgesia.

Are there universal mechanisms for all general anesthetics that can explain in which way they cause the shift between the conscious and unconscious state? Beyond other components of anesthesia and their already known sites of action, what are the molecular and neuronal mechanisms that underlie this extraordinary phenomenon of consciousness? Anesthesia gives an exclusive and intense opportunity to understand consciousness because it is quite selective, and in fact many brain activities continue during anesthesia while conscious awareness vanishes. Consciousness is scantily defined and cannot be measured but understanding anesthetic mechanisms require considering consciousness, and vice versa.

Notwithstanding considerable advances in our comprehension of the molecular properties of anesthetics, much remains to be studied about the deep and complex changes which occur at the level of neural structures during general anesthesia.

2 Hypotheses and Theories in the History of Anesthesia

The possibility of a communication between inhalational anesthetics and proteins was first suggested by Claude Bernard. He presented anesthetic gases as the cause of reversible cessation of the organized movement of cytoplasm within the cell interior (protoplasmic streaming). Exposing amoeboid cells to the anesthetic gas chloroform, Bernard found that the streaming was ended [11]. He firmly considered the "protoplasm" as the life-harboring substance for all living forms. Bernard, in the nineteenth century, was responsible for establishing a unified paradigm of anesthetic action across life forms. The "protoplasm coagulation" theory and the following unified theory of narcosis proclaimed by Claude Bernard were received with enthusiasm by the scientific community.

In the twentieth century, the next major research into anesthetic mechanisms occurred when Meyer and Overton found a conspicuous correlation between potency of anesthetics, and their solubility in a nonpolar, lipid-like, "hydrophobic" milieu [12, 13]. Since neuronal membranes carry signals, and are mostly lipid, anesthetics were assumed to act in lipid regions of brain neuronal surface membranes. Claude Bernard's view was forgotten, and a new archetype was believed. Charles Ernst Overton (1865–1933) and Hans Horst Meyer (1853–1939) coined the Meyer–Overton hypothesis, celebrated as a foundation stone for anesthesia understanding [14].

In 1915, Harvey stated that n-alkanols gases reversibly depressed the luminescence of certain marine bacteria and that inhibitory potency correlated with n-alkanol chain length [15]; this observation supported again the hypothesis of anesthetic interactions with proteins. Because the light from marine bacteria derives from the activity of specific enzymes called "luciferases," Harvey deducted that anesthetics bind to proteins and their binding cause an alteration in protein activity.

Going from protein activity to protein structure, Östergren hypothesized that general anesthetics exert their effects on the lipophilic portion of proteins [16]. In the twentieth century, the scientific community began to reunite the observations of Overton and Meyer with those of protein target.

The turning point came from Nick Franks and Bill Lieb. They discovered that anesthetics act straight on proteins, via weak London force interactions in lipid-like, intraprotein nonpolar "hydrophobic pockets" [17]. More recently, Eckenhoff (1915–1996) [18] considered allosteric mechanisms as the key of the anesthetic effects: the simple existence of a general anesthetic in hydrophobic pockets or cavities exercises its effects in the whole protein, by fixing it in a certain conformation. In fact, normal proteins occupy an ensemble of functional conformations, some of which contain hydrophobic pockets or cavities large enough to bind anesthetics. The consequences of this drug–protein binding are a change in protein conformation and structural effects. Nevertheless, a variety of gases bind in hydrophobic pockets of different neural dendritic proteins, giving a variety of measurable effects but confusing the scientific background of molecular interactions between inhaled anesthetics and proteins. In fact, there are not anesthetic gases that follow the Meyer–Overton correlation and bind in hydrophobic pockets but do not cause immobility or loss of consciousness [19] and other Meyer–Overton gases which are predominantly excitatory and cause convulsions [20].

Up to now, mainstream anesthesia research in the early twenty-first century is not just looking for functional and molecular target or to conceive the most plausible theory about the action of anesthetics but is aimed to shift the attention on a different neurobiological view. The mechanism by which general anesthetics stop consciousness is unknown because the mechanism by which brain keeps on consciousness is still not completely explained. The base to understand the act of anesthetics is to return to the concept of consciousness. Neuroscience has no solid theory of consciousness but during the last decade, progress in neuroscience and neuroimaging led to real advances in the understanding of the neural correlates of consciousness. From a neurophysiological point of view, consciousness and anesthetic gases both act through extremely weak London forces (a class of van der Waals forces) in hydrophobic pockets within dendritic proteins. Straightening out this common pathway may reveal how anesthetics operate and also how we keep the vigilance state. We nowadays know that anesthetics do not shut down all cerebral activities, but they induce a shift in the brain state. This condition is a specific and complex state, which can be studied by modern neuroimaging techniques.

In this scientific context, another unitary theory to explain anesthetic action is emerging: the quantum hypothesis. London forces are weak instantaneous couplings between pairs of electron induced dipoles (e.g., between adjacent nonpolar amino acid groups). In normal vigilance state, endogenous van der Waals London forces occur among nonpolar amino acid groups in hydrophobic pockets of neural proteins and facilitate the regulation of their conformation and then function. Quantum effects (*see* description in Box 1) mediated by endogenous London forces in hydrophobic pockets of select neural proteins may be necessary for consciousness. Based on the new and current scientific knowledge, Stuart Hemeroff [21] strongly supports this theory and suggests that "the mechanism of anesthetics may be to inhibit, by exogenous London forces, the necessary quantum states," in contrast to Eckenhoff allosteric theory. Which proteins are interested? Not the membrane receptor proteins, but proteins of cytoskeletal microtubules, which allow the normal function of synapses, primarily dendritic ones.

Box 1 Quantum State

The term "quantum" refers to a discrete element of energy in a system, related to a fundamental frequency of its oscillation. This relation between energy levels and frequencies of oscillation underlies the particle or wave duality inherent in quantum phenomena. Neither the word "particle" nor the word "wave" adequately conveys the true nature of a basic quantum entity, but both provide useful and reproducible graphic representations. Because of their oscillations, quantum particles can exist in two or more states simultaneously, and in a multiple coexisting superposition of alternatives of these states. These quantum superpositions of states are called "quantum bits" or "qubits" (*see* Fig. 1). These particles with their "qubits" have a dynamic behavior that contributes to the collective constitution of a macroscopic state, with continuous jumping from a "unitary state," that represents the equilibrium of the macroscopically visible system, and "reduction states," in which this state collapses. Hameroff and Penrose [22] proposed that consciousness is the product of a fundamental and universal physical feature that is the quantum state, described above. They suggested that the oscillations of this system derived from electron cloud in a series of hydrophobic regions in the proteins and their electric and magnetic waves determinate the quantum superpositions. The protein site is the microtubule, involved in the neural information processing, which regulates dendritic–somatic integrations, influences axonal firing, controls synaptic plasticity. By its composition in tubulin dimers and polarity characteristics, microtubule networks determinate quantum states well "orchestrated," resulting in an appropriately organized system able to develop and maintain quantum coherent superposition, which is adequately organized to coordinate cognitive information, capable to integration and computation, that is to say, the consciousness state.

Is this a new acceptable paradigm or is it another theory conditioned by the main scientific discoveries of our years? (Table 1).

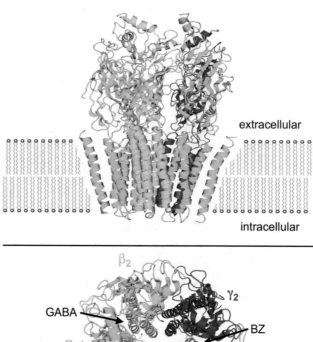

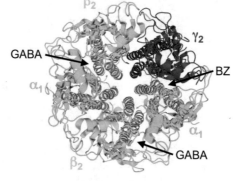

Fig. 1 Three-dimensional structure of GABA$_A$ receptor. Top section of the figure: the receptor associated with the lipid bilayer; bottom: the receptor constituted by the assembly of five subunits illustrated in different colors and consisted of two α (α1/α1), two β (β2/β2) and one γ2 subunit. The extracellular portion of the receptor is directly involved in binding drug (binding site) while the chloride ion channel across the plasma membrane is implicated in the process of gating. Two binding sites for GABA (γ aminobutyric acid) in to the interface between the α and β subunits and the site for benzodiazepines that bind to the interface between the α and γ2 subunit

Table 1
Hypotheses and theories in the history of anesthesia

Authors	Date	Hypothesis
Bernard C.	1875	Protoplasm coagulation theory—anesthetic's interaction with protoplasm
Overton C. E. and Meyer H. H.	1901	Lipid solubility theory—anesthetic lipid potency correlation
Harvey E.N.	1915	Interaction with soluble proteins
Östergren G.	1944	Interaction with lipophilic protein domains
Franks N. and Lieb B.	1984	Anesthetic binding in hydrophobic pockets of soluble proteins
Eckenhoff R. G.	1997	Allosteric binding theory
Hameroff S.	1998	Unitary quantum hypothesis—microtubules theory

3 Molecular Interaction Between Anesthetics and Proteins

3.1 The Task of Binding Forces

Proteins function depends on their conformation: the linear chains of amino acids that constitute a protein "fold" into its three-dimensional conformation. The "shape" depends on attractive and repellent forces among various amino acids: uncharged nonpolar amino acid groups join together, repelled by solvent water. These "hydrophobic" groups attract each other by not covalent force, and cover themselves within the protein interior. Intra protein hydrophobic pockets are composed of side groups of nonpolar, but polarizable, amino acids. Noncovalent forces in hydrophobic pockets determine the protein shape and its dynamic conformational changes. These types of forces, operating among amino acid side groups within a protein, include charged interactions such as ionic forces and hydrogen bonds, as well as interactions between dipoles, separated charges in electrically neutral groups, the above-mentioned van der Waals forces. van der Waals forces are dipole couplings among close atoms or molecules. The weakest type of van der Waals forces, known as London dispersion force, is a dynamic interaction between two neutral, nonpolar atoms or molecules. Adjacent nonpolar electron clouds polarize each other, inducing temporary, instantaneous, dipoles, which attract each other. London force attractions depend on the distance between electron clouds since they are extremely weak but, acting collectively and coherently, they are strong enough to regulate protein conformation in the not polar regions as hydrophobic pockets.

Anesthetic gas molecules occupy hydrophobic pockets by forming London force interactions with not polar amino acid groups, altering protein conformational dynamics and neuronal functions. Franks and Lieb [17] suggested that anesthetics following the Meyer–Overton correlation, by their mere presence in hydrophobic pockets, prevent conformational changes. Indeed, this explanation is not satisfactory since other molecules, which follow the Meyer–Overton correlation and occupy the same pockets, are not anesthetic [19] or they are even convulsant [23], while, on the other hand, not all the target proteins have endogenous hydrophobic ligands [23].

Another explanation of how anesthetics occupancy of hydrophobic pockets may change protein conformational dynamics derives from the allosteric mechanism [17]. As already reported, according to Eckenhoff, the presence of anesthetics in hydrophobic cavities blocks the proteins in one particular conformation. To bind the anesthetic, the protein should be in a conformation which has large hydrophobic pockets. These particular states with large pockets, according to Eckenhoff, are inactive and stabilized by anesthetic occupancy in an inactive state. However, anesthetic-induced structural changes are not proved remarkable [24] and the simple presence of molecules in hydrophobic pockets of proteins is insufficient to explain anesthesia.

Another possibility the scientists are now verifying is that anesthetics somehow disrupt London force interactions in some critical hydrophobic pockets. As suggested by Hameroff and Watt in the early 1980s [25], by forming their own London force attractions in hydrophobic pockets, anesthetics may interfere on electron mobility required for protein dynamics while non-anesthetics may occupy hydrophobic pockets without altering electron mobility. On the contrary convulsant drugs may form cooperative van der Waals interactions that increase electron mobility in excitatory proteins [26].

4   Sites of Anesthetic Action

4.1   The Role of Membrane Proteins

Starting from the mid-twentieth century, when Hodgkin and Huxley gave a *"quantitative description of membrane current and its application to conduction and excitation in nerve"* (1952) [27], researchers started to look for one postsynaptic membrane protein receptor or channel responsible for anesthetic action. The conclusion of many studies, with conflicting results, is that by interfering with global neuronal networks, anesthesia causes a controlled and reversible loss of consciousness *by* potentiating inhibitory proteins and inhibiting excitatory ones [28, 29].

Nevertheless, it should be understood how this mechanism works. The concept that anesthetics disrupt lipid bilayers, or act on some other nonspecific mechanism, has been put aside, and the research is going toward the concept that [1, 21, 30, 31] the inhaled anesthetics work only on few receptors.

We may now list the main results about the most accredited membrane targets of general anesthetics:

- GABA and glycine protein channels [32].
- Glutamate and acetylcholine protein channels [23].
- Potassium channels [33].

Most anesthetics modulate receptors for GABA and other neurotransmitter-gated ion channels at both synaptic and extrasynaptic sites [34]. $GABA_A$ receptors stimulation gives a fast inhibitory postsynaptic current. Volatile gases prolong this inhibitory postsynaptic potential but also increase tonic inhibition mediated by extrasynaptic $GABA_A$ receptors [35].

The most important cerebral target of anesthetics is historically recognized as gamma-amino butyric acid (GABA) receptor, mainly placed in the cortex, thalamus, striatum, and brainstem [36]. Surely, $GABA_A$ receptors have an important role in the anesthetic-induced loss of consciousness, but they are more important for some general anesthetics than for others. Almost all general anesthetics, inhaled gases as well as intravenous agents, play with cellular protein channels, controlling synaptic transmission, in particular to potentiate GABA-induced Cl^- currents and, at higher concentrations, directly activate $GABA_A$ receptors in the absence of GABA itself. Of relevance, xenon has little or no effect on $GABA_A$ receptors [37, 38].

The binding site of the halogenated agents on the ionotropic receptor of GABA is an allosteric site arranged in the lipid bilayer of membrane cell. The more lipophilic molecules reach this allosteric site more easily and the variations in structure of the halogens are responsible for a different affinity for this binding site. Lipophilic agents, in agreement with Overton–Meyer's correlation, can cross the lipid membrane and join the binding sites within hydrophobic pockets or even intracellular domains of the receptor or others cytoplasmatic protein how described above. The allosteric interaction explains the additive effect of the various hypnotics among themselves. In fact, the halogenated agents improve the sensitivity of GABA receptors to their agonists; that is, evident and clinically relevant is the strengthening of inhaled anesthetics by benzodiazepines, agonists of GABA [39, 40] (Fig. 1).

Glycine receptors are inhibitory, homologous to and often colocalized with $GABA_A$ receptors, particularly in the lower brainstem and spinal cord, where they might mediate the action of volatile anesthetics [41–43].

Anesthetics act more powerfully on $GABA_A$ and glycine inhibitory receptors than on other target proteins. Some anesthetics potentiate $GABA_A$ inhibition at concentrations below 1 mM but inhibit $GABA_A$ effects at higher concentrations [18]. Genetic studies and evidence from brain-imaging have excluded an unique role of $GABA_A$ receptors in producing anesthesia, even if they support the hypothesis that the effects of anesthetics are mediated at least in part through GABAergic system.

Current research has made possible to point out that other protein targets are implicated in the anesthetic effect. Halogenated anesthetics stabilize the acetylcholine receptors in an inactive conformational stage [44] but halogenated agents without anesthetic properties receptors do not do the same effects [45]. This was observed for nicotinic receptors, particularly the innately copious *Torpedo* nAChR, that was among the first ligand-gated ion channel models used to probe general anesthetic mechanisms [46]. Up to now, we are far from understanding how these ubiquitous receptors are involved in the mechanisms off loss of consciousness during anesthesia.

Volatile anesthetics have effects on axons and dendrites and presynaptic and postsynaptic membranes as well as on the somatic membranes of neurons and glia. In this last structure, it was shown that some volatile anesthetics caused the augmentation of glutamate uptake into astrocytes, suggesting a rule of this neurotransmitter on decreasing excitatory transmission in the brain [47].

All these functional modifications of cortical and glial neurons are associated with the action of ion channels at the level of the neuronal membrane. Over the years, attention has shifted to ion channels, first of all potassium channels. The conformational modification of these channels in an open form induces a membrane hyperpolarization, which results in a reduction in excitatory responses and alteration in the synchronization of neuronal networks [48].

The biochemical research of recent decades has focused on the anesthetic interaction with the voltage-gated ion channels. The opening of K^+ channels by anesthetics was already mentioned in the 1980s, and there is emergent evidence that K^+ channels mediate some of the effects of volatile agents [49]. They are two-pore-domain K^+ (2PK) channels that provide "background" modulation of neuronal excitability. There are many different 2PK subunits and little is known about the determinant of their anesthetic-sensitivity; five members of this channel family (TREK1, TREK2, TASK1, TASK3 and TRESK) can be straight activated via volatile general anesthetics [50]. Patel et al. in 1999 proved that baseline K^+ channels can be activated by various volatile general anesthetics, establishing 2PK channels as possible targets for general anesthetics [51]. Previously, in the late 1980s Franks and Lieb [52] were the promoters of the first characterization of an anesthetic-activated K+ current which, later was shown to be mediated by a 2PK channel [53]. Anesthetic activation of 2PK channels inhibit neuronal activity by either hyperpolarizing the membrane and/or increasing the membrane conductance, thus reducing the effects of excitatory currents.

Other ion channels that are sensitive to anesthetics are voltage-gated Na^+ and Ca^{2+} channels, very important for excitability and synaptic transmission. Various researchers identify presynaptic Na^+ channels as relevant anesthetic targets for volatile anesthetics [54, 55]. From current literature, come to light that inhibition of Ca^{2+} channels by inhaled anesthetics might be relevant but it is still unclear.

But we returned to the starting point. Is there any single interaction that can explain the global neuronal effect of inhaled anesthetics on consciousness? The same question applies to all interactions between anesthetics molecules and specific neural receptors.

After the Franks and Lieb's discover that anesthetic action in lipid-like hydrophobic pockets of proteins, the prevalence of evidence indicate the hydrophobic pockets included in many brain proteins as primary targets of anesthetic effects. Relatively few proteins have hydrophobic pockets large enough for anesthetics, approximately 15% of neural proteins as described by Eckenhoff et al. in their "*Multiple specific binding targets for inhaled anesthetics in the mammalian brain*" (2002) [56]. Furthermore, drugs effect is not equal to drug binding.

4.2 Interaction with Soluble Proteins

Since Bernard experimented that anesthesia resulted from revocable "coagulation" of cellular proteins, the research attempted to demonstrate which intracellular proteins were involved in the complex mechanics induced by anesthetics. Protoplasmic streaming depends on polymerization cycles of the cytoskeletal protein actin and we now know that anesthetic gases depolymerize actin in dendritic spines in neurons [57]. Actual evidence, from genomics and proteomics investigations, points to anesthetic action in dendritic interiors proteins [58, 59] after many years of focusing on neuronal membrane proteins. Anesthetics act inside the dendrites, via both metabotropic receptors, including Glutamate and $GABA_B$ receptors, and directly on cytoplasmic proteins [60, 61]. Volatile anesthetics bind cytoplasmic protein kinase C, adenylate cyclase, second messenger G proteins, postsynaptic density proteins, actin in dendritic spines, and tubulin in microtubules [18, 62]. In this contest, great importance is given to the role of cytoskeleton, manly constituted by microtubules and actin. Along with actin and other cytoskeletal structures, microtubules establish cell shape, direct growth, and organize neuron functions.

It is important to consider that microtubules are made of tubulin; anesthetics weakly bind to tubulin, which is indeed extremely represented inside the cells [63]. Proteomic analysis of genetic expression following binding demonstrated that some gases functionally act through protein involved in neuronal growth, proliferation, division and communication [64], all microtubule-dependent functions. Studies about genetic expressions of neural proteins in animals show that expression of some soluble proteins, including tubulin, changed following gases anesthetics exposure, while genetic expression of membrane proteins was not affected [65].

Murine brain studies showed alterations in tubulin genetic expression for several days after halogenated anesthetics exposure [58, 59]. Recent discoveries have shown that microtubule instability is associated with postoperative cognitive dysfunction and separation from protein tau (same as in Alzheimer's disease) and microtubules [22, 66, 67]. Microtubules can influence axonal firings to implement behavior, they can regulate synaptic plasticity and act as guides for motor proteins transporting synaptic precursors from cell body to distal synapses [68, 69].

Using computer modeling, it has been demonstrated the anesthetic bind in hydrophobic channels within tubulin [48]. These hydrophobic channels associate with those in adjacent tubulins, to create macroscopic "quantum channels" and collective dipoles through microtubules and neurons [48, 70]. In such quantum channels, anesthetics can inhibit electron mobility and disperse dipoles [25, 71, 72], and these activities could stop cognitive activities essential to consciousness [73] (Fig. 2).

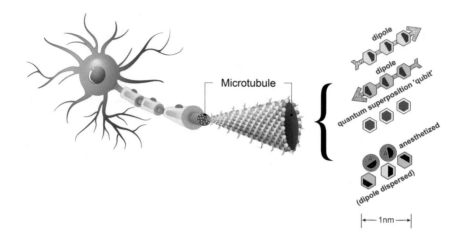

Fig. 2 Tubulin dipoles and anesthesia. Internal cytoskeleton structure of a neuron with section of a microtubule, cylindrical polymer built on thousands of tubular dimers, constituting a polarizable electron dipole with electric properties. Electron dipoles (schematized at the right top as hexagonal structures) are coupled by van der Waals London forces (instantaneous dipole-induced dipole attractions between electron clouds) and oscillate in relation to their neighbors, in each direction (left and right in the figure). Superposition of all states results in quantum balance that yields stable, definite biomolecular coherence (Quantum state). Anesthetic gas molecules (A) form their own (van der Waals London force) dipole couplings, dispersing jointly dipoles and disrupting their collective effects

5   Further Research Perspectives

My brain? It's my second favorite organ

—Woody Allen

Multiple molecular pathways to anesthesia are present in our brain; different anesthetics can use one or more of these pathways to produce anesthesia and particularly the loss of consciousness.

Researchers are still actively searching a unitary mechanism of anesthesia. Altering collective oscillations in microtubules has been suggested as a profitable research pathway by Craddock and Hemeroff, in order to understand how inhaled anesthetics suppress consciousness and then the mechanism by which brain determines consciousness. No one of the molecular targets recognized up to now can, by itself, explain the loss of consciousness. Nevertheless, microtubules appear a promising field of study: modeling and measuring their electrostatic and conductive properties has been proposed by biophysics scientists. Quantum computers and algorithmic models to elucidate the role of quantum mechanics are the future steps in the study of anesthesia.

References

1. Campagna JA, Miller KW, Forman SA (2003) Mechanisms of actions of inhaled anesthetics. N Engl J Med 348:2110–2124

2. Dwyer R, Bennett HL, Eger EI II et al (1992) Effects of isoflurane and nitrous oxide in sub-anesthetic concentrations on memory and responsiveness in volunteers. Anesthesiology 77:888–898

3. Ghoneim MM, Block RI (1997) Learning and memory during general anesthesia: an update. Anesthesiology 87:387–410

4. Eger EI 2nd (2001) Age, minimum alveolar anesthetic concentration, and minimum alveolar anesthetic concentration-awake. Anesth Analg 93:947–953

5. Roizen MF, Horrigan RW, Frazer BM (1981) Anesthetic doses blocking adrenergic (stress) and cardiovascular responses to incision: MAC BAR. Anesthesiology 54:390–398

6. Rampil IJ, Mason P, Singh H (1993) Anesthetic potency (MAC) is independent of forebrain structure in rats. Anesthesiology 78:707–712

7. Caraiscos VB, Newell JG, You-Ten KE et al (2004) Selective enhancement of tonic GABAergic inhibition in murine hippocampal neurons by low concentrations of the volatile anesthetic isoflurane. J Neurosci 24 (39):8454–8458

8. Alkire MT, Nathan SV (2002) Does the amygdala mediate anesthetic-induced amnesia? Basolateral amygdala lesions block sevoflurane-induced amnesia. Anesthesiology 102:754–760

9. Alkire MT, Haier RJ, Fallon JH (2000) Toward a unified theory of narcosis: brain imaging evidence for a thalamocortical switch as the neurophysiologic basis of anesthesia-induced unconsciousness. Conscious Cogn 9:370–386

10. John ER, Prichep LS (2005) The anesthetic cascade: a theory of how anesthesia suppresses consciousness. Anesthesiology 102:447–471

11. Bernard C (1875) Leçons sur les anesthésiques et sur l'asphyxie. Librairie J-B Baillière et Fils, Paris

12. Meyer HH (1901) Zur Theorie der Alkoholnarkose III. Der Einfluss wechselnder Temperatur auf Wikungs-starke and Teilungs Koefficient der Nalkolicka. Arch Exp Pathol Pharmakol 154:338–346

13. Overton CE (1901) Studien uber die narkose: zugleich ein beitrag zur allgemeinen pharmakologie. Gustav Fischer, Jena

14. Perouansky M (2015) The Overton in Meyer–Overton: a biographical sketch commemorating the 150th anniversary of Charles Ernest Overton's birth. Br J Anaesth 114 (4):537–541

15. Harvey EN (1915) The effect of certain organic and inorganic substances upon light production by luminous bacteria. Biol Bull 29:308–311

16. Ostergren G (1984) Colchicine mitosis, chromosome contraction, narcosis and protein chain folding. Hereditas 30:429–467

17. Franks NP, Lieb WR (1984) Do general anaesthetics act by competitive binding to specific receptors? Nature 310:599–601

18. Eckenhoff RG (2001) Promiscuous ligands and attractive cavities: how do the inhaled anesthetics work? Mol Interv 1(5):258–268

19. Koblin DD, Chortkoff BS, Laster MJ et al (1994) Polyhalogenated and perfluorinated compounds that disobey the Meyer-Overton hypothesis. Anesth Analg 79:1043–1048

20. Fang ZX, Sonner J, Laster MJ et al (1996) Anesthetic and convulsant properties of aromatic compounds and cycloalkanes: implications for mechanisms of narcosis. Anesth Analg 83:1097–1104

21. Hameroff SR (1998) Anesthesia, consciousness and hydrophobic pockets—a unitary quantum hypothesis of anesthetic action. Toxicol Lett 100-101:31–39

22. Le Freche H, Brouillette J, Fernandez-Gomez FJ et al (2012) Tau phosphorylation and sevoflurane anesthesia: an association to postoperative cognitive impairment. Anesthesiology 116:779–787

23. Hameroff SR (2006) The entwined mysteries of anesthesia and consciousness. Is there a common underlying mechanism? Anesthesiology 105:400–412

24. Raines DE (2000) Perturbation of lipid and protein structure by general anesthetics: how little is too little? Anesthesiology 92:1492–1494

25. Hameroff SR, Watt RC (1983) Do anesthetics act by altering electron mobility? Anesth Analg 62(10):936–940

26. Hameroff SR, Nip A, Porter M (2004) Conduction pathways in microtubules, biological quantum computation, and consciousness Stuart. Biosystems 64:149–168

27. Hodgkin AL, Huxley AF (1952) A quantitative description of membrane current and its application to conduction and excitation in nerve. J Physiol 117(4):500–544

28. Evers AS, Steinbach JH (1999) Double-edged swords: volatile anesthetics both enhance and inhibit ligand-gated ion channels. Anesthesiology 90:1–3

29. Uhrig L, Dehaene S, Jarra B (2014) Cerebral mechanisms of general anesthesia. Ann Fr Anesth Reanim 33(2):72–78

30. Sonner JM et al (2003) Inhaled anesthetics and immobility: mechanisms, mysteries, and minimum alveolar anesthetic concentration. Anesth Analg 97:718–740

31. Rudolph U, Antkowiak B (2004) Molecular and neuronal substrates for general anaesthetics. Nat Rev Neurosci 5:709–720

32. Franks NP, Lieb WR (1994) Molecular and cellular mechanisms of general anaesthesia. Nature 367:607–614

33. Bertaccini EJ, Dickinson R, Trudell JR et al (2014) Molecular modeling of a tandem two pore domain potassium channel reveals a putative binding site for general anesthetics. ACS Chem Neurosci 5:1246–1252

34. Hemmings HC Jr et al (2005) Emerging molecular mechanisms of general anesthetic action. Trends Pharmacol Sci 26(10):503–510

35. Hemmings HC Jr (2009) Molecular Targets of General Anesthetics in the Nervous System. In: Hudetz A, Pearce R (eds) Suppressing the mind. Springer. Chapter 2

36. Nakahiro M, Yeh JZ, Brunner E et al (1989) General anesthetics modulate GABA receptor channel complex in rat dorsal root ganglion neurons. FASEB J 3:1850–1854

37. De Sousa SL, Dickinson R, Lieb WR et al (2000) Contrasting synaptic actions of the inhalational general anesthetics isoflurane and xenon. Anesthesiology 92:1055–1066

38. Raines DE, Claycomb RJ, Scheller M et al (2001) Nonhalogenated alkane anesthetics fail to potentiate agonist actions on two ligand-gated ion channels. Anesthesiology 95:470–477

39. Goetz T, Arslan A, Wisden W et al (2007) GABAA receptors: structure and function in the basal ganglia. Prog Brain Res 160:21–41

40. Hemmings HC Jr, Egan TD et al (2013) Chapter 1, Mechanisms of drug action. In: Proekt A, Hemmings HC Jr (eds) pharmacology and physiology for anesthesia: foundations and clinical application. Elsevier Health Science, Portland, OR

41. Harrison NL, Kugler JL, Jones MV et al (1993) Positive modulation of human GABAA and glycine receptors by the inhalation anesthetic isoflurane. Mol Pharmacol 44(3):628–632

42. Downie DL, Hall AC, Lieb WR et al (1996) Effects of inhalational general anaesthetics on native glycine receptors in rat medullary neurones and recombinant glycine receptors in Xenopus oocytes. Br J Pharmacol 118:493–502

43. Mascia MP, Machu TK, Harris RA (1996) Enhancement of homomeric glycine receptor function by long-chain alcohols and anaesthetics. Br J Pharmacol 119:1331–1336

44. Stuart A, Formana B, David C et al (2015) Anesthetics target interfacial transmembrane sites in nicotinic acetylcholine receptors. Neuropharmacology 96(Pt B):169–177

45. Douglas E, Raines MD (1996) Anesthetic and nonanesthetic halogenated volatile compounds have dissimilar activities on nicotinic acetylcholine receptor desensitization kinetics. Anesthesiology 84:663–671

46. Sobel A, Heidmann T, Changeux JP (1977) Purification of a protein binding quinacrine and histrionicotoxin from membrane fragments rich in cholinergic receptors in Torpedo marmorata. C R Acad Sci Hebd Seances Acad Sci D 285:1255–1258

47. Miyazaki H, Nakamura Y, Arai T et al (1997) Increase of glutamate uptake in astrocytes a possible mechanism of action of volatile anesthetics. Anesthesiology 86:1359–1366

48. Ries CR, Puil E (1999) Ionic mechanism of isoflurane's actions on thalamocortical neurons. J Neurophysiol 81(4):1802–1809

49. Sonner JM, Cantor RS (2013) Molecular mechanisms of drug action: an emerging view. Annu Rev Biophys 42:143–167

50. Franks NP, Honoré E (2004) The TREK K2P channels and their role in general anaesthesia and neuroprotection. Trends Pharmacol Sci 25(11):601–608

51. Patel AJ, Honoré E, Lesage F et al (1999) Inhalational anesthetics activate two-pore-domain background K + channels. Nat Neurosci 2:422–426

52. Franks NP, Lieb WR (1988) Volatile general anaesthetics activate a novel neuronal K+ current. Nature 333(6174):662–664

53. Andres-Enguix CA, Yustos R et al (2007) Determinants of the anesthetic sensitivity of two-pore domain acid-sensitive potassium channels: molecular cloning of an anesthetic-activated potassium channel from *Lymnaea stagnalis*. J Biol Chem 282(29):20977–20990

54. Schlame M, Hemmings HC Jr (1995) Inhibition by volatile anesthetics of endogenous glutamate release from synaptosomes by a presynaptic mechanism. Anesthesiology 82(6):1406–1416

55. Westphalen RI, Hemmings HC Jr (2003) Selective depression by general anesthetics of glutamate versus GABA release from isolated cortical nerve terminals. J Pharmacol Exp Ther 304(3):1188–1196

56. Eckenhoff MF, Chan K, Eckenhoff RG (2002) Multiple specific binding targets for inhaled anesthetics in the mammalian brain. J Pharmacol Exp Ther 300:172–179

57. Kaech S, Brinkhaus H, Matus A (1999) Volatile anesthetics block actin-based motility in dendritic spines. Proc Natl Acad Sci U S A 96:10433–10437

58. Fütterer CD, Maurer MH, Schmitt A et al (2004) Alterations in rat brain proteins after desflurane. Anesthesiology 100:302–308

59. Kalenka A, Hinkelbein J, Feldmann RE Jr et al (2007) The effects of sevoflurane anesthesia on rat brain proteins: a proteomic time-course analysis. Anesth Analg 104:1129–1135

60. Franks NP, Lieb WR (1998) Which molecular targets are most relevant to general anaesthesia? Toxicol Lett 100–101:1–8

61. Sugimura M, Kitayama S, Morita K et al (2002) Effects of GABAergic agents on anesthesia induced by halothane, isoflurane, and thiamylal in mice. Pharmacol Biochem Behav 72 (1–2):111–116

62. Xi J, Liu R, Asbury GR et al (2004) Inhalational anesthetic- binding proteins in rat neuronal membranes. J Biol Chem 279:19628–19633

63. Hameroff S, Penrose R (2004) Consciousness in the universe: a review of the 'OrchOR' theory. Phys Life Rev 11:39–78

64. Pan JZ, Xi J, Tobias JW, Eckenhoff MF et al (2007) Halothane binding proteome in human brain cortex. J Proteome Res 6(2):582–592

65. Pan JZ, Xi J, Eckenhoff MF, Eckenhoff RG (2008) Inhaled anesthetics elicit region-specific changes in protein expression in mammalian brain. Proteomics 8(14):2983–2992. https://doi.org/10.1002/pmic.200800057

66. Eckenhoff RG, Planel E (2012) Postoperative cognitive decline: where art tau? Anesthesiology 116:751–752

67. Craddock TJA, St. George M, Freedman H, Barakat KH, Damaraju S, Hameroff S, Tuszynski JA (2012) Computational predictions of volatile anesthetic interactions with the microtubule cytoskeleton: implications for side effects of general anesthesia. PLoS One 7(6):e37251. https://doi.org/10.1371/journal.pone.0037251

68. Sánchez C, Díaz-Nido J, Avila J (2000) Phosphorylation of microtubule-associated protein 2 (MAP 2) and its relevance for the regulation of the neuronal cytoskeleton function. Prog Neurobiol 61(2):133–168

69. Kneussel M (2010) Tubulin post-translational modifications: encoding functions on the neuronal microtubule cytoskeleton. Rev Trends Neurosci 33(8):362–372

70. Hameroff S, Nip A, Porter M et al (2002) Conduction pathways in microtubules, biological quantum computation and microtubules. Biosystems 64(13):149–168

71. Hameroff SR, Watt RC, Borel JD et al (1982) General anesthetics directly inhibit electron mobility: dipole dispersion theory of anesthetic action. Physiol Chem Phys 14(3):183–187

72. Hameroff S (1998) Anesthesia, consciousness and hydrophobic pockets – a unitary quantum hypothesis of anesthetic action. Toxicol Lett 100(101):31–39

73. Hameroff S (2006) The entwined mysteries of anesthesia and consciousness. Anesthesiology 105:400–441

Chapter 7

Intravenous Hypnotic Agents: From Binding Sites to Loss of Consciousness

Daniela Baldassarre, Filomena Oliva, and Ornella Piazza

Abstract

All the intravenous hypnotic drugs important for clinical anesthesiology reversibly unsettle functional brain networks, in order to undermine the information transfer on which consciousness depends. Three classes of intravenous hypnotic drugs are the most used nowadays: the carboxylated imidazole derivate propofol, the short-acting benzodiazepine midazolam, and the barbiturates, which show action on $GABA_A$ Receptors, potentiating gamma-aminobutyric acid (GABA) action. The dissociative agent ketamine, instead, mainly exerts its effects by reversibly blocking the activity of N-methyl-D-aspartate receptors while the most recent dexmedetomidine is an alpha-2 adrenergic receptor agonist. Nevertheless, other receptors are also involved in anesthesia determining, that is voltage-gated and ligand-gated ion channels and it is probable that each intravenous hypnotic agent alters neuronal activity acting at different levels and at multiple sites in a way not yet entirely clear.

Key words Propofol, Midazolam, GABA, $GABA_A$ receptors, Ketamine, Dexmedetomidine

1 Introduction

Several intravenous hypnotic agents, as well as volatile compounds, lead to the common expression of unconsciousness but only partially through the same molecular action. Intravenous hypnotic agents currently in use include the carboxylated imidazole derivate propofol, which replaced the old etomidate, the short-acting benzodiazepine midazolam, and the barbiturates. These three classes of hypnotic drugs show a prevalent and demonstrated action on gamma-aminobutyric acid class A ($GABA_A$) receptors, potentiating GABA action, while the still used dissociative agent ketamine, unlike the other intravenous agents, do not act primarily through the potentiation of GABA transmission but mainly exerts its effects by reversibly blocking the activity of N-methyl-D-aspartate (NMDA) receptors. To the aforementioned drugs, the most recent alpha 2 adrenergic receptor agonist dexmedetomidine was recently added. In spite of structural, pharmacological and neurobiological

Marco Cascella (ed.), *General Anesthesia Research*, Neuromethods, vol. 150, https://doi.org/10.1007/978-1-4939-9891-3_7,
© Springer Science+Business Media, LLC, part of Springer Nature 2020

differences, all the hypnotic drugs important for clinical anesthesiology, reversibly unsettle functional brain networks in order to undermine the information transfer on which consciousness seems to depend [1]. It is not effectively clear how outstanding complex and dynamic cognitive functions actually come out from the interactions of these brain networks, but the reduced efficiency in these information neural integrations has been proposed as the main feature of intravenous hypnotic drugs induced unconsciousness [2], even if it is unclear whether this conclusion can be generalized to anesthetics with different molecular targets. Many anesthetic mechanisms have been proposed to be responsible for generating the state of general anesthesia, but none of them actually can explain alone how molecular actions modify brain function.

2 The Molecular Mechanism Underlying Intravenous Hypnotics Action

2.1 The Development of the Receptor Binding Concept

All components of the most important families of ligand and voltage-gated ion channels and their many subtypes studied until now are involved in the bond with the intravenous general anesthetics, including the most well know and important $GABA_A$ receptor [3] and Glycine receptor channels [4], NMDA Glutamate receptors [5], Nicotinic acetylcholine receptor (AChR) [6] and the novel ATP-gated P2X receptor cation channel (P2X) receptors modulating presynaptic release of some neurotransmitters [7]. A part of clinically important general anesthetics like propofol, midazolam, and barbiturates positively modulate the function of $GABA_A$ receptors through interactions with their transmembrane domains but there are also hypnotic drugs that have no or little effect on GABA receptor and they produce unconsciousness as well. The thing even more interesting is that none of the current clinical general anesthetics are selective for a single receptor or voltage-gated ion channel and at clinical concentrations, every anesthetic modulates the function of more types of protein components in the central nervous system. Many of other systems are also influenced by intravenous anesthetics agents with no involvement of cerebral functions designate to produce unconsciousness but it could be not otherwise considering that the presence of ion channels for example, that is most known molecular target of an intravenous anesthetic, is widely diffused in human cellulose. The diversity of the binding sites involved reflects the enormous structural variability of the anesthetic compounds and the effects produced by each of the bonds at the molecular level on unconsciousness establishment are still not clear. The actual knowledge steer towards a convenient conclusion that all of the sites of action implicated, to a greater or lesser extent, leading to an extended disruption in the exchange of substantial brain information as demonstrated for some of these drugs [8]. Understanding which mechanisms underlie the interaction between intravenous hypnotic agents and their molecular targets is a noteworthy current challenge that can give an important contribution to the knowledge of the mechanism underlying the still obscure constitution of consciousness but it also can give the possibility to improve the specificity of the drugs used or to synthesize other drugs in this sense. Current research oriented to discover the crucial binding site that determinate only the effect which the hypnotic agents are used for, take into consideration the development of the receptor concept in the last few years. The capacity to bind a macromolecule by a drug does not allow automatically classify as a receptor because of drug binding receptor is not equivalent to drug effect. Propofol, for example, shows a pronounced binding for serum albumin following intravenous administration but there is no direct effect on consciousness from this interaction despite affinity [9]. The receptor is defined as a binding site with a functional correlate [10] and it corresponds to a small and specific fraction of the total amino acid sequence of the protein. Diverse hypnotic drugs have the capacity to modulate the function of the same receptor through allosteric interactions because of the multiple distinct binding sites on the same macromolecules. It has already been known that on $GABA_A$ receptor there are binding sites in transmembrane interfaces that are a one of a kind modulator sites without established ligands or rather an exclusive site for some precise hypnotic agents [11]. The study on the structure of the Glycine receptor showed that specific amino acid residues known to bind some anesthetic drugs were in close proximity each other [12] forming a distinctly evident cavity within each of the four-helix bundle transmembrane domain subunits of this receptor [13]. The aim of many electrophysiology, photolabeling, and molecular modeling studies is to identify the exact molecular structure of these folding sequences and the alignment of the amino acid that assembles the allocations for any anesthetic drugs inside the three-dimensional structure of macromolecules. These cavities could be located in either intra-subunit or inter-subunit surface within the transmembrane part of the receptor but also in the extracellular domain or in the pore as for the cationic ion channels. These types of receptors have shown a link with anesthetic drugs too but many technical difficulties for their isolation in different conformational states make less clear their rule. The argument is even more difficult because of their wide distribution practically in every human tissue.

2.2 The Nature of the Binding Sites

The physicochemical properties of these performed protein cavities in which anesthetic compounds accommodated revealed a very complex binding structure. As the inhaled anesthetics, described elsewhere, also the intravenous agents show weak binding interactions with their specific receiving pockets [14]. These weak chemical interactions have in themselves the nature of reversible bond forming hydrogen bonds and other intermolecular forces [15], as reversible is the anesthetic effect. Some actual molecular models clarify the amphiphilic nature of the anesthetic binding sites. This means that the bonds of the anesthetic compounds with these three-dimensional cavities, otherwise called crevices, are determined by hydrophobic and hydrophilic types of interactions and not just hydrophobic as described for many years. Recognizing the real nature of the anesthetic binding sites gives reason to those who have tried to explain, for more than 20 years, the action and potency of several agents that not obey the Meyer–Overton correlation [16]. The access of hypnotic agents to the amphiphilic cavities does not happen only by dissolution in the phospholipid bilayer correlated with the compound's lipid solubility, as supposed by Meyer–Overton hypothesis, but also by other nonlipophilic accesses [17]. Actually, one of two well-known binding sites of propofol on GABAA receptor is located in the transmembrane region [18] and the other is a pocket quite near the extracellular domain [19] whereas the benzodiazepines act exclusively on a binding site on the extracellular side [20]. This knowledge, though still incomplete, on the one side complicate an already heterogeneous scene, but on the other side slowly materialize the concept of specificity of action of each of these drugs at the molecular level as unique and specific is the clinical effects that they produced.

From multiple molecular targets to neurobiological effect. The current results point out the great variability of molecular targets implicated in a bond with the intravenous hypnotic agents [21]. The GABA receptors have been studied over the years and the role of receptors type A in the anesthetic effects is indisputable.

Propofol bind $GABA_A$ receptor directly activating GABAergic inhibitory activity in the brain and at lower intravenous concentrations it can enhance the receptor action when GABA or other ligands are bonded on it [22]. The ability to bind this type of receptors is not possessed by all the hypnotic drugs used indeed some of them have no GABAergic action at all, and besides, others act on more than only one target. Binding sites on different glycine receptor subunits have been identified for propofol [4], as well as on the nicotinic acetylcholine receptors (nAChRs) [23] despite controversies about the role of these molecular targets on brain functions at the base of loss of consciousness. The review of years of investigations confirms the molecular bond of propofol also on voltage-gated ion channels, first of all, and confirmed by molecular modeling, on subfamilies of sodium channels that are crucial for setting membrane neuronal potential and the initiation and propagation of action potentials in the brain [24]. The recent discovery of the action of propofol on kinesins, fundamental motor proteins expressed in the brain on the microtubules, refers to the hypothesis already advanced for volatile anesthetics of mechanistic support for the anesthetic state [25]. As is common knowledge ketamine binds more than one binding sites on many other neuronal receptors. It has distinct pockets on the NMDA receptor reducing neural excitation of this neurotransmitter circuits and therefore modulates the inhibitory interneurons in a reversible way [26]. Among its many effects on other several molecular targets, the role on NMDA is still the most investigated and it takes into account as the determiner in controlling consciousness. It is not negligible the action of ketamine on subtypes of cationic channels that have an important role in governing neuronal excitability and synaptic transmission among brain networks, first of all, some family of hyperpolarization-activated cyclic nucleotide-gated (HCN) channels that are expressed in nervous system implicated into hypnotic action [27]. Currently, anesthesiologists use drugs that can induce unconsciousness in a reversible way without truly knowing how they act at the neuronal level. The example of this paradox is the use in the clinical practice of barbiturates drugs for more than a century, now widely substituted by propofol and benzodiazepines with less probability of overdose and addiction, without really knowing their molecular targets. Only recently some data acknowledge the nature of the bond between barbiturates and a cavity on some macromolecules through electrostatic interaction, hydrogen bonds and van der Waals forces whit with reversibility feature [28]. The same uncertainty overshadows dexmedetomidine in use since the 90 years when was described its role on a subtype of alpha 2 receptors in the central nervous system and his alteration in transmembrane potassium conductance resulting in indirectly suppressing neural firing [29, 30]. Its therapeutic use for the prevention of postoperative and intensive care unit delirium has aroused the interest toward the action of dexmedetomidine on $GABA_A$ receptors and indirectly the inhibitory pathways [31]. Based on a study conducted with magnetic resonance imaging, Hashmi et al. highlighted how dexmedetomidine severely influences the efficiency of information exchange between neuronal networks [32]. They support the hypothesis now most influences the scientific opinion that unconsciousness is depended on a downturn in the ability for efficient information transmission in the brain [33]. Comparable results have been obtained for various hypnotic drugs with diverse molecular action as ketamine [34] and propofol [2, 35]. This view complies with the concept that consciousness is the expression of a global organization of the central nervous system and their neural circuits [36], and seems to agree with the multiple features of the hypnotic agents used to induce in the end the same pharmacological alteration on it. It is as if each compound, with its distinctive molecular action, weakens the complexity of cerebral interactions or the neural connectivity in specific brain regions depending on the circuits that they interrupt [37]. The topic is very complex and the debate on the global state of consciousness is uncertain and not yet completely convincing because of state of consciousness is composed of different higher brain functions with various degree of multidimensional states [38]. Sanders and his colleagues [39] supposed that some brain functions as consciousness, connectedness with environment and responsiveness unfastened during general intravenous anesthesia and depended on drugs used or different doses of the same compound. It has been observed that low concentration of propofol-induced unresponsiveness to verbal command and subcortical structures become functionally disconnected but the consciousness might remain operative in some degree [40]. At higher concentrations of propofol, the cortical connectivity was interrupted as though a global impairment of neural information integrations [35, 41]. The responsiveness is not a measuring of consciousness in the same manner as the disconnection consciousness from the environment is not equivalent to unconsciousness. It has been proposed for each of these neuronal abilities the involvement of own neuronal circuits and molecular targets during the use of hypnotic drugs at different doses. The high cholinergic tone in the cortex, for example, may underlie dreaming, a state of disconnected consciousness as Sanders defines it [42], in both sleep and anesthesia [43]. A cholinesterase inhibitor, physostigmine, provokes the return of consciousness in patients sedated with propofol [44]. As well as administration of a muscarinic antagonist, scopolamine, seems to prevent dreaming in patients under propofol anesthesia [45]. The correlation between the degree of involvement of the various molecular targets and the modulation of different neuronal circuits producing unconsciousness remains unclear. A lot of data exists and there are many advanced technologies that allow the identification of binding sites of intravenous hypnotic agents but the physiology of consciousness still remains without concrete explanation as well as the actions of these drugs on the brain.

3 Conclusion

Discussions of the molecular mechanism of intravenous hypnotic drugs have been dominated for decades by the view that enhancement of the activity of $GABA_A$ receptors is the most important component, perhaps the only component, of the mechanism of unconsciousness. It is clear that some intravenous general anesthetics and, as described elsewhere, inhaled agents too, have little or no effect on GABA receptors. It is commonly known that other receptors are also involved, as a variety of voltage-gated and ligand-gated ion channels connected with neuronal conductance and influencing functions of brain networks. Each different intravenous hypnotic agent alters neuronal activity acting at different levels and at multiple sites. Current models describe cavities on the macromolecules with amphiphilic features that bind the hypnotic agents and act in a way that is not yet entirely clear.

References

1. Mashour GA (2017) Network inefficiency: a Rosetta Stone for the mechanism of anesthetic-induced unconsciousness. Anesthesiology 126(3):366–368

2. Monti MM, Lutkenhoff ES, Rubinov M et al (2013) Dynamic change of global and local information processing in propofol-induced loss and recovery of consciousness. PLoS Comput Biol 9(10):e1003271. https://doi.org/10.1371/journal.pcbi.1003271

3. Shin DJ, Germann AL, Johnson AD, Forman SA, Steinbach JH, Akk G (2018) Propofol is an allosteric agonist with multiple binding sites on concatemeric ternary GABAA receptors. Mol Pharmacol 93(2):178–189

4. Yevenes GE, Zeilhofer HU (2011) Allosteric modulation of glycine receptors. Br J Pharmacol 164(2):224–236

5. Li L, Vlisides PE (2016) Ketamine: 50 years of modulating the mind. Front Hum Neurosci 10:612

6. Jayakar SS, Dailey WP, Eckenhoff RG, Cohen JB (2013) Identification of propofol binding sites in a nicotinic acetylcholine receptor with a photoreactive propofol analog. J Biol Chem 288(9):6178–6189

7. Stojilkovic SS, Leiva-Salcedo E, Rokic MB, Coddou C (2014) Regulation of ATP-gated P2X channels: from redox signaling to interactions with other proteins. Antioxid Redox Signal 21(6):953–970

8. Hamilton C, Ma Y, Zhang N (2017) Global reduction of information exchange during anesthetic-induced unconsciousness. Brain Struct Funct 222(7):3205–3216

9. Bhattacharya AA, Curry S, Franks NP (2000) Binding of the general anesthetics propofol and halothane to human serum albumin. High resolution crystal structures. J Biol Chem 275(49):38731–38738

10. (1994) Receptor and ion channel nomenclature. Trends Pharmacol Sci (Suppl):1–51

11. Nourmahnad A, Stern AT, Hotta M et al (2016) Tryptophan and cysteine mutations in M1 helices of α1β3γ2L γ-aminobutyric acid type a receptors indicate distinct intersubunit sites for four intravenous anesthetics and one orphan site. Anesthesiology 125(6):1144–1158

12. Bertaccini EJ, Shapiro J, Brutlag DL et al (2005) Homology modeling of a human glycine alpha 1 receptor reveals a plausible anesthetic binding site. J Chem Inf Model 45(1):128–135

13. Bertaccini EJ, Wallner B, Trudell JR et al (2010) Modeling anesthetic binding sites within the glycine alpha one receptor based on prokaryotic ion channel templates: the problem with TM4. J Chem Inf Model 50(12):2248–2255

14. Fahrenbach VS, Bertaccini EJ (2018) Insights into receptor-based anesthetic pharmacophores and anesthetic-protein interactions. Methods Enzymol 602:77–95

15. Sandorfy C (2004) Weak intermolecular associations and anesthesia. Anesthesiology 101(5):1225–1227

16. Koblin DD, Chortkoff BS, Laster MJ et al (1994) Polyhalogenated and perfluorinated compounds that disobey the Meyer-Overton hypothesis. Anesth Analg 79(6):1043–1048

17. Trudell JR, Bertaccini E (2002) Molecular modelling of specific and non-specific anaesthetic interactions. Br J Anaesth 89(1):32–40

18. Jayakar SS, Zhou X, Chiara DC et al (2014) Multiple propofol-binding sites in a γ-aminobutyric acid type A receptor (GABAAR) identified using a photoreactive propofol analog. J Biol Chem 289(40):27456–27468

19. Yip GM, Chen ZW, Edge CJ (2013) A propofol binding site on mammalian GABAA receptors identified by photolabeling. Nat Chem Biol 9(11):715–720

20. Chiara DC, Jounaidi Y, Zhou X (2016) General anesthetic binding sites in human α4β3δ γ-aminobutyric acid type a receptors (GABAARs). J Biol Chem 291(51):26529–26539

21. Yamakura T, Bertaccini E, Trudell JR et al (2001) Anesthetics and ion channels: molecular models and sites of action. Annu Rev Pharmacol Toxicol 41:23–51

22. Shin DJ, Germann AL, Johnson AD et al (2018) Propofol is an allosteric agonist with multiple binding sites on concatemeric ternary GABAA receptors. Mol Pharmacol 93(2):178–189

23. Jayakar SS, Dailey WP, Eckenhoff RG et al (2013) Identification of propofol binding sites in a nicotinic acetylcholine receptor with a photoreactive propofol analog. J Biol Chem 288(9):6178–6189

24. Tang P, Eckenhoff R (2018) Recent progress on the molecular pharmacology of propofol. F1000Res 7:123

25. Bensel BM, Guzik-Lendrum S, Masucci EM et al (2017) Common general anesthetic

propofol impairs kinesin processivity. Proc Natl Acad Sci U S A 114(21):E4281–E4287

26. Mion G (2017) History of anaesthesia: The ketamine story - past, present and future. Eur J Anaesthesiol 34(9):571–575

27. Zhou C, Douglas JE, Kumar NN et al (2013) Forebrain HCN1 channels contribute to hypnotic actions of ketamine. Anesthesiology 118 (4):785–795

28. Oakley S, Vedula LS, Bu W et al (2012) Recognition of anesthetic barbiturates by a protein binding site: a high resolution structural analysis. PLoS One 7(2):e32070. https://doi.org/10.1371/journal.pone.0032070

29. Hayashi Y, Maze M (1993) Alpha 2 adrenoceptor agonists and anaesthesia. Br J Anaesth 71 (1):108–118

30. Nacif-Coelho C, Correa-Sales C, Chang LL et al (1994) Perturbation of ion channel conductance alters the hypnotic response to the alpha 2-adrenergic agonist dexmedetomidine in the locus coeruleus of the rat. Anesthesiology 81(6):1527–1534

31. Wang DS, Kaneshwaran K, Lei G et al (2018) Dexmedetomidine prevents excessive γ-aminobutyric acid type a receptor function after anesthesia. Anesthesiology 129 (3):477–489

32. Hashmi JA, Loggia ML, Khan S et al (2017) Dexmedetomidine disrupts the local and global efficiencies of large-scale brain networks. Anesthesiology 126(3):419–430

33. Hudetz AG, Mashour GA (2016) Disconnecting consciousness: is there a common anesthetic end-point? Anesth Analg 123 (5):1228–1240

34. Bonhomme V, Vanhaudenhuyse A, Demertzi A et al (2016) Resting-state network-specific breakdown of functional connectivity during ketamine alteration of consciousness in volunteers. Anesthesiology 125(5):873–888

35. Boveroux P, Vanhaudenhuyse A, Bruno MA et al (2010) Breakdown of within- and between-network resting state functional

magnetic resonance imaging connectivity during propofol-induced loss of consciousness. Anesthesiology 113(5):1038–1053

36. Berlucchi G, Marzi CA (2019) Neuropsychology of consciousness: some history and a few new trends. Front Psychol 10:50. https://doi.org/10.3389/fpsyg.2019.00050

37. Pappas I, Adapa RM, Menon DK et al (2019) Brain network disintegration during sedation is mediated by the complexity of sparsely connected regions. NeuroImage 186:221–233

38. Bayne T, Hohwy J, Owen AM (2016) Are there levels of consciousness? Trends Cogn Sci 20(6):405–413

39. Sanders RD, Tononi G, Laureys S et al (2012) Unresponsiveness ≠ unconsciousness. Anesthesiology 116(4):946–959

40. Mhuircheartaigh RN, Rosenorn-Lanng D, Wise R et al (2010) Cortical and subcortical connectivity changes during decreasing levels of consciousness in humans: a functional magnetic resonance imaging study using propofol. J Neurosci 30(27):9095–9102

41. Backman SB, Fiset P, Plourde G et al (2004) Cholinergic mechanisms mediating anesthetic induced altered states of consciousness. Prog Brain Res 145:197–206

42. Sanders RD, Raz A, Banks MI, Boly M, Tononi G (2016) Is consciousness fragile? Br J Anaesth 116(1):1–3

43. Meuret P, Backman SB, Bonhomme V et al (2000) Physostigmine reverses propofol-induced unconsciousness and attenuation of the auditory steady state response and bispectral index in human volunteers. Anesthesiology 93(3):708–717

44. Toscano A, Pancaro C, Peduto VA (2007) Scopolamine prevents dreams during general anesthesia. Anesthesiology 106(5):952–955

45. Hugel S, Schlichter R (2000) Presynaptic P2X receptors facilitate inhibitory GABAergic transmission between cultured rat spinal cord dorsal horn neurons. J Neurosci 20(6):2121–2130

Chapter 8

Pharmacological Considerations for the Use of General Anesthetics in the Elderly

Francesca Guida, Enza Palazzo, Serena Boccella, Livio Luongo, Giulio Scala, Francesca Gargano, Gorizio Pieretti, Ida Marabese, Mariantonietta Scafuro, Vito de Novellis, and Sabatino Maione

Abstract

Aging is a physiological condition involving progressive degenerative modifications and loss of functionality in all organ systems. In general, aged patients are frail and sensitive to anesthesia and surgery. Clinical practice indicates that being more sensitive to anesthetic drugs and more susceptible to the side effects, elderly requires usually lower doses to reach the clinical anesthesia. This is mainly associated with the significant changes in pharmacokinetics (and pharmacodynamics) occurring with advancing age. However, how the dose–response relationship is affected by age should be better defined. This chapter focuses on the main changes that physiologically occur with age, how these changes affect organs and systems, and their impact on anesthetic care. The agents used in general anesthesia are discussed in detail with particular emphasis on aspects concerning the dose, safety, and efficacy of their use in the elderly. Moreover, we highlight some recent preclinical evidence on the use of general anesthetics in aged animals.

Key words Pharmacology, Elderly, General anesthetics, Side effects, Pharmacokinetics

1 Introduction: *Age-Related Physiological Changes*

It is expected that the elderly population, which includes people aged 65 and over, is destined to increase progressively. In Europe the current population of elderly will increase from 20% to 25% in 2030 and 30% in 2050.

Aging is a physiological process associated with progressive degeneration of the structure and function of tissues and organs over time [1]. The aging process includes changes of physiology on three types of processes: (a) homeostatic mechanisms, such as body temperature and fluid volume regulation, (b) decrease in organ mass, and (c) functional reserves of the systems of the body. These functional reserves are extremely important for the individual's ability to face challenges such as trauma or surgery.

Marco Cascella (ed.), *General Anesthesia Research*, Neuromethods, vol. 150, https://doi.org/10.1007/978-1-4939-9891-3_8,
© Springer Science+Business Media, LLC, part of Springer Nature 2020

The aging process is highly complex and variable. It depends on numerous factors such as lifestyle, inheritable factors, and chronic diseases and is influenced by a wide range of external factors, such as microorganism exposure, exercise, pollutants, diet, and ionizing radiation. Indeed, elderly patient population ages differently, a large proportion of them is functionally independent suffering of milder chronic diseases, appropriately treated. A consistent part of elderly individuals instead shows coexisting diseases, such as atherosclerosis, heart failure, diabetes, chronic obstructive lung disease, kidney or liver impairments and cognitive decline. Normally, a physiological reduction in function of organs and systems occurs with age, although the amount and the rate of the progression differ between body systems within the same individual and between individuals.

1.1 Cardiovascular System

Changes in the heart functionality with aging are generally caused by collagen proliferation affecting myocardial layers and vessels. Arterial wall thickening, smooth muscle tone increases and matrix composition changes with increased glycosylation reactions induce vessel stiffening. These anatomical changes decrease diastolic compliance as well as increases pulse pressure. There is a reduced responsiveness to the stimulation of adrenergic receptors (notwithstanding an increase in circulating catecholamines) and a reduced response of baroreceptors and chemoreceptors. These changes lead to functional alterations that determine pathological conditions such as atrioventricular conduction defects, systolic hypertension, calcification of the aortic valve, diastolic dysfunction, up to heart failure. Moreover, the compensatory response to erect posture mediated by plasma renin activity and the response of sodium restriction mediated by aldosterone are reduced, or even absent.

1.2 Nervous System

The amount of brain neurons is remarkably reduced in elderly with a decrease which is an approximately of 20%. As a result, the size of the brain tends to shrink and brain areas such as substantia nigra, striatum, hippocampus, locus coeruleus, and the cerebral cortex tend to atrophy. The concentration of main central nervous system neurotransmitters, including serotonin, catecholamines, and acetylcholine, decreases in elderly with consequent effects on mood, memory, and motor function. The loss of autonomic, sensory, and motor fibers reduces afferent and efferent conduction rates and causes denervation and muscle atrophy. Parasympathetic system activity decreases and sympathetic system activity increases with aging. The increase in sympathetic nervous system activity augments in turn vascular resistance and, paradoxically, reduces the response to stimulation the β-adrenergic receptors.

1.3 Renal and Urinary System

Aging reduces renal cortex affecting the volume weight and whole size. The number of glomeruli of kidney is reduced together with glomerular filtration rate (GFR) which results decreased until to 50% in 80 years old subjects. The hardening of the renal vessels, the accumulation of increased quantities of nephrotoxic substances and ischemic damage increase the risk of renal failure. Dysuresia, pollakisuria, and urinary incontinence are main urinary system impairments associated with elderly.

1.4 Respiratory System

The deterioration of collagen and elastic fibers of the lung tissue that occurs with age reduces the elasticity and the ability of the lungs to spread. The loss of elastic fibers at the alveolar level favors the collapse of the alveoli and terminal bronchioles, reducing effective lung volumes [2]. Aging also reduces PCO_2 in arterial blood, expiratory volume and velocity, and airway cilia motility, thus compromising defense mechanisms. An increase in dead space, weakening of respiratory muscles, and diaphragm stiffening, which altogether contribute to reduce respiratory efficiency, are also observed in the elderly.

1.5 Skeletal System

Bone mineral density decline is a consequence of calcium and protein level decrease in elderly. The reduction in bone mass is mainly observed in female due to the fall in estrogen levels. Osteoblast degeneration and irregular bone resorption or deposition are due to biochemical changes. Cell degeneration obstacles the cell communication and consequent lipofuscin deposition induces necrosis. The reduction of bone mineral density affects mainly the bottom of the radial bones, femoral neck, and spine which may easily fracture. Furthermore, the deterioration of joint cartilage causes wear and loss of elasticity compromising the motor capacity of joints.

1.6 Gastrointestinal System

Age-related alterations in the structure of gastrointestinal tract cause changes in neuromuscular, absorptive and secretory function of the bowel. Functional alterations in absorption and secretion are mainly found in the stomach and small bowel. Reflux and achalasia are primarily associated with neuromuscular impairment of the upper gastrointestinal tract. Structural changes distally in the colon are responsible of age-related diverticula.

1.7 Hepatic System

Aging also reduces the volume and blood perfusion of the liver. Interestingly, there is a loss of hepatocytes and an increase in the mean cell volume as an adaptive mechanism of the remaining cells. The synthesis of coagulation factors and several proteins can be reduced without altering, however, the baseline function. A decrease in cytochrome P450 enzymes can instead reduce the ability of the liver to metabolize drugs increasing the risk of adverse reactions. A lowering in immune responses against pathogenic microorganisms or cancer cells in the elderly may predispose to viral hepatitis and development of hepatocellular carcinoma. The impairment of dendritic and regulatory T cell differentiation may instead predispose to autoimmune liver diseases [3].

Apart from the several physiological changes in main body systems (Table 1) geriatric population assumes several drugs, often more than 5 drugs a day. The intake of several drugs a day is called polypharmacy. Polypharmacy is associated with high risk of drug interactions and adverse reactions. These in turn cause a reduction in the compliance.

Table 1
Physiological changes in elderly in the main body systems

System	Changes
Cardiovascular	Reduced cardiac output Reduced baroreceptor activity Reduced peripheral resistance
Renal	Reduced renal blood flow Reduced glomerular filtration Reduced tubular secretion
Nervous	Reduced neuron density Reduced reflexes Reduced sympathetic responses
Respiratory	Reduced lung capacity Reduced peak airflow and gas exchange Weakening of respiratory muscle Reduced vital capacity Reduced tidal volume Increased residual volume
Gastroenteric	Reduced gastric emptying Reduced gastrointestinal motility Increased gastric pH Reduced intestinal blood flow
Hepatic	Decreased liver volume Decreased blood flow Decreased cytochrome P450 activity
Skeletal	Loss of calcium Reduced collagen synthesis Decreased bone mass Increased bone fragility

2 Pharmacology in the Elderly

Age increases sensitivity to anesthetics and the risk of adverse reactions. Thus usually lower doses are required to reach the desired clinical outcomes. This is mainly due to the significant pharmacokinetic (what the body makes to the drug) and pharmacodynamic (what the drug makes to the body) changes occurring with advancing age (Fig. 1).

2.1 Pharmacokinetic Implications

2.1.1 Biometric Parameters

Generally, for a given dose of drug, the blood concentration and the volume of distribution (Vd) are inversely related. Drug pharmacokinetic is strongly affected by the aging. Compared with young, aged patients show an increase and a decrease of adipose tissue and body water, respectively. The Vd of lipid drugs, which is augmented in fat patients, generates an enhancement of their elimination half-life with a significant pharmacological impact specially after repeated or continuous dosing (i.e., midazolam). On the other hand, old patients show reduced total body water, and this effect may be something aggravated by the usage of diuretics. Hence, hydrophilic drugs which have a small Vd and high plasma concentration can generate a greater pharmacologic effect with aging (i.e., morphine). Anesthetic drugs are often bound to plasma proteins, in particular albumin, whose concentration is fairly decreased in the elderly. Thus, free fraction of drugs with highly protein binding, that is, propofol, can be significantly higher, though its clinical relevance is probably limited.

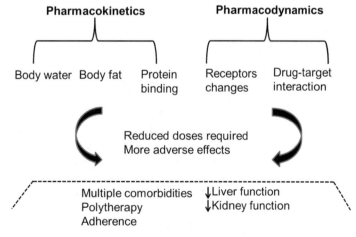

Fig. 1 Pharmacokinetic and pharmacodynamic modifications occurring with advancing age. Aging-associated biometric and physiological modifications result in altered distribution volume and clearance. Quantitative and functional alterations of receptors may be responsible for altered pharmacological effects and more adverse effects

2.1.2 Hepatic Metabolism	Metabolism of the drugs in the liver is divided into two different phases (I and II). The phase I reactions consist of hydrolysis, oxidation, reduction which are catalyzed by cytochrome P450 (CYP) enzymes, while the reactions of phase II consist of chemical conjugation with substrates which change the phase I metabolites into a more polar products. Only the phase I seems to be affected by aging. Reductions in blood perfusion and liver mass decreasing the metabolic capacity of drugs are observed with age. The clearance of drugs such as ketamine, flumazenil, morphine, and fentanyl is decreased up to 30–40% in older patients [4].
2.1.3 Renal Elimination	During elimination, many drugs, including anesthetics, undergo glomerular filtration and reach the renal tubules. Lipophilic compounds can be reabsorbed whereas hydrophilic metabolites are excreted. The glomerular filtration rate is affected by the age, together with the renal blood flow, by impacting on the pharmacological effects (Table 1). As a consequence, the half-life of drugs mainly eliminated through renal system increase promoting the risk of toxicity.
2.2 Pharmacodynamic Implications	Quantitative and functional alterations of receptors, which are targets of anesthetic drugs, in the brain, may be responsible for altered pharmacological effects, together with toxicity, in elderly patients. Specific changes in N-methyl-D-aspartate (NMDA) receptors, which is the target of several anesthetics, have been associated with advancing age [5, 6]. Moreover, alterations in γ-aminobutyric acid (GABA)$_A$ receptors could at least partly, explain the augmented effects at given dose of benzodiazepines (i.e., diazepam or midazolam) or propofol [7–9] in the elderly.

3 Anesthetics Management

Elderly patients require special consideration for anesthesia. Since the pharmacokinetic and pharmacodynamic, as well as the response to drugs, change with aging, the choice of anesthetic drugs should reflect such modifications. Diminished cardiac output together with a late clearance in aged patients contributes to the slow drugs onset and prolonged clinical effects. Thus, it is essential to titrate drug dosage and to use short-acting drugs, also to avoid possible anesthetic overdose. Moreover, elderly patients assume usually several drugs which can complicate the anesthetic management. The possibility of drug interactions affecting metabolism and the end effects must be considered when arranging the anesthetic strategy.

3.1 Benzodiazepines

They are commonly used in premedication to obtain anxiolysis, anterograde amnesia and sedation; however, long-acting compounds (i.e., diazepam) are considered potentially inappropriate in geriatric patients because of possible secondary effects, including respiratory depression and hypotension. Because of their increased activity in the elderly, low doses of short- or intermediate-acting benzodiazepines are considered safer and more effective (i.e., midazolam). The perioperative use of midazolam (and other benzodiazepines) is also associated with post-intervention sedation and confusion until postoperative delirium [10–12], which may promote the risk of falls and ensuing fractures [13].

3.2 Muscle Relaxants

Muscle relaxants in elderly patients have generally a delayed onset of action, owing to a decrease in cardiac output and muscle blood perfusion. Because of the decrease in body water, muscle relaxants which are highly hydrophilic, exert greater effects in elderly than that in younger patients. The duration of action and recovery index are increased in elderly (i.e., atracurium, mivacurium, rocuronium). For above reasons, the onset time of succinylcholine is prolonged, on the other hands, pseudocholinesterase expression might be altered by aging. Drugs which undergo hepatic or renal excretion (i.e., vecuronium) may show an increase in the half-life. The use of sugammadex, able to neutralize the rocuronium-mediated neuromuscular blockade, is particularly useful in the elderly [14, 15].

3.3 Opioids

In the elderly hypersensitivity to morphine is mainly due to altered pharmacokinetics. Compared with young patients, morphine Vd is decreased by 50%, together with a reduced elimination of its active metabolites, morphine-3- and morphine-6-glucuronide, due to the decrease in glomerular filtration. Thus, the initial dose of morphine should be reduced. Fentanyl is commonly used perioperatively. It shows a rapid onset in part due to its lipophilicity. Aging decreases fentanyl dose requirements up to 50% [16]. Remifentanil undergoes a rapid metabolization by nonspecific esterases, resulting in an ultrashort duration of action. Mild or severe hypotension associated with bradycardia may be observed after bolus injection of remifentanil in the elderly, who are also more prone to the hypnotic effects of remifentanil.

3.4 Inhalation Agents

Sevoflurane and desflurane are the low-soluble inhalation anesthetics which offer as clinical advantages for long surgical procedures a faster induction and recovery over isoflurane [17]. There are not studies indicating geriatric-specific problems that would limit the usefulness of both, desflurane and sevoflurane, in the elderly. In particular, desflurane shows a more rapid wash in and washout than the other inhalation agents. From a clinical point of view, the washout rate translates into a shorter reawakening time of 50% compared to that observed with isoflurane, which is an important advantage for a faster recovery [14, 18].

3.5 Propofol   Propofol is the medication most commonly used for inducing anesthesia in surgical patients in the USA. Propofol dose requirement in the elderly population should be reduced by as much as 50%. Indeed, according to the manufacturer's recommendation its dose used for induction must be decreased in elderly patients from a recommended dose of 2–2.5 to 1–1.5 mg/kg [19]. Due to its narrow therapeutic index, the use of propofol in old patients is potentially associated with severe adverse reactions, such as the risk of bradycardia, hypotension, and severe apnea. However, a recent large size retrospective study ($n = 17,540$ patients aged >65 years) showed that despite this finding, the dose of propofol for induction was not independently associated with a greater 30-day mortality rate [20]. On the contrary, Reich et al. [21] suggested that propofol induction should be avoided in the geriatric population, especially in patients with baseline low blood pressure and hemodynamic instability.

4 Preclinical Evidence

The term postoperative cognitive decline or dysfunction refers to a wide spectrum of clinical conditions featuring a decline in a variety of neuropsychological domains including memory, executive functioning, and speed of processing emerging week to months after anesthesia and surgery [10]. Although it represents a fascinating field of study, the clinical studies on the use of anesthetics in the old individuals are conflicting, due either to the fact that anesthetics are rarely administered in the absence of surgery or dissociation between neuropathology and manifesting symptoms [22–24]. Therefore, the preclinical evaluations are necessary to better understand the molecular mechanisms underlying the problems related to general anesthetics in the elderly.

Small animals (murine and canine) are commonly used in preclinical research to reproduce pathological diseases because of their high similarity in anatomy and physiology with humans [25–29]. The anesthesia assures paralysis, analgesia and reduces stress during several experimental procedures. In laboratory animals the anesthesia is problematic for different reasons, such as small body size, fast metabolic rate, and the hypoglycemia and hypothermia states. It is more important considering, as it happens among humans, that individual variation plays a key role in pharmacokinetics and therapeutic response.

In general, genotype, age, sex, body composition, and diseases must be well evaluated for the choice of the anesthetic procedure. The age of the animals can affect the anesthetic metabolism. For example, mice older than 18 months metabolize anesthetics less efficient than adults due to senescence-related pathologies such as cardiac, hepatic or renal alterations. A similar condition is observed in mice younger than 8 weeks that present immature hepatic enzymatic system and reduced homeostatic function [30]. Obesity, frequent in old rodents, also raises the risk of surgical and anesthetic complications. Indeed, obese mice display altered distribution of lipophilic drugs and hepatic dysfunctions that are responsible for hypoventilation and hypoxia post-anesthesia [31]. Animal models that reproduce several human pathologies, such as myocardial ischemia, diabetes mellitus, or cancer, have shown that specific anesthetic protocols are required. For example, transgenic mice that lack the gene for neuropeptide Y receptor type 2 (NY2), that is involved in cognition, show high body weight and increased sensitivity to pentobarbital [32].

4.1 Main Anesthetics in Preclinical Research

Anesthetics used in laboratory can be either injectable or inhaled, based on the administered drug property and on experimental method. Although inhaled anesthetics allow rapid recovery, fast adjustments, and ensure a simple maintenance of a steady anesthetic depth, injectable anesthetics are preferred for small animals, due to the easy administration, the minimal equipment and the safer profile than inhalator anesthetics. Moreover, injectable anesthetics agents have the potential of reversibility by pharmacological antagonism. Generally, pre-anesthetic care is applied to reduce the occurrence of complications that can take place during anesthesia and for avoiding stress-during induction and recovery. In particular, atropine has been used in pre-anesthesia to minimize bronchial secretions and to prevent vagal inhibition of the cardiac rate. In mice a single administration of the drug is recommended to prevent stress caused by multiple injections. The injection volume should be carefully considered based on the route of administration [33]. The most used injectable anesthetics in rodents include tribromoethanol (TBE), ketamine and pentobarbital, which are often combined with other agents, such as xylazine, acepromazine, or diazepam. The anesthetic effect of TBE, which lasts for 15–30 min and is associated with a good muscular relaxation, is unpredictable in old mice, in obese or diabetic mouse models, and in strains with genetic vulnerability to hyperglycemia state such as the C57Bl/6J [34]. Pentobarbital together with thiopental represent the two most popular short-acting oxybarbiturate. They inhibit the release of noradrenaline and glutamate by binding to $GABA_A$ receptor [35]. Augmented responses to pentobarbital administration, in terms of sleep period and loss of reflexes, have been recorded between selected strains of mice [36, 37].

An increasing number of studies describe the age-related functional changes in the body systems that may interfere with drug effect. These include lack of drug responses, mainly due to the altered GABAergic tone. Indeed, extracellular GABA concentration and presynaptic levels of the glutamic acid decarboxylase (GAD), the GABA synthesizing enzyme, resulted downregulated across the auditory cortex (AC) in aged rodents [38–40]. Changes in NMDA receptor subunits have been also found in sub-regions of the cerebral cortex and hippocampus in old mice [41]. Propofol is the most used intravenous anesthetic for long-lasting sedation in veterinary medicine and basic research. It provides rapid, smooth recovery and shows little analgesic effects. As in humans, the initial dose of propofol should be reduced in aged animals. The successful use of propofol and remifentanil combinations, administered by the intraperitoneal route, has been described for anesthesia in rats [42].

Drugs as ketamine and tiletamine induce a pronounced analgesic effect with a wide safety margin. Ketamine is often combined with α_2 agonist xylazine for improving the quality of anesthesia while reducing adverse reactions [43].

In spite of the larger use of injectable anesthetics, inhalation anesthesia is usually preferred for prolonged procedures. The most commonly used inhalation anesthetics for laboratory animals are nitrous oxide and halogenated anesthetics (halothane, isoflurane and sevoflurane, etc.). Isoflurane is the most used for surgical operations regardless of the time required for surgery because it is characterized by short induction and recovery times. Sevoflurane is characterized by an induction and recovery of anesthesia even faster than isoflurane, but like isoflurane or halothane, it presents the risk of dose-dependent respiratory depression and hypotension [30]. Such effects are even exacerbated in old animals. Emerging evidence show that inhalant agents beside typical side effects (respiratory depression, myocardial depression, vasodilation, and hypotension), also exhibit a negative impact on the development of neurodegenerative diseases. In particular, preclinical studies suggest that inhaled anesthetics may facilitate the risk of Alzheimer's disease [44, 45]. Transgenic mice have been developed to reproduce clinical markers of Alzheimer's disease such as beta-amyloid accumulation (plaque), neurofibrillary tangles and cognitive dysfunction. Previous studies on the neurotoxicity induced by anesthesia [44, 46] found that in old transgenic Tg2576 mice, the exposure of subminimum (0.8%) isoflurane or halothane concentration in the alveoli exhibited an increase in amyloidopathy, greater in halothane-exposed mice than in isoflurane-exposed mice [47]. Moreover, significant behavioral changes in learning and memory were observed in isoflurane exposed mice respect to the aged littermate mice [48]. Although these findings clearly demonstrate that isoflurane may be neurotoxic, there have been many reports about the neuroprotective effects of isoflurane [49, 50].

It is possible that isoflurane, and perhaps other anesthetics, depending on the dose- and time of exposure can have dual effects: induction/potentiation or attenuation, of neurotoxicity. More recent results have reported a greater safety of desflurane and nitrous oxide anesthesia compared to isoflurane and sevoflurane. In vitro experiments showed that both two anesthetics failed to induce APP processing and Aβplaques generation [51]. These data are consistent with behavioral studies showing that desflurane did not induce cognitive impairment in mice, differently from isoflurane [52].

Overall these findings indicate that minimizing anesthetic exposure in old animals might be important. At the moment, none of the animal models reproduces the human age-related functional changes with high fidelity, but at least they permit to establish less dangerous anesthetic, the vulnerability of critical age and to draw mechanistic hypothesis. Possible interactions between anesthetics and other drugs or pathophysiological situations are confirmed in preclinical studies. Therefore, further investigations need to be addressed for new advances in anesthetic techniques to achieve the maximum effect with minimal side effects related to the old age.

5 Conclusions

Aging is a physiological process characterized by structural changes associated with progressive degenerative modifications and loss of functionality in all organ systems. Advancing age causes pharmaco-kinetic and pharmacodynamic changes which increase the sensitivity of old patients to anesthetic drugs. In elderly, less doses are usually required to reach clinical effects. Moreover, the comorbidity of pathologies in elderly requiring pharmacological therapies adds the risk of drug interaction with anesthetics. Therefore, the choice of anesthesia in the elderly should be individual, considering that the changes in organ systems, coexisting diseases and drugs taken on a daily basis altogether can alter the response to drugs, as well as endanger the life of the patient. Therefore, a better definition of the effects of aging on the clinical pharmacology of anesthetics, together with the preclinical evidence that assesses the risks and benefits of each anesthetic in elderly animals, would certainly increase the quality of anesthesia management minimizing the risks of the procedure.

References

1. Nigam Y, Knight J, Bhattacharya S, Bayer A (2012) Physiological changes associated with aging and immobility. J Aging Res 2012:468469. https://doi.org/10.1155/2012/468469

2. Zaugg M, Lucchinetti E (2000) Respiratory function in the elderly. Anaesthesiol Clin North Am 18:47–58

3. Tajiri K, Shimizu Y (2013) Liver physiology and liver diseases in the elderly. World J Gastro-enterol 19(46):8459–8467

4. Le Couteur DG, McLean AJ (1998) The aging liver: drug clearance and an oxygen diffusion barrier hypothesis. Clin Pharmacokinet 34:359–373

5. Magnusson KR (1998) The aging of the NMDA receptor complex. Front Biosci 3: e70–e80

6. Magnusson KR, Nelson SE, Young AB (2002) Age-related changes in the protein expression of subunits of the NMDA receptor. Brain Res Mol Brain Res 99(1):40–45

7. Reidenberg MM, Levy M, Warner H et al (1978) Relationship between diazepam dose, plasma level, age, and central nervous system depression. Clin Pharmacol Ther 23:371–374

8. Swift CG, Ewen JM, Clarke P, Stevenson IH (1985) Responsiveness to oral diazepam in the elderly: relationship to total and free plasma concentrations. Br J Clin Pharmacol 20:111–118

9. Castleden CM, George CF, Marcer D, Hallett C (1977) Increased sensitivity to nitrazepam in old age. Br Med J 1:10–12

10. Cascella M, Bimonte S (2017) The role of general anesthetics and the mechanisms of hippocampal and extra-hippocampal dysfunctions in the genesis of postoperative cognitive dysfunction. Neural Regen Res 12 (11):1780–1785

11. Cascella M (2015) Anesthesia awareness. Can midazolam attenuate or prevent memory consolidation on intraoperative awakening during general anesthesia without increasing the risk of postoperative delirium? Korean J Anesthesiol 68(2):200–202

12. Cascella M, Schiavone V, Muzio MR, Cuomo A (2016) Consciousness fluctuation during general anesthesia: a theoretical approach to anesthesia awareness and memory modulation. Curr Med Res Opin 32(8):1351–1359

13. Fredman B, Lahav M, Zohar E, Golod M, Paruta I, Jedeikin R (1999) The effect of midazolam premedication on mental and psychomotor recovery in geriatric patients undergoing brief surgical procedures. Anesth Analg 89:1161–1166

14. Kruijt Spanjer M, Bakker NA, Absalom AR (2011) Pharmacology in the elderly and newer anaesthesia drugs. Best Pract Res Clin Anaesthesiol 25:355–365

15. Kirmeier E, Eriksson LI, Lewald H, et al; POPULAR Contributors (2019) Post-anaesthesia pulmonary complications after use of muscle relaxants (POPULAR): a multicentre, prospective observational study. Lancet Respir Med 7 (2):129–140

16. Scott JC, Stanski DR (1987) Decreased fentanyl and alfentanil dose requirements with age: a simultaneous pharmacokinetic and pharmacodynamic evaluation. J Pharmacol Exp Ther 240:159–166

17. Sakai EM, Connolly LA, Klauck JA, Parker CJ, Hunter JM, Snowdon SL (1992) Effect of age, sex and anaesthetic technique on the pharmacokinetics of atracurium. Br J Anaesth 69:439–434

18. Heavner JE, Kaye AD, Lin BK et al (2003) Recovery of elderly patients from two or more hours of desflurane or sevoflurane anaesthesia. Br J Anaesth 91(4):502–506

19. Claeys MA, Gepts E, Camu F (1988) Haemodynamic changes during anaesthesia induced and maintained with propofol. Br J Anaesth 60:3–9

20. Phillips AT, Deiner S, Mo Lin H et al (2015) Propofol use in the elderly population: prevalence of overdose and association with 30-day mortality. Clin Ther 37(12):2676–2685

21. Reich DL, Hossain S, Krol M (2005) Predictors of hypotension after induction of general anesthesia. Anesth Analg 101(3):622–628

22. Avidan MS, Searleman AC, Storandt M, Barnett K, Vannucci A, Saager L et al (2009) Long-term cognitive decline in older subjects was not attributable to noncardiac surgery or major illness. Anesthesiology 111:964–970

23. Ehlenbach WJ, Hough CL, Crane PK, Haneuse SJ, Carson SS, Curtis JR, Larson EB (2010) Association between acute care and critical illness hospitalization and cognitive function in older adults. JAMA 303:763–770

24. Cascella M, Muzio MR, Bimonte S, Cuomo A, Jakobsson JG (2018) Postoperative delirium and postoperative cognitive dysfunction: updates in pathophysiology, potential translational approaches to clinical practice and further research perspectives. Minerva Anestesiol 84(2):246–260

25. Erden IA, Altinel S, Saricaoglu F et al (2012) Effect of intra-articular injection of levobupivacaine on articular cartilage and synovium in rats. Anaesthesist 61:420–423

26. Maione S, Palazzo E, Guida F et al (2013) New insights on neuropathic pain mechanisms as a source for novel therapeutical strategies. In: Souayah N (ed) Peripheral neuropathy – a new insight into the mechanism, evaluation and management of a complex disorder. IntechOpen. https://doi.org/10.5772/55276

27. Chemonges S, Shekar K, Tung KP et al (2014) Optimal management of the critically ill: Anaesthesia, monitoring, data capture, and point-of-care technological practices in ovine models of critical care. Biomed Res Int 2014:468309. https://doi.org/10.1155/2014/468309

28. D'Aniello A, Luongo L, Romano R et al (2017) D-aspartic acid ameliorates painful and neuropsychiatric changes and reduces β-amyloid Aβ1-42peptide in a long lasting model of neuropathic pain. Neurosci Lett 651:151–158

29. Boccella S, Guida F, Palazzo E et al (2018) Spared nerve injury as a long-lasting model of neuropathic pain. Methods Mol Biol 1727:373–378

30. Paddleford R (2000) Small animals anesthesia. Masson, Milano-Cremona

31. Kushi A, Sasai H, Koizumi H, Takeda N, Yokoyama M, Nakamura M (1998) Obesity and mild hyper-insulinemia found in neuropeptide Y-Y1 receptor-deficient mice. Proc Natl Acad Sci U S A 26:15659–15664

32. Naveilhan P, Canals JM, Arenas E, Ernfors P (2001) Distinct roles of the Y1 and Y2 receptors on neuropeptide Y-induced sensitization to sedation. J Neurochem 78(6):1201–1207

33. Flecknell PA (1989) Laboratory animal anaesthesia. Academic Press, San Diego

34. Zeller W, Meier G, Bürki K, Panoussis B (1998) Adverse effects of tribromoethanol as used in the production of transgenic mice. Lab Anim 32(4):407–413

35. Weinberger J, Nicklas WJ, Berl S (1976) Mechanism of action of anticonvulsants. Role of the differential effects on the active uptake of putative neurotransmitters. Neurology 26 (2):162–166

36. Bennett B (2000) Congenic strain developed for alcohol- and drug-related phenotypes. Pharmacol Biochem Behav 67:671–681

37. Christensen SC, Johnson TE, Markel PD et al (1996) Quantitative trait locus analyses of sleep-times induced by sedative-hypnotics in LSXSS recombinant inbred strains of mice. Alcohol Clin Exp Res 20:543–550

38. Burianova J, Ouda L, Profant O, Syka J (2009) Age-related changes in GAD levels in the central auditory system of the rat. Exp Gerontol 44 (3):161–169

39. de Villers-Sidani E, Alzghoul L, Zhou X, Simpson KL, Lin RC, Merzenich MM (2010) Recovery of functional and structural age-related changes in the rat primary auditory cortex with operant training. Proc Natl Acad Sci U S A 107(31):13900–13905

40. Ling LL, Hughes LF, Caspary DM (2005) Age-related loss of the GABA synthetic enzyme glutamic acid decarboxylase in rat primary auditory cortex. Neuroscience 132 (4):1103–1113

41. Magnusson KR et al (2002) Age-related changes in the protein expression of subunits of the NMDA receptor. Brain Res Mol Brain Res 99(1):40–45

42. Alves HN, da Silva AL, Olsson IA, Orden JM, Antunes LM (2010) Anesthesia with intraperitoneal propofol, medetomidine, and fentanyl in rats. J Am Assoc Lab Anim Sci 49 (4):454–459

43. Kiliç N, Henke J (2004) Comparative studies on the effects of S(+)-ketamine-medetomidine and racemic ketamine-medetomidine in mice. YYÜ Vet Fak Derg 15:15–17

44. Xie Z, Culley DJ, Dong Y, Zhang G, Zhang B, Moir RD, Frosch MP, Crosby G, Tanzi RE (2008) The common inhalation anesthetic isoflurane induces caspase activation and increases amyloid beta-protein level in vivo. Ann Neurol 64(6):618–627

45. Bianchi SL, Tran T, Liu C, Lin S, Li Y, Keller JM et al (2008) Brain and behavior changes in 12-month-old Tg2576 and nontransgenic mice exposed to anesthetics. Neurobiol Aging 29:1002–1010

46. Eckenhoff RG, Johansson JS, Wei H et al (2004) Inhaled anesthetic enhancement of amyloid-beta oligomerization and cytotoxicity. Anesthesiology 110(2):427–430

47. Chen G, Chen KS, Knox J, Inglis J, Bernard A, Martin SJ et al (2000) A learning deficit related to age and betaamyloid plaques in a mouse model of Alzheimer's disease. Nature 408:975–979

48. Rosenholm M, Paro E, Antila H, Võikar V (2017) Repeated brief isoflurane anesthesia during early postnatal development produces negligible changes on adult behavior in male mice. PLoS One 12(4):e0175258. https://doi.org/10.1371/journal.pone.0175258

49. de Klaver MJ, Manning L, Palmer LA, Rich GF (2002) Isoflurane pretreatment inhibits cytokine-induced cell death in cultured rat smooth muscle cells and human endothelial cells. Anesthesiology 97:24–32

50. Gray JJ, Bickler PE, Fahlman CS, Zhan X, Schuyler JA (2005) Isoflurane neuroprotection in hypoxic hippocampal slice cultures involves increases in intracellular Ca2+ and mitogen-activated protein kinases. Anesthesiology 102 (3):606–615

51. Xie Z, Xu Z (2013) General anesthetics and β-amyloid protein. Prog Neuropsychopharmacol Biol Psychiatry 47:140–146

52. Loop T, Dovi-Akue D, Frick M et al (2005) Volatile anesthetics induce caspase-dependent, mitochondria-mediated apoptosis in human T lymphocytes in vitro. Anesthesiology 102:1147–1157

Chapter 9

Propofol Effects in Breast Cancer Cell Progression: Evidences from In Vitro Studies

Sabrina Bimonte, Marco Cascella, Aldo Giudice, Francesca Bifulco, Stefan Wirz, and Arturo Cuomo

Abstract

Propofol (2,6 diisopropylphenol), belonging to the class of intravenous anesthetics, is largely used as a sedative-hypnotic agent in humans, outlined by lower toxicity and fast resumption from anesthesia. This anesthetic possesses multiple properties by which it exerts many biological functions. Particularly, due to its antioxidant and anti-inflammatory features, many in vitro studies suggested that propofol could have direct inhibitory or promoting effects on cancer cells proliferation, by influencing their activities, especially for breast cancer cells.

In this chapter, we summarize and describe these studies trying to elucidate the molecular mechanisms underlying the roles of propofol in breast cancer progression. Data emerged from these research works suggest that this anesthetic has different effects on breast cancer cell proliferation depending mainly on breast cancer heterogeneity.

Key words Propofol, General anesthesia, General anesthetic, Intravenous anesthetic, Breast cancer, Apoptosis, Proliferation, Cancer progression, Cancer migration

1 Introduction

Breast cancer is the most prevalent form of tumor in women and a cause of their death worldwide, due to its higher invasiveness and metastatic potential [1]. Independently from the early (I or II) or the advanced stage of disease (III or IV), surgery represents the first line of treatment for breast malignancy, although the survival rate is lower especially after a metastatic or local recurrence [2, 3]. Different clinical studies proved evidence that some anesthetics, intravenous or volatile, applied in surgery on patients with breast cancer, could affect their fallout [4–6]. Importantly, compared to volatile anesthetics such as isoflurane, desflurane, and sevoflurane, propofol may have a better effect on oncological patients' outcomes. Moreover, many preclinical studies supported the hypothesis that propofol is able to affect breast cancer cell migration and invasion, by

Marco Cascella (ed.), *General Anesthesia Research*, Neuromethods, vol. 150, https://doi.org/10.1007/978-1-4939-9891-3_9,
© Springer Science+Business Media, LLC, part of Springer Nature 2020

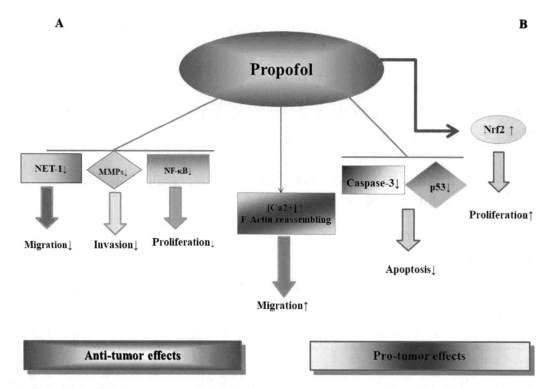

Fig. 1 Molecular mechanisms underlying the effects of propofol on breast cancer. (**a**) The cartoon shows the antitumor effects of propofol, thought an impairment of cell migration ($\downarrow$NET-1), proliferation ($\downarrow$NF-κB), and invasion ($\downarrow$MMPs) and (**b**) the protumor effects of propofol by an enhancement of proliferation ($\uparrow$Nrf2) and migration ($\uparrow$Ca^{2+}) and F-actin rearrangement) and apoptosis inhibition ($\downarrow$p53, and $\downarrow$caspase-3). Abbreviations: *NET-1* neuroepithelial cell transforming gene 1, *MMP* metalloproteinase, *NF-κB* nuclear factor kappa-light-chain-enhancer of activated B cells, *Nrf2* nuclear factor erythroid 2, $\uparrow$ upregulated, $\downarrow$ downregulated. caspase-3$\downarrow$)

acting on different molecular pathways [7–14] (Fig. 1). In this chapter, we summarize and describe these studies trying to elucidate the molecular mechanisms underlying the roles of propofol in breast cancer progression. Data emerged from these research works suggest that this anesthetic has different effects on breast cancer cell proliferation, depending mainly on the heterogeneity of breast cancer. Furthermore, the other aim of this work is to offer suggestions for translational research and perspectives.

2   Propofol: An Intravenous Anesthetic with Multiple Biological Proprieties

Propofol represents one of the principal intravenous anesthetics. Experimental pieces of evidence suggested that it may have many potential biological roles in the regulation of breast cancer progression [15]. Interestingly, propofol is also able to affect the immune system, mainly by regulating the functions of natural killer cells

(NK) which represent significant players in the regulation of cellular resistance to cancer, particularly at metastatic stage [16]. Specifically, it has been proved that propofol is able to improve the activities of NK cells in cardiomyocytes [17] and breast cancer cells especially after surgical procedures [18], thus potentially reducing tumor invasion and metastasis formation, through regulation of the beta (β)-adrenoreceptor-mediated signal transduction. Moreover, propofol could enhance the cytotoxic T lymphocytes activity, and, in turn, potentiate the antitumor immunity [19]. Additionally, it has been shown that propofol is able to activate the T-helper cells and to promote the differentiation of T-helper 1 cells, by increasing the ratio of interferon c (IFN-c)/interleukin-4 (IL-4) and the expression of CD28 in lung cancer patients subjected to pulmonary lobectomy [20]. Regarding breast cancer cells, a recent study suggested that propofol may suppress tumor development and metastasis in MCF-7 cells by inhibiting the expression of IL-6, IL-8 and Cyclooxygenase 2 (COX2) (detected by Western blotting analysis) [21].

Most interesting findings concerning propofol effects other than its anesthetic proprieties come even from clinical investigations. Several clinical retrospective studies [22, 23] proved that compared to sevoflurane anesthesia, propofol administration was associated with a lower incidence of chronic pain after breast cancer surgery, although the underlying mechanisms remain still unclear. Concerning cancer outcomes, as reported by several retrospectives [5, 6, 24] and perspective [25–28] clinical studies in surgical breast cancer patients, propofol (compared to volatile anesthetics, e.g., sevoflurane) could prolong the survival rate and diminish the cancer recurrence rate, although these shreds of evidence need to be strongly potentiated and verified.

Again, a growing body of evidence from preclinical studies demonstrated that propofol is able to modulate the malignancy of many types of cancers [29] including those affecting the colon [30, 31], the pancreas [32, 33], and the lung [34–36] by acting on different molecular pathways. Altogether, these findings strongly suggest that propofol could have different antitumor effects on cancer cells.

3   The Effects of Propofol on Breast Cancer Cells Progression: A Controversial Issue

Different preclinical studies indicate that propofol, at high concentrations, has an inhibitory or a promoting role in breast cancer cells progression, depending mainly on the diversity of breast cancer [7–14]. Notably, breast cancer can be classified in different subtypes at histopathological and molecular levels. Specifically, on the basis of the histological features of breast cancer cells, such as growth, and cell morphology, this disease can be classified in up to

21 distinct histological subtypes [37, 38]. This diversity has significant prognostic outcomes. On the other side, four principal subtypes have been identified at a molecular level from the presence or not of the progesterone receptor/PR, or the estrogen receptor/ER, or by higher levels of human epidermal growth factor receptor 2 (HER2+/HER2-) [39, 40]. Moreover, breast tumor represents a complex pathology including different disorder outlined by distinct pathological and biological features, clinical appearance, reply to treatments, and fallout [41].

As reported by Li et al. [15] many in vitro studies have been conducted to establish breast cancer cell lines (about 51), which have been used to study the principal antitumor effects of different substances (as a drug, natural compounds) on cancer cells [42]. Depending on the type of breast cancer cells used and on the different experimental conditions, several preclinical studies highlighted a direct effect (inhibitory or promoting) of propofol on breast cancer cells, by acting on different molecular pathways. Due to the clinical relevance of the matter, the discrepancy on these results represents a controversial issue which will be necessarily solved.

3.1 The Promoting Role of Propofol on Breast Cancer Cell Proliferation

Few preclinical evidence highlighted a promoting role of propofol in breast cancer initiation and progression (Table 1). The first study on this topic was reported by Garib et al. [12] on MDA-MB-468 cells treated with anesthetics at clinical relevant concentrations (propofol 3, 6, 9 mg/L, etomidate 2, 3, 4 mg/L, and lidocaine 1.25, 2.5, 5 mg/L for more than 10 h). Specifically, the authors

Table 1
The promoting effects of propofol on breast cancer cell progression

Breast cancer cells	Dose	Time of incubation (h)	Effects, molecular mechanism	References
MDA-MB-468	3, 6, 9 mg/L	Up to 10	Propofol enhanced migration and morphology of breast cancer cells	[12]
MDA-MB-468	6 µg/L		Propofol activated GABA-A receptor leading to an increased intracellular calcium concentration, to actin reorganization and to an increased migration of breast cancer cells; $[Ca^{2+}]\uparrow$	[13]
MDA-MB-231	2–10 µg/mL	Up to 12	Propofol enhanced proliferation and apoptosis of breast cancer cells; Nrf2$\uparrow$, p53$\downarrow$, caspase-3$\downarrow$	[14]

GABA gamma-aminobutyric acid, *Ca²⁺* calcium ions, *Nrf2* nuclear factor erythroid 2, $\uparrow$ upregulated, $\downarrow$ downregulate studied the effects of different anesthetics on the migration of MDA-MB-468 cells. Data showed that propofol, as lidocaine but not etomidate, enhanced the migratory behavior (detected with a three-dimensional collagen matrix using time-lapse video microscopy and computer-assisted cell-tracking) of breast cancer cells in a dose-dependent manner. Moreover, in a subsequent investigation, the authors demonstrated that propofol affected the architecture pattern of the F-actin cytoskeleton (detected by Confocal scanning microscopy) of these cells and increased the intracellular level of calcium, an effect reversed by verapamil (an L-type calcium channel blocker) [13]. Probably, propofol was able to enhance the locomotion of breast cancer cells, by activating the gamma-aminobutyric acid (GABA)-A receptor chloride channels, although this remains still to be proved by other in vitro and in vivo studies. Accordingly, Meng et al. [14] demonstrated that propofol (2–10 μg/mL) enhanced cell proliferation (detected by Cell proliferation assay) and migration (detected by Wound healing assays) of MDA-MB-231. These effects were associated with inhibition of p53 expression and cell apoptosis (detected by Western blotting analysis and TUNEL assay) and activation of the nuclear factor erythroid 2-related factor 2 (Nrf2) pathway. Taken together, these findings suggest that propofol, at clinically relevant doses, has protumor effects on breast cancer, by acting on different molecular mechanisms.

3.2 The Inhibitory Role of Propofol on Breast Cancer Cell Proliferation

On the contrary to those previously reported, accumulated shreds of evidence showed that propofol has antitumor effects on breast cancer, mainly by inhibiting cell proliferation and migration (Table 2). The first evidence was reported by Siddiqui et al. [10] on MDA-MB-231 propofol-treated (25 μM) cells in combination with fatty acids docosahexaenoic (DHA) or eicosapentaenoic acid (EPA). Data emerged from this study showed that the lipid compounds formed by the conjugation of propofol and DHA (propofol-DHA), and propofol and EPA, (propofol-EPA) were able to inhibit the cell migration (detected by cell migration assay) and adhesion (detected by cell adhesion assay) and to induce significant apoptosis (detected by apoptosis and cytochrome c release assays) within MDA-MB-231 breast cancer cells compared to the unconjugated compounds DHA, EPA or propofol. Overall, these findings suggest that the propofol-DHA and propofol-EPA compounds may be used for research on breast cancer treatment.

Later on, Li et al. [7] demonstrated that propofol played an antitumor role in breast cancer cells. Specifically, MDA-MB-231 cells were treated (from 0 to 24 h) with propofol (2–10 μg/mL) and then tested in vitro for migration (detected by Wound-healing assay) and invasion (assessed by transwell assay). Data showed that propofol was able to restrain the locomotion of MDA-MB-231 cells, mainly by inhibiting the production and the expression of

Table 2
The inhibitory effects of propofol on breast cancer cell progression

Breast cancer cells	Dose	Time of incubation (h)	Effects and molecular mechanism	References
MDA-MB-231	Propofol (25 μM) DHA (25 μM) EPA (25 μM) Propofol-DHA (25 μM) Propofol-EPA (25 μM)	4–24	Propofol-DHA and Propofol-EPA induced the apoptosis and inhibited the proliferation in breast cancer cells more than are the unconjugated parent compounds DHA, EPA, or propofol	[10]
MDA-MB-231	Propofol (2–10 mg/ mL) TNF-a (50 μM)	0–24	Propofol inhibited the migration and the invasion of breast cancer cells MMP-2↓, MMP-9↓, NF-κ B↓, IKK-β↓	[7]
MDA-MB-231, MCF-7	Propofol (1–10 μg/ mL); bupivacaine (0.5–100 μg/mL)		Propofol inhibited the invasion of breast cancer cells NET-1↓	[8]
MDA-MB-231	Propofol (25, 50, and 100 μmol/L)	24	Propofol inhibited the migration and the invasion of breast cancer cells H19↓	[43]
MDA-MB-435	Propofol (10 μM)	6	Propofol enhanced the apoptosis of breast cancer cells miR-24↓, p27↑, cleaved caspase-3↑	[44]
MCF-7	Propofol (10 μM)	0–48	Propofol inhibited the proliferation and epithelial-mesenchymal transition of breast cancer cells. Moreover, promoted the apoptosis miR-21, p53↑, Wnt/β-catenin↓, PI3K/AK↓	[45]

DHA fatty acids docosahexaenoic, *EPA* eicosapentaenoic acid, *MMP-2* metalloproteinase-2, *MMP-9* metalloproteinase-9, *NF-κB* nuclear factor kappa-light-chain-enhancer of activated B cells, *IKK-β* inhibitor of nuclear factor kappa-B kinase subunit beta, *miR-21* micro-RNA 21, *miR-24* microRNA-24, *PI3K/AK* phosphoinositide 3-kinase/protein kinase B, ↑ upregulated, ↓ downregulated

matrix metalloproteinase (MMP)-2 and -9 (detected by Western blotting and ELISA analysis) which are involved in the development of metastasis. Moreover, the authors showed that propofol downregulated the expression of MMPs by affecting the nuclear factor kappa-light-chain-enhancer of activated B cells (NF-κB) pathways. Specifically, propofol reduced the phosphorylation of inhibitor of nuclear factor kappa-B kinase subunit beta IKK-β (detected by Western blotting analysis) and the activation of NF-κB (detected by EMSA). Finally, the downregulation of MMP-9 (detected by Western blotting analysis) induced by treating MDA-MB-231 cells with propofol (5 μg/mL for 4 h) was lowered by the pretreatment with tumor necrosis factor-α (TNF-α, 50 μm) which commonly activates NF-κB. Altogether these findings strongly suggest that because propofol is capable to act on the principal pathways involved in the regulation of proliferation and migration of cancer cells, it could be considered a valuable therapeutic agent for the treatment of breast cancer. Similar data were reported by Ecimovic et al. [8] in a fascinating study on the inhibitory role of propofol on the migration of breast cancer cells. In this investigation MDA-MB-231 (ER−−) and MCF-7 (ER+) cells were treated with propofol (1–10 μg/mL) and bupivacaine (0.5–100 μg/mL), and then were tested for their functions (cell proliferation assay, cell migration, and invasion assay). Results indicated that propofol was able to inhibit only the migration and not the proliferation of both cancer cell lines; the effect was obtained through a downregulation of neuroepithelial cell transforming gene 1 (NET-1), mainly in MDA-MB-231 cells. Of note, NET-1 is expressed in breast adenocarcinoma cell line, enhancing excessively their migration and progression. The propofol-induced inhibition on cell migration was counteracted by silencing NET-1, which in this way can be considered a mediator between the propofol and the breast cancer cell function. Interestingly, propofol inhibited invasion of MCF-7 but not of MDA-MB-231 cells, while no effects on functions of both cell lines or NET-1 expression were detected in cells treated with bupivacaine.

Subsequently, Bai et al. [43] illustrated that propofol (25, 50, and 100 μmol/L for 24 h) inhibited the migration and invasion (determined by wound-healing and transwell assays, respectively) of MDA-MB-231 cells by altering the expression of H19 (detected by RT-PCR). Recently, Yu et al. [44] reported that propofol (10 μM for 6 h) promoted the apoptosis (determined by TUNEL staining and flow cytometry) of MDA-MB-435 cells by acting on micro-RNA (miR)-24/p27 signal pathway (determined by RT-PCR and Western blotting analysis). Specifically, through the downregulation of miR-24, the upregulation of p27 and of cleaved caspase-3 induced the cell death in breast cancer cells. Similar results have been recently reported by Du et al. [45]. In this study, the authors investigated the effect of propofol on proliferation and epithelial–mesenchymal transition (EMT)—a crucial biological process involved in the embryogenesis and in the cancer progression in which epithelial cell assume a mesenchymal cell phenotype and, in turn, acquire enhanced migratory capacity, invasiveness, and elevated resistance to apoptosis—of MCF-7 cells. Data showed that propofol (0–10 μg/mL) inhibited the proliferation and migration (determined by Proliferation and Transwell assays) of MCF-7 cells while induced apoptosis (determined by apoptosis assay). In particular, propofol reduced EMT (determined by Western blotting analysis and qRT-PCR), through the deregulation of miR-21 which, in turn, influenced phosphoinositide

3-kinase/protein kinase B (PI3K/AKT) and Wnt3a/β-catenin pathways (determined by Western blotting analysis and qRT-PCR). Overall, these findings suggest that propofol could be considered an innovative therapeutic agent for breast cancer treatment.

3.3 Propofol and Its Effect on Pretreated Breast Cancer Cells

Several experiments were conducted by using the serum from breast cancer patients subjected to different types of anesthesia in order to test its effects on breast cancer cell lines [46–48].

In a fascinating study, Degan et al. [46] assessed the effects of different anesthetic techniques on breast cancer cell proliferation and migration by treating MDA-MB-231 cells with serum (10%) from breast cancer surgery patients subjected to either sevoflurane/opioid anesthesia–analgesia (GA) or propofol and paravertebral block anesthesia (PPA). Data showed a significant decrease in cellular proliferation (not detected in cell migration) of MDA-MB-231 cells treated with serum from PPA patients compared to the groups of cells treated with serum from GA. This underlines that differences in anesthetic procedure could alter the proliferation of breast cancer cells by altering the serum molecular profile of breast cancer patients, although the underlying mechanism remains to be explored.

In a subsequent study Jaura et al. [47] investigated the effect of serum (2–10%) from breast cancer surgery patients (24 h post-surgery) randomized to receive distinct anesthetic techniques (GA or PPA), on apoptosis of MDA-MB-231 cells. Data showed that serum from GA patients reduced apoptosis in MDA-MB231 cells more than the serum from PPA patients. Later on, Buckley et al. [48] tested the effects of anesthetics on NK cells anticancer function of serum from patients receiving PPA or GA anesthesia. Data showed that the serum from the PPA group significantly induced HCC1500 cells apoptosis, form the serum of the GA group. Moreover, an upregulation in NK cells in CD107a expression was detected in the serum from the PPA group.

Altogether these findings suggest that regional anesthesia could be preferred to opioids and volatile agents in the surgical procedure to diminish perioperative residual disease. Additional research will be necessary to shed light on this important issue.

4   Future Perspectives and Outcomes

In this chapter, we highlight the roles of propofol on breast cancer biology. Despite few studies reported protumor effects in breast cancer, mainly depending by the blown heterogeneity of the disease, the majority of the literature suggests that this intravenous general anesthetic, at high concentrations, has antitumor effects on breast cancer cells. Although these findings come from preclinical investigations, their potential clinical impact in term of cancer patients' outcomes is of huge interest. Propofol may exert its anticancer activity by modulating different molecular pathways that to date are not completely dissected. However, despite encouraging, these results lack of consistency and cannot be translated into clinical practice. Thus, there is a need for more preclinical and clinical studies (a) to dissect the molecular mechanisms underlying the effect of propofol on breast cancer taking into account of the large heterogeneity of this disease; (b) to evaluate the anesthetic in on other cell lines; (c) to elucidate the effects of anesthetics (e.g., on the immunity) by using in vivo models; (d) to cover the unavailability of prospective clinical trials.

Results from these research works will be helpful to clinicians to select the most appropriate anesthetic for breast cancer treatment, with the aim of reducing the risk of worsening its progression.

References

1. American Cancer Society (2017) Cancer facts & figures 2017. American Cancer Society, Atlanta. https://www.cancer.org/content/dam/cancer-org/research/cancer-facts-and-statistics/annual-cancer-facts-and-figures/2017/cancer-facts-and-figures-2017.pdf. Accessed 20 June 2018

2. DeSantis CE, Lin CC, Mariotto AB et al (2014) Cancer treatment and survivorship statistics. CA Cancer J Clin 64(4):252–271

3. Rafferty EA, Park JM, Philpotts LE et al (2013) Assessing radiologist performance using combined digital mammography and breast tomosynthesis compared with digital mammography alone: results of a multicenter, multireader trial. Radiology 266(1):104–113

4. Wigmore TJ, Mohammed K, Jhanji S et al (2016) Long-term survival for patients undergoing volatile versus IV anesthesia for Cancer surgery: a retrospective analysis. Anesthesiology 124(1):69–79

5. Enlund M, Berglund A, Andreasson K et al (2014) The choice of anaesthetic—sevoflurane or propofol—and outcome from cancer surgery: a retrospective analysis. Ups J Med Sci 119(3):251–261

6. Lee JH, Kang SH, Kim Y et al (2016) Effects of propofol-based total intravenous anesthesia on recurrence and overall survival in patients after modified radical mastectomy: a retrospective study. Korean J Anesthesiol 69(2):126–132

7. Li Q, Zhang L, Han Y et al (2012) Propofol reduces MMPs expression by inhibiting NF-kappaB activity in human MDA-MB-231 cells. Biomed Pharmacother 66(1):52–56

8. Ecimovic P, Murray D, Doran P et al (2014) Propofol and bupivacaine in breast cancer cell function in vitro - role of the NET1 gene. Anticancer Res 34(3):1321–1331

9. Yu B, Gao W, Zhou H et al (2017) Propofol induces apoptosis of breast cancer cells by downregulation of miR-24 signal pathway. Cancer Biomark 21(3):513–519

10. Siddiqui RA, Zerouga M, Wu M et al (2005) Anticancer properties of propofol-docosahexaenoate and propofol-eicosapentaenoate on breast cancer cells. Breast Cancer Res 7(5):R645

11. Harvey KA, Xu Z, Whitley P et al (2010) Characterization of anticancer properties of 2,6-diisopropylphenol-docosahexaenoate and analogues in breast cancer cells. Bioorg Med Chem 18(5):1866–1874

12. Garib V, Niggemann B, Zänker KS et al (2002) Influence of non-volatile anesthetics on the migration behavior of the human breast cancer cell line MDA-MB-468. Acta Anaesthesiol Scand 46(7):836–844

13. Garib V, Lang K, Niggemann B et al (2005) Propofol-induced calcium signalling and actin reorganization within breast carcinoma cells. Eur J Anaesthesiol 22(8):609–615

14. Meng C, Song L, Wang J et al (2017) Propofol induces proliferation partially via downregulation of p53 protein and promotes migration via activation of the Nrf2 pathway in human breast cancer cell line MDA-MB-231. Oncol Rep 37(2):841–848

15. Li R, Liu H, Dilger JP et al (2018) Effect of Propofol on breast cancer cell, the immune

system, and patient outcome. BMC Anesthesiol 18(1):77

16. Pross HF, Lotzová E (1993) Role of natural killer cells in cancer. Nat Immun 12:279

17. Kurokawa H, Murray PA, Damron DS (2002) Profol attenuates β-Adrenoreceptor–mediated signal transduction via a protein kinase C–dependent pathway in Cardiomyocytes. Anesthesiology 96(3):688–698

18. Desmond F, McCormack J, Mulligan N et al (2015) Effect of anaesthetic technique on immune cell infiltration in breast cancer: a follow-up pilot analysis of a prospective, randomised, investigator-masked study. Anticancer Res 35(3):1311–1319

19. Kushida A, Inada T, Shingu K (2007) Enhancement of antitumor immunity after propofol treatment in mice. Immunopharmacol Immunotoxicol 29(3–4):477–486

20. Ren XF, Li WZ, Meng FY et al (2010) Differential effects of propofol and isoflurane on the activation of T-helper cells in lung cancer patients. Anaesthesia 65(5):478–482

21. Li X, Li LL, Liang F et al (2018) Anesthetic drug propofol inhibits the expression of interleukin-6, interleukin-8 and cyclooxygenase-2, a potential mechanism for propofol in suppressing tumor development and metastasis. Oncol Lett 15(6):9523–9528

22. Cho AR, Kwon JY, Kim KH et al (2013) The effects of anesthetics on chronic pain after breast cancer surgery. Anesth Analg 116(3):685–693

23. Abdallah FW, Morgan PJ, Cil T et al (2015) Comparing the DN4 tool with the IASP grading system for chronic neuropathic pain screening after breast tumor resection with and without paravertebral blocks: a prospective 6-month validation study. Pain 156(4):740–749

24. Kim MH, Kim DW, Kim JH et al (2017) Does the type of anesthesia really affect the recurrence-free survival after breast cancer surgery? Oncotarget 8(52):90477–90487

25. Royds J, Khan AH, Buggy DJ (2016) An update on existing ongoing prospective trials evaluating the effect of anesthetic and analgesic techniques during primary cancer surgery on cancer recurrence or metastasis. Int Anesthesiol Clin 54(4):e76–e83

26. Subramani S, Poopalalingam R (2014) Bonfils assisted double lumen endobronchial tube placement in an anticipated difficult airway. J Anaesthesiol Clin Pharmacol 30(4):568–570

27. Weng H, Xu ZY, Liu J et al (2010) Placement of the Univent tube without fiberoptic bronchoscope assistance. Anesth Analg 110(2):508–514

28. Schuepbach R, Grande B, Camen G et al (2015) Intubation with VivaSight or conventional left-sided double-lumen tubes: a randomized trial. Can J Anaesth 62(7):762–769

29. Song J, Shen Y, Zhan J et al (2014) Mini profile of potential anticancer properties of propofol. PLoS One 9(12):e114440

30. Tat T, Jurj A, Selicean C et al (2019) Antiproliferative effects of propofol and lidocaine on the colon adenocarcinoma microenvironment. J BUON 24(1):106–115

31. Xu YJ, Li SY, Cheng Q et al (2016) Effects of anaesthesia on proliferation, invasion and apoptosis of LoVo colon cancer cells in vitro. Anaesthesia 71(2):147–154

32. Zhang Z, Zang M, Wang S et al (2018) Effects of propofol on human cholangiocarcinoma and the associated mechanisms. Exp Ther Med 17(1):472–478

33. Du QH, Xu YB, Zhang MY et al (2013) Propofol induces apoptosis and increases gemcitabine sensitivity in pancreatic cancer cells in vitro by inhibition of nuclear factor-κB activity. World J Gastroenterol 19(33):5485–5492

34. Freeman J, Crowley PD, Foley AG et al (2019) Effect of perioperative lidocaine, propofol and steroids on pulmonary metastasis in a murine model of breast cancer surgery. Cancers (Basel) 11(5). pii: E613

35. Kang FC, Wang SC, So EC (2019) Propofol may increase caspase and MAPK pathways, and suppress the Akt pathway to induce apoptosis in MA-10 mouse Leydig tumor cells. Oncol Rep 41(6):3565–3574

36. Qin Y, Ni J, Kang L et al (2019) Sevoflurane effect on cognitive function and the expression of oxidative stress response proteins in elderly patients undergoing radical surgery for lung cancer. J Coll Physicians Surg Pak 29(1):12–15

37. Tavassoli FA, Devilee P (2003) World Health Organization classification of tumors: pathology and genetics of tumors of the breast and female genital organs. IARC Press, Lyon

38. Lakhani SR, Ellis IO, Schnitt SJ et al (2012) WHO classification of tumors of the breast. IARC, Lyon

39. Perou CM, Sorlie T, Eisen MB et al (2000) Molecular portraits of human breast tumours. Nature 406(6797):747–752

40. Network CGA (2012) Comprehensive molecular portraits of human breast tumours. Nature 490(7418):61–70

41. Dieci MV, Orvieto E, Dominici M et al (2014) Rare breast cancer subtypes: histological,

molecular, and clinical peculiarities. Oncologist 19(8):805–813

42. Neve RM, Chin K, Fridlyand J et al (2006) A collection of breast cancer cell lines for the study of functionally distinct cancer subtypes. Cancer Cell 10(6):515–527

43. Bai JJ, Lin CS, Ye HJ et al (2016) Propofol suppresses migration and invasion of breast cancer MDA-MB-231 cells by down-regulating H19. Nan Fang Yi Ke Da Xue Xue Bao 36(9):1255–1259

44. Yu B, Gao W, Zhou H et al (2018) Propofol induces apoptosis of breast cancer cells by downregulation of miR-24 signal pathway. Cancer Biomark 21(3):513–519

45. Du Q, Zhang X, Zhang X et al (2019) Propofol inhibits proliferation and epithelial-mesenchymal transition of MCF-7 cells by suppressing miR-21 expression. Artif Cells Nanomed Biotechnol 47(1):1265–1271

46. Deegan CA, Murray D, Doran P et al (2009) Effect of anaesthetic technique on oestrogen receptor-negative breast cancer cell function in vitro. Br J Anaesth 103(5):685–690

47. Jaura AI, Flood G, Gallagher HC et al (2014) Differential effects of serum from patients administered distinct anaesthetic techniques on apoptosis in breast cancer cells in vitro: a pilot study. Br J Anaesth 113(Suppl 1):i63–i67

48. Buckley A, McQuaid S, Johnson P et al (2014) Effect of anaesthetic technique on the natural killer cell anti-tumour activity of serum from women undergoing breast cancer surgery: a pilot study. Br J Anaesth 113:56–62

Chapter 10

Postoperative Cognitive Function Following General Anesthesia in Children

Maiko Satomoto

Abstract

Research into the developmental effects of anesthetic use began from the study of the fetal and postnatal effects of maternal alcohol consumption. Broadly speaking, alcohol is an anesthetic of sorts, and its toxicity as such has been firmly established: there is a consensus on the danger of alcohol to children as the warning label on every can of beer reminds us. Furthermore, studies using rodents have largely confirmed the toxicity of anesthetics on the developing brain. Although some human retrospective observational studies have found anesthetics to be toxic while others have not, as mentioned earlier the difficulty of assembling cohorts with matching backgrounds has rendered interpretation of the results difficult. To date, two large-scale human studies have been conducted; the preliminary findings of one of these demonstrated no apparent effect of the short-term use of anesthetics on healthy children. Against this, we have the warning issued by the FDA and based on the results of the animal studies and retrospective observation studies, which applies to all anesthetics and sedatives besides alpha-2 agonists, as previously described: the extended or multiple use of these agents in children younger than 3 years of age or in fetuses during the third trimester may adversely affect brain development. In light of the foregoing information, the clinician is left to ponder the very serious question of which anesthetic procedures to use with pediatric patients. Given the state of our current knowledge, the answer to this question must be to shorten exposure as much as possible and to limit the dosage to strictly appropriate quantities. The prevailing view at present is that short-term exposure of healthy children to anesthetics has no adverse effects on brain development.

Key words Anesthetics, Neurodegeneration, Brain growth spurt, Brain development, Cognitive impairment, Children

1 Introduction

Research in this field began with the study of fetal alcohol syndrome. Previous human studies have reported characteristic cognitive and facial features of infants resulting from excessive maternal consumption of alcohol during pregnancy [1, 2]. The brain is primarily affected, and an association between excessive maternal alcohol consumption and depression and mental disorder in offspring has also been observed [3]. The neurotoxic effects of alcohol consumption appear leave their mark during the period of brain

Marco Cascella (ed.), *General Anesthesia Research*, Neuromethods, vol. 150, https://doi.org/10.1007/978-1-4939-9891-3_10,
© Springer Science+Business Media, LLC, part of Springer Nature 2020

development corresponding to synaptogenesis. This period is also known as the brain growth spurt (BGS) and is generally thought to occur between the third trimester of pregnancy and 3 years of age in humans, and in the first 2 postnatal weeks in rodents [4]. Not only maternal alcohol consumption but also exposure to toxins, external trauma, and hypoxia are known to produce similar neurological sequelae on brain development in offspring [5]. Administering MK801, an NMDA antagonist, to 7-day-old rats reportedly induced apoptosis in brain cells [6]. Similarly, administering ethanol 2.5 g/kg twice to 7-day-old rats also reportedly induced widespread apoptosis of brain cells [7].

Many anesthetic agents work either as NMDA antagonists, GABA agonists, or as both [8]. It is also well known that alcohol has NMDA-antagonistic and GABA agonistic effects [7, 9]. This similarity instigated research into the effects of anesthetic agents on brain development, a topic which continues to be of considerable interest to clinicians.

2   Results of Animal Experiments

Animal experiments investigating the neurotoxic effects of anesthetic agents on developing brains using mice, rats, guinea pigs, and rhesus monkeys began to be conducted with increasing frequency from 2000 [10–14]. Differences due to variations in concentration and exposure time were found in the effects of the anesthetic agents tested using animal models; however, with the exception of dexmedetomidine, an alpha2-agonist, neurotoxic effects on developing brains were established for all inhalation anesthetics including nitrous oxide, isoflurane, sevoflurane, and desflurane, as well as for intravenous anesthetics including midazolam, propofol, ketamine, and thiopental [8, 10–34]. Moreover, a number of these studies established the long-term effects of neurotoxicity on learning by testing the affected offspring after rearing them to adulthood [8, 16, 27, 35–40].

3   Human Retrospective Observational Studies

Here we will consider a number of human retrospective observational studies which have also investigated the effects of anesthetic agents on developing brains. The first of these was a retrospective birth cohort study which conducted by the Mayo Clinic from 1976 to 1982 and enrolling 5357 cases [41], which found that in 593 children who received an anesthetic agent before 4 years of age, the hazard ratio of the risk of cognitive impairment resulting from anesthetic administration did not increase among subjects who received an anesthetic only once, but rose to 1.59 for subjects

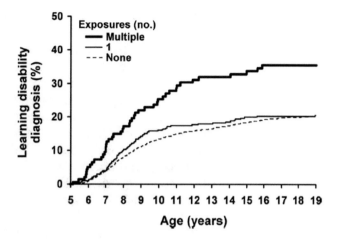

Fig. 1 The ratio of the number of children who underwent anesthesia when 4 years old or younger to subsequent learning impairment. The bold line indicates children who underwent anesthesia two times or more. The rate of learning impairment is clearly higher for this group than for those who never underwent anesthesia or did so once only (*see* ref. 41)

who received an anesthetic two times, and to 2.6 for subjects receiving an anesthetic three or more times (Fig. 1). Of those who received an anesthetic only once, the rate of leaning impairment was 20% (showing no significant difference with the non-anesthesia group); for those who received it two or more times, the rate rose to 35% with a corresponding change in the hazard ratio to 1.56 after 2 h under anesthesia. The second study enrolling 243 patients who underwent urological surgery before 6 years of age found that the rate of aberrant behavior was higher among those who underwent surgery when 2 years old or younger than among those who underwent surgery when 2 years old or older [42]. The third study enrolling 383 subjects who had undergone inguinal hernia surgery when 3 years old or younger showed a 2.3 times higher rate of developmental behavioral abnormalities than the background-matched control group [43].

Among studies reporting no difference, a Dutch study enrolling 1143 monozygous twin pairs found a decrease in learning ability until 3 years of age among subjects who had received an anesthetic when 3 years old or younger compared to the non-anesthesia group. However, the same study found no significant difference between twins [44]. A Danish study of 2547 subjects who had undergone inguinal hernia surgery when 1-year-old or younger showed equivalent learning test results between the anesthesia and non-anesthesia groups when the subjects were 15 years old [45]. In the fourth study [46] enrolling healthy subjects who had received general anesthesia when 1-year-old or younger for one of each of the following procedures: inguinal hernia repair and orchiopexy (withorwithoutherniarepair), pyloromyotomy, and circumcision. The study compared the results of the Iowa State scholastic aptitude test scores of the subjects and the non-surgery group when the subjects were 7–10 years of age and found that those who had received anesthesia and an operation when 1 year of age or younger (287 subjects) more frequently scored in the lower fifth percentile than those who had not received either anesthesia or surgery. The same study conducted a subgroup analysis of 58 subjects after excluding those with confirmed or suspected central nervous system anomalies and found that the same tendency persisted. Furthermore, in the same group lower scores on the scholastic aptitude test correlated with longer anesthesia and surgery times.

Several caveats are in order when considering the findings of such retrospective observational studies as these. First, the patient backgrounds vary; for instance, the academic performance of children is known to vary depending on factors like the ethnicity of their parents and level of education. However, matching the subject group and control groups for the ethnicity of the children's parents and educational level is difficult in practice. Second, the comparison would ideally be based on healthy subjects without any medical complications, but in practice assembling such a cohort is difficult, as low birth weight infants tend to have a higher rate of sedation or surgery with consequent variations in the proportion of low birth weight infants who are included.

4   Cohort Studies

Two cohort studies are of special relevance to the present discussion and deserve mentioning. The first is the General Anaesthesia and Awake-regional Anaesthesia in Infancy (GAS) Study [47] and the second is the Pediatric Anesthesia and NeuroDevelopment Assessment (PANDA) study [48]. The GAS study was an international prospective study conducted from 2007 to 2013 in Australia, New Zealand, the USA, and the UK with 28 participating institutions enrolling healthy children who had undergone hernia repair surgery. The aim of this study was to determine whether there were any developmental effects on children who had received either general anesthesia or local anesthesia inducing unconsciousness. The primary outcome, the results of follow-up assessment at 5 years of age, will be measured when the subjects reach the stipulated age; the results of the 2-year follow up, however, have already been published in an article. The subjects were children born later than 26 weeks of gestation who were 60 weeks old or younger at the commencement of the study. At the time of their surgery, the average age of the subjects was 10 months, the average body weight was 4.2 kg, and 85% were male. Of the 363 subjects allocated to the local anesthesia group, 140 had required either sedation or general anesthesia as additional anesthesia, while 287 subjects had received only local anesthesia. The study found that blood pressure failed to decrease at the same rate in the local anesthesia group as in the general anesthesia group, but that there was no significant difference in neurological assessment results between the two groups at 2 years of age.

The PANDA study, which examined the effect of family environment on the development of children [48], was conducted between 2009 and 2015 at four universities led by Columbia University, and enrolled children born at more than 36 weeks of gestation who had undergone planned hernia repair. The control group consisted of their siblings, who together constituted 105 pairs. The neurological assessments were conducted from age 8 to 15 years. Of the anesthesia group 9.5% were females. The average age of the group was 17.3 months (about 1-year-old), and the average anesthesia time was 80 min. The average age of the subjects at the time of their assessment was 10.6 years. The authors assessed a broad range of items including not only the subjects' IQ but also their memory/learning, motor/processing speed, visuospatial function, attentiveness, executive functions, speech, and behavior but found no significant difference between groups.

5   Topics

Let us now examine some more recent studies. The findings of the Mayo clinic cohort study conducted from 1976 to 1982 [41], which, as mentioned earlier, produced two significant findings—a significant difference in learning impairment between children who had undergone anesthesia two or more times and those who had never undergone anesthesia or only did so only once and long-term learning impairment among the former group—were astonishing. A current birth cohort study of children born between 1996 and 2000 has yet to publish its results, but the preliminary data appear to indicate that, as in Fig. 1, poorer learning correlates with multiple anesthetic procedures [49, 50]. A study conducted at the University of California at San Francisco which assessed 28 ASA 1–2 children between the ages of 6 and 11 years who had been exposed to a volatile anesthetic for 2 h or longer at age 1-year-old or younger demonstrated no difference in IQ compared to the control group, but had decreased recall memory, that is, contextual memory consisting of the ability to remember the details of time and place, for example, was poorer. On the other hand, familiarity, or the feeling of having experienced an event, as opposed to the ability to recall the specifics of the event, remained intact. Furthermore, the same study includes a figure demonstrating a similar deficit in recall memory in adult rats which had been exposed at 7 days of age for 4 h to a 1 minimum alveolar concentration (MAC) of volatile anesthetic [51]. The study, led principally by Columbia University [52], used magnetic resonance imaging (MRI) findings to determine whether the local anesthetic administered to 60% of women at delivery had any long-term effects on offspring. Thirty-seven children born between 2005 and 2010 to a healthy mother were assessed within 6 weeks of birth by MRI. Of these, 24 had been delivered under local anesthesia while 13 had not. Subjects whose mother had received local anesthesia had a larger bilateral frontal and occipital lobes and cingulated gyrus. Furthermore, the correlation between maternal epidural anesthesia and occipital lobe volume in the offspring strengthened with increasing anesthesia time. The study concludes that long-term follow up and neurological assessment will be necessary.

6   The FDA's Perspective

In April 2017, the US Food and Drug Administration published safety information concerning anesthetic agents on its official website. This information also provides an exhaustive list of anesthetic agents besides alpha-2 agonists, including inhalational anesthetics (desflurane, halothane, isoflurane, and sevoflurane) and intravenous anesthetics (etomidate, ketamine, lorazepam injection, methohexital, midazolam injection and syrup, pentobarbital, and propofol) in addition to the warning that multiple or extended use of these agents for children under 3 years of age or fetuses in the third trimester may adversely affect brain development. At the same time the FDA advises that necessary surgery should not be postponed or omitted due to concerns about the safety of anesthetics.

References

1. Jones KL, Smith DW (1973) Recognition of the fetal alcohol syndrome in early infancy. Lancet 302:999–1001

2. Jones KL, Smith DW, Ulleland CN et al (1973) Pattern of malformation in offspring of chronic alcoholic mothers. Lancet 1:1267–1271

3. Clarren SK, Smith DW (1978) The fetal alcohol syndrome. N Engl J Med 298:1063–1067

4. Dobbing J, Sands J (1979) Comparative aspects of the brain growth spurt. Early Hum Dev 3:79–83

5. Rice D, Barone S Jr (2000) Critical periods of vulnerability for the developing nervous system: evidence from humans and animal models. Environ Health Perspect 108S:511–533

6. Ikonomidou C, Bosch F, Miksa M et al (1999) Blockade of NMDA receptors and apoptotic neurodegeneration in the developing brain. Science 283:70–74

7. Ikonomidou C, Bittigau P, Ishimaru MJ et al (2000) Ethanol-induced apoptotic neurodegeneration and fetal alcohol syndrome. Science 287:1056–1060

8. Fredriksson A, Pontén E, Gordh T et al (2007) Neonatal exposure to a combination of N-methyl-D-aspartate and gamma-aminobutyric acid type A receptor anesthetic agents potentiates apoptotic neurodegeneration and persistent behavioral deficits. Anesthesiology 107:427–436

9. Hemmings HC Jr, Akabas MH, Goldstein PA et al (2005) Emerging molecular mechanisms of general anesthetic action. Trends Pharmacol Sci 26:503–510

10. Paule MG, Li M, Allen RR et al (2011) Ketamine anesthesia during the first week of life can cause long-lasting cognitive deficits in rhesus monkeys. Neurotoxicol Teratol 33:220–230

11. Rizzi S, Carter LB, Ori C et al (2008) Clinical anesthesia causes permanent damage to the fetal guinea pig brain. Brain Pathol 18:198–210

12. Shen X, Dong Y, Xu Z et al (2013) Selective anesthesia-induced neuroinflammation in developing mouse brain and cognitive impairment. Anesthesiology 118:502–515

13. Shen X, Liu Y, Xu S et al (2013) Early life exposure to sevoflurane impairs adulthood spatial memory in the rat. Neurotoxicology 39:45–56

14. Zou X, Liu F, Zhang X et al (2011) Inhalation anesthetic-induced neuronal damage in the developing rhesus monkey. Neurotoxicol Teratol 33:592–597

15. Istaphanous GK, Howard J, Nan X et al (2011) Comparison of the neuroapoptotic properties of equipotent anesthetic concentrations of desflurane, isoflurane, or sevoflurane in neonatal mice. Anesthesiology 114:578–587

16. Jevtovic-Todorovic V, Hartman RE, Izumi Y et al (2003) Early exposure to common anesthetic agents causes widespread neurodegeneration in the developing rat brain and persistent learning deficits. J Neurosci 23:876–882

17. Rizzi S, Ori C, Jevtovic-Todorovic V (2010) Timing versus duration: Determinants of anesthesia-induced developmental apoptosis in the young mammalian brain. Ann N Y Acad Sci 1199:43–51

18. Young C, Jevtovic-Todorovic V, Qin YQ et al (2005) Potential of ketamine and midazolam, individually or in combination, to induce apoptotic neurodegeneration in the infant mouse brain. Br J Pharmacol 146:189–197

19. Nikizad H, Yon JH, Carter LB et al (2007) Early exposure to general anesthesia causes significant neuronal deletion in the developing rat brain. Ann N Y Acad Sci 1122:69–82

20. Sanders RD, Xu J, Shu Y et al (2008) General anesthetics induce apoptotic neurodegeneration in the neonatal rat spinal cord. Anesth Analg 106:1708–1711

21. Straiko MMW, Young C, Cattano D et al (2009) Lithium protects against anesthesia-induced developmental neuroapoptosis. Anesthesiology 110:662–668

22. Sanders RD, Sun P, Patel S et al (2010) Dexmedetomidine provides cortical neuroprotection: impact on anaesthetic-induced neuroapoptosis in the rat developing brain. Acta Anaesthesiol Scand 54:710–716

23. Yon JH, Carter LB, Jevtovic-Todorovic V (2006) Melatonin reduces the severity of anesthesia-induced apoptotic neurodegeneration in the developing rat brain. Neurobiol Dis 21:522–530

24. Cattano D, Young C, Olney JW (2008) Sub-anesthetic doses of propofol induce neuroapoptosis in the infant mouse brain. Anesth Analg 106:1712–1714

25. Ma D, Williamson P, Januszewski A et al (2007) Xenon mitigates isoflurane-induced neuronal apoptosis in the developing rodent brain. Anesthesiology 106:746–753

26. Johnson SA, Young C, Olney JW (2008) Isoflurane-induced neuroapoptosis in the developing brain of non-hypoglycemic mice. J Neurosurg Anesth 20:21–28

27. Sanders RD, Xu J, Shu Y et al (2009) Dexmedetomidine attenuates isoflurane-induced neurocognitive impairment in neonatal rats. Anesthesiology 110:11077–11085

28. Zhang X, Xue Z, Sun A (2008) Subclinical concentration of sevoflurane potentiates neuronal apoptosis in the developing C57BL/6 mouse brain. Neurosci Lett 447:109–114

29. Cattano D, Williamson P, Fukui K et al (2008) Potential of xenon to induce or to protect against neuroapoptosis in the developing mouse brain. Can J Anesth 55:429–436

30. Brambrink AM, Evers AS, Avidan MS et al (2010) Isoflurane-induced neuroapoptosis in the neonatal rhesus macaque brain. Anesthesiology 112:834–841

31. Brambrink AM, Evers AS, Avidan MS et al (2012) Ketamine-induced neuroapoptosis in the fetal and neonatal rhesus macaque brain. Anesthesiology 116:372–384

32. Brambrink AM, Back SA, Riddle A et al (2012) Isoflurane-induced apoptosis of oligodendrocytes in the neonatal primate brain. Ann Neurol 72:525–535

33. Slikker W Jr, Zou X, Hotchkiss CE et al (2007) Ketamine-induced neuronal cell death in the perinatal rhesus monkey. Toxicol Sci 98:145–158

34. Zou X, Patterson TA, Divine RL et al (2009) Prolonged exposure to ketamine increases neurodegeneration in the developing monkey brain. Int J Dev Neurosci 27:727–731

35. Fredriksson A, Archer T (2004) Neurobehavioural deficits associated with apoptotic

neurodegeneration and vulnerability for ADHD. Neurotox Res 6:435–456

36. Satomoto M, Satoh Y, Terui K et al (2009) Neonatal exposure to sevoflurane induces abnormal social behaviors and deficits in fear conditioning in mice. Anesthesiology 110:628–637

37. Stratmann G, Sall JW, May LD et al (2009) Isoflurane differentially affects neurogenesis and long-term neurocognitive function in 60- and 7-day-old rats. Anesthesiology 110:834–848

38. Wozniak DF, Hartman RE, Boyle MP et al (2004) Apoptotic neurodegeneration induced by ethanol in neonatal mice is associated with profound learning/memory deficits in juveniles followed by progressive functional recovery in adults. Neurobiol Dis 17:403–414

39. Sun Z, Satomoto M, Adachi YU et al (2016) Inhibiting NADPH oxidase protects against long-term memory impairment induced by neonatal sevoflurane exposure in mice. Br J Anaesth 117:80–86

40. Satomoto M, Sun Z, Adachi YU et al (2016) Neonatal sevoflurane exposure induces adulthood fear-induced learning disability and decreases glutamatergic neurons in the basolateral amygdala. J Neurosurg Anesthesiol. https://doi.org/10.1097/ANA.0000000000000387

41. Wilder RT, Flick RP, Sprung J et al (2009) Early exposure to anesthesia and learning disabilities in a population-based birth cohort. Anesthesiology 110:796–804

42. Kalkman CJ, Peelen L, Moons KG et al (2009) Behavior and development in children and age at the time of first anesthetic exposure. Anesthesiology 110:805–812

43. DiMaggio C, Sun LS, Kakavouli A et al (2009) A retrospective cohort study of the association of anesthesia and hernia repair surgery with behavioral and developmental disorders in young children. J Neurosurg Anesthesiol 21:286–291

44. Bartels M, Althoff RR, Boomsma DI (2009) Anesthesia and cognitive performance in children: no evidence for a causal relationship. Twin Res Hum Genet 12:246–253

45. Hansen TG, Pedersen JK, Henneberg SW et al (2011) Academic performance in adolescence after inguinal hernia repair in infancy: a nationwide cohort study. Anesthesiology 114:1076–1085

46. Block RI, Thomas JJ, Bayman EO et al (2012) Are anesthesia and surgery during infancy associated with altered academic performance during childhood? Anesthesiology 117:494–503

47. Davidson AJ, Disma N, de Graaff JC et al (2016) Neurodevelopmental outcome at 2 years of age after general anaesthesia and awake-regional anaesthesia in infancy (GAS): an international multicentre, randomised controlled trial. Lancet 387:239–250

48. Sun LS, Li G, Miller TL et al (2016) Association between a single general anesthesia exposure before age 36 months and neurocognitive outcomes in later childhood. JAMA 315:2312–2320

49. Hu D, Flick RP, Gleich SJ et al (2016) Construction and characterization of a population-based cohort to study the association of anesthesia exposure with neurodevelopmental outcomes. PLoS One 11:e0155288

50. Pinyavat T, Warner DO, Flick RP et al (2016) Summary of the update session on clinical neurotoxicity studies. J Neurosurg Anesthesiol 28:356–360

51. Stratmann G, Lee J, Sall JW et al (2014) Effect of general anesthesia in infancy on long-term recognition memory in humans and rats. Neuropsychopharmacology 39:2275–2287

52. Spann MN, Serino D, Bansal R et al (2015) Morphological features of the neonatal brain following exposure to regional anesthesia during labor and delivery. Magn Reson Imaging 33:213–221

Chapter 11

The Challenge of Opioid-Free Anesthesia

Maher Khalife, Graziela Biter, Marco Cascella, and Raffaela Di Napoli

Abstract

The introduction of opioids in the clinical practice of anesthesia was a revolution. By blocking the sympathetic response to surgical stimuli and obtaining a reduced requirement of hypnotic agents, a safer and a more stable hemodynamically perioperative period was made possible. However, administration of opioids can be associated with several side effects that can be responsible for delayed patient recovery and hospital discharge, as well as leading to increased health service costs. Furthermore, opioid use is related to a wide range of side effects including gastrointestinal (nausea, vomiting, ileus), respiratory (decreased central respiratory drive impacting respiratory rate, and tidal volume), central nervous system effects (sedation, delirium, dysphoria, catalepsy, hallucinations) effects, as well as urinary retention, pruritus, bradycardia, and dizziness. Another undesired effect of opioids as primary pain therapy in the perioperative period, is the development of "acute tolerance" to the analgesic effect of these drugs. This diminished analgesic effect may also be the result of the "opioid-induced hyperalgesia" (OIH) phenomenon. Again, some evidence has suggested that opioids may interfere with the immune system. Numerous studies have shown that opioids can influence the progression of cancer, metastasis, and cancer recurrence. In the 1990s, in the light of experience and studies conducted on opioids and their effects within the broader context of multimodal analgesia, their use came into question. New anesthesia techniques started to be developed that aimed to achieve the sparing use of opioids. These approaches culminated, in the 2000s, with the development of opioid-free anesthesia (OFA) pathways. The strategy of OFA is a realistic alternative that can lead to enhanced recovery and increased patient satisfaction by reducing important opioid related side effects; it can also facilitate the use of lower doses of opioids postoperatively in order to achieve a pain-free recovery and reduce pain scores while providing faster and safer mobilization and rehabilitation. The drugs used are hypnotics, N-methyl-D-aspartate (NMDA) antagonists (ketamine, magnesium sulfate), sodium channel blockers (local anesthetics), anti-inflammatory drugs [nonsteroidal anti-inflammatory drugs (NSAIDs), dexamethasone, local anesthetic], and alpha-2 agonists (dexmedetomidine, clonidine). The association of OFA and locoregional anesthetic techniques is very common. Several types of patients can benefit from this technique including narcotic history patients, obese patients with obstructive sleep apnea, patients with hyperalgesia and history of chronic pain, immune deficiency individuals, patients undergoing oncologic surgery as well as those affected by inflammatory conditions, chronic obstructive pulmonary disease, and asthma. While different OFA protocols have been reported in the literature, the publications rely mostly on case reports and small size investigations. More studies are necessary to assess what the interactions between these drugs are. Clinical researchers must design studies with rigorous methodology in order to correctly assess the risks and benefits of OFA for patients in different surgical settings.

Key words Opioid-free anesthesia, Opioids, Opioid-related side effects, Ketamine, Dexmedetomidine, Magnesium, Lidocaine

Marco Cascella (ed.), *General Anesthesia Research*, Neuromethods, vol. 150, https://doi.org/10.1007/978-1-4939-9891-3_11,
© Springer Science+Business Media, LLC, part of Springer Nature 2020

1   Introduction

Opioids have been used to relieve pain for millennia, in the form of opium extracted from poppies. The oldest reference to opium poppy growth comes from Sumerians of Mesopotamia, who cultured the plant around 5000 years ago and described it in an ideogram as the "joy plant" [1]. In 1805, the German pharmacist Friedrich Wilhelm Adam Ferdinand Sertürner (1783–1841) isolated morphine from poppy. Subsequently, in 1874, Charles Adler Wright (1844–1894) synthesized diacetylmorphine, also known as heroin, in an attempt to obtain a less addictive substance, and it was marketed by the Bayer Laboratories (1888) as a pain reliever, sedative, and cough suppressant. Further derivatives of morphine and codeine were developed at the beginning of the twentieth century, including hydromorphone, dihydrocodeine, hydrocodone, oxymorphone, meperidine, and oxycodone [2]. Fentanyl was developed by Paul A.J. Jansen, in 1953, and became available as of 1960. It opened the gate to developing other fentanyl analogs, such as carfentanil (1974), sufentanil (1974), lofentanil (1975), and alfentanil (1976) [3]. At the beginning of the 1990s, remifentanil became available for clinical use [4].

The introduction of opioids in the clinical practice of anesthesia was a revolution. By blocking the sympathetic response to surgical stimuli and obtaining a reduced requirement of hypnotic agents, a safer and a more stable hemodynamically perioperative period was allowed. However, administration of opioids can be associated with several side effects that can be responsible for delayed patient recovery and hospital discharge, as well as leading to increased health service costs. Another undesired effect of opioids as primary pain therapy in the perioperative period is the development of "acute tolerance" to the analgesic effect of these drugs [5, 6]. This diminished analgesic effect may also be the result of the "opioid-induced hyperalgesia" (OIH) effect. The two phenomena are very different from a pharmacological point of view but can both lead to significant opioid dose increases over time. Overprescription of opioids for postoperative pain control has now led to the opioid epidemic, with increased addiction and overdose-related deaths worldwide.

2   The Rationale for Opioid-Free Anesthesia

While opioids have been one of the pillars of the anesthetic practice for decades, anesthesiologists must deal with their side effects, some of which are well known and studied, and others whose mechanisms still need investigation (Table 1). Pain and paralytic ileus is often the cause of delayed hospital release. Furthermore,

Table 1
Opioids side effects

Opioids side effects
Gastrointestinal: nausea and vomiting, ileus
Respiratory: decreased central respiratory drive (respiratory rate, and tidal volume)
Central nervous system: sedation, delirium, dysphoria, catalepsy, hallucinations
Urinary retention
Bradycardia
Dizziness
Pruritus
Chronic effects
Opioid-induced tolerance
Physical dependence

opioids can exacerbate obstructive sleep apnea and increase its severity. Studies have shown that 50–80% of patients that are administered opioids will have at least one side effect [7]. At the same time, the efficacy of pain relief during movement is only moderate. In the era of fast-track surgery with concerns for the quality of care and economic efficiency in health care, the avoidance of these common side effects in the postoperative period is essential. Physical dependence and opioid-induced tolerance represent a special issue, while the relationship between opioid use and cancer should be better investigated. For this reason, alternative anesthetic protocols aimed at reducing or totally avoiding opioids in the perioperative setting have been developed [8].

In the 1990s, in the light of experience and studies conducted on opioids and their effects within the broader context of multimodal analgesia, their use came into question. New anesthesia techniques started to be developed that aimed to achieve the sparing use of opioids, which culminated in the 2000s with the development of opioid-free anesthesia (OFA) pathways. It was in 1993 that Kehlet and Dahl introduced the term "multimodal analgesia" for management of postoperative pain, along with the term "balanced analgesia" [9]. However, it is also possible to replace opioids completely in the perioperative setting with other drugs, providing perioperative hemodynamic stability and improving postoperative pain management. The strategy of OFA is a realistic alternative that can lead to enhanced recovery and increased patient satisfaction by reducing important opioid-related side effects; it can also facilitate the use of lower doses of opioids postoperatively in order to achieve a pain free recovery and reduce pain scores while providing faster and safer mobilization and rehabilitation [10]. Furthermore, it is established that opioid-induced hyperalgesia and chronic pain syndromes are more frequent when high-dose opioids are used perioperatively [11].

2.1 Anesthesia and Opioid Dependence

Although pharmaceutical companies stated that patients would not become addicted to opioid pain relievers, their consumption has risen in the last three decades and a rise in prescription medication then followed, after letters to medical journals had underlined the safety and low addictive potential of opioid use [12]. Consequently, there was an increase in the misuse of opioid prescriptions. In 2016, there were more than 130 deaths per day due to opioid-related drug overdoses in the USA and a total of 11.4 million people who misused prescription opioids, from which two million people for the first time. (www.hhs.gov/opioids). The US government has declared the opioid epidemic a public health emergency.

In the perioperative period, patients that receive opioids as primary pain therapy will most likely require increased doses to maintain the same analgesic effects. This is due to the phenomenon of opioid-induced hyperalgesia and tolerance. The opioid paradox is that the more opioids that are used intraoperatively, the more opioids will be required postoperatively and has been measured in surgical patients up to two days postoperatively [13]. In a study of over 36,000 opioid-naïve patients undergoing elective surgery between 2013 and 2014 in the USA, an incidence of 6% of chronic opioid use after surgery, major or minor, was found (opioid use lasting more than 90 days postoperatively) [14]. Shah et al. [15] found that every day of opioid therapy after the third postoperative day increases the risk of the patient becoming a chronic opioid consumer. Dramatic increases are seen after the 5th and the 31st day of therapy. The incidence of chronic opioid consumption was found to be approximately 15% in patients that had their first opioid-based treatment for a period longer than 8 days and as high as 30% in patients whose first treatment was 31 days or more.

The Australian and New Zealand College of Anesthetists and its Faculty of Pain Medicine released the following statement in 2018: "Slow-release opioids are not recommended for use in the management of patients with acute pain" [16]. They also made recommendations that, when opioids are indicated and if the oral route is available, then the most appropriate treatment is an age-based dose of immediate-release oral opioid. Controlled release opioids should only be used in patients that are already on such treatment. (Australia and New Zealand College of Anesthesia. Position statement on the use of slow-release opioid preparations in the treatment of acute pain.)

2.2 Opioid-Induced Tolerance and Hyperalgesia

Tolerance is defined as a decrease in a pharmacologic response following repeated or prolonged drug administration [17]. In the case of opioid tolerance, the phenomenon is linked to opioid receptor desensitization, as well as down-regulation, phosphorylation, and endocytosis [18].

The concept of differential tolerance development states that varying opioid targets will develop tolerance at different degrees and at different speeds. For example, patients will become rapidly tolerant to the analgesic effect, but when it comes to gastrointestinal side effects, tolerance will develop more slowly and to a lesser degree. The opioid signaling system has a remarkable ability for tolerance development. Opioid tolerance can develop in a short time frame, possibly within hours, if patients are exposed to high doses. In principle, tolerance can be overcome by increasing opioid dose, but this will expose patients to opioid-induced hyperalgesia (OIH) that will be worse with dose increment [19].

Opioid-Induced tolerance and OIH phenomena have a significant clinical impact. In a recent meta-analysis, Fletcher et al. addressed the issue of OIH in patients undergoing surgery. They included 27 studies with a total of 1494 patients and concluded that those who received a high intraoperative opioid dose had higher pain scores and increased morphine consumption in the first 24 h postoperatively. These findings were particularly evident with the use of remifentanil [20].

2.3 Opioids and Cancer

Some evidence has suggested that opioids may interfere with the immune system. Numerous studies have shown that opioids can influence the progression of cancer, metastasis and cancer recurrence through numerous and complex mechanisms such as improved angiogenesis and overall impairment of host immunity [21–23]. Furthermore, opioids may directly stimulate cancer growth by overexpression of mu opioid receptors in the neoplastic tissue [24–26].

However, most of these findings come from preclinical studies; results regarding opioids and cancer outcomes can be contradictive. For instance, although Bimonte et al. [27] demonstrated that naloxone may counteract the promotion of tumor growth, as induced by morphine in an animal model of triple-negative breast cancer, the translational perspectives of these findings seem to be a complex.

Again, in the clinical context, the strength of these studies is limited because of their retrospective design and the quality of data. To date, no conclusion can be drawn on a potential link between opioids and the immune function in patients with cancer.

3 Features of Opioid-Free Anesthesia (OFA)

The OFA techniques avoid the use of giving systemic intraoperative opioids, as well as those given by neuraxial and intracavitary methods. Instead, it promotes an association of nonopioid multimodal analgesics, with the aim of ensuring adequate pain control and at the same time, ensuring the hemodynamic stability of the patient. The drugs used are hypnotics, NMDA antagonists (ketamine, magnesium sulfate), sodium channel blockers (local anesthetics), anti-inflammatory drugs (NSAID, dexamethasone, local anesthetic), and alpha-2 agonists (dexmedetomidine, clonidine). The association between OFA and locoregional anesthetic techniques is very common.

Normal pain transmission is a complex, dynamic process. Originating from the peripheral receptors, the transduced noxious stimuli is transmitted via afferent fibers to the dorsal horn of the spinal cord, where the signal can be extensively modulated by intrinsic spinal interneurons, glia, and descending pathways. It is here that NMDA receptors, GABA, enkephalins, and other neurotransmitters intervene. The sympathetic–parasympathetic reactions to noxious stimuli during anesthesia are equally complex. Opioid-free anesthesia can offer an alternative to the current opioid-inclusive balanced anesthesia, especially for obese and oncologic patients or opioid addicts.

The most frequent side effects of OFA are hypotension and bradycardia. A meta-analysis by Frauneknecht et al. [28] investigated the analgesic impact of intraoperative opioids versus opioid-free anesthesia. It included 1304 patients undergoing different types of surgeries and concluded that there is high-quality evidence that opioid-inclusive anesthesia when compared to opioid-free anesthesia, did not reduce the level of pain or opioid consumption postoperatively. However, the authors found increased postoperative nausea and vomiting in the opioid inclusive regimen.

3.1 Drugs and Pharmacological Considerations

3.1.1 Ketamine

Ketamine is a water-soluble phencyclidine derivative. It was first used in 1964 for veterinary purposes and was FDA approved in 1970. It has been used mainly for induction in hemodynamically unstable patients or as an adjunct for analgesia or sedation. It is a rapid-acting general anesthetic that produces profound analgesia, cardiovascular and respiratory stimulation and produces normal or slightly increased skeletal muscle tone. It does not alter the normal pharyngeal-laryngeal reflexes and can lead to transient and minimal respiratory depression. By selectively interrupting association pathways of the brain, it produces a "dissociative anesthesia" at doses more than 1 mg/kg. Ketamine enhances descending inhibiting serotoninergic pathways and can exert antidepressive effects (with concentrations as low as ten times the dose needed for obtaining anesthesia). Its analgesic effect is produced by preventing central sensitization in the dorsal horn neurons and through inhibition of the synthesis of nitric oxide [29]. Again, this medication seems to be useful for attenuating the development of opioid tolerance and opioid hyperalgesia [30, 31]. Analgesia is obtained with subanesthetic doses and it can reverse tolerance to opioids. Low-dose ketamine reduces pain and hyperalgesia, and there is an interest in using it for the management of depression [32, 33].

Ketamine acts as a noncompetitive antagonist of the central nervous system (CNS) N-methyl-D-aspartate (NMDA) receptors. It blocks sensory input and impairs limbic functions. It is an agonist of α- and β-adrenergic receptors and an antagonist of muscarinic receptors in the brain. This medication blocks the reuptake of catecholamines. Finally, it acts as an agonist for the kappa-type opioid receptor, and binds to the delta and mu-type receptor.

With regard to pharmacokinetics, after intravenous administration, the onset time is 30 s. The drug has a distribution half-life of 11–16 min whereas the elimination half-life is 2–3 h. It is metabolized in the liver by microsomal cytochrome P450 enzymes and excreted in the form of metabolites, mainly in the urine.

Relative contraindications to sub-anesthetic ketamine doses are a high-risk coronary or vascular disease, uncontrolled hypertension, elevated intracranial pressure, elevated intraocular pressure, globe injuries, history of psychosis, sympathomimetic syndrome, hepatic dysfunction, recent liver transplantation, and porphyria. A risk–benefit assessment should be done when the patient presents one or more of the abovementioned pathologies.

Regarding doses, the most common in published studies are in the range of 0.15–0.5 mg/kg given as a bolus and 0.1–0.2 mg/kg/h as an infusion. Psychosensory effects increase with boluses above 0.3 mg/kg and the effects dissipate after 30–45 min [34]. Furthermore, the ideal body weight for obese patients must be considered. According to Gorlin et al. [34], the dose regimes are dependent on the operating time (Table 2).

3.1.2   Magnesium

Magnesium is the fourth most common cation in the body. It acts as a noncompetitive antagonist of NMDA receptors in the CNS and regulates calcium influx into the cell. These are believed to be the mechanisms by which it promotes analgesia. It has been reported that it suppresses neuropathic pain [35], potentiates analgesia after opioid administration and attenuates opioid tolerance [36]. Some studies have shown that its use imparts an important reduction in opioid consumption in the first 24 h postoperatively. A smaller effect was observed on pain scores over the same time interval.

Adverse side effects from magnesium sulfate administration are a circulatory collapse, respiratory paralysis, hypothermia, pulmonary edema, depressed reflexes, hypotension, flushing, drowsiness, depressed   cardiac   function,   diaphoresis,   hypocalcemia,

Table 2
Ketamine dose regimens [adapted from [34]]

Timing	Dose regimen
Operating time	
Less than 60 min	0.1–0.3 mg/kg iv bolus at induction
More than 60 min	0.1–0.3 mg/kg iv bolus at induction
Intraoperative to postoperative	
No postoperative infusion	0.1–0.3 mg/kg iv bolus at induction that can be repeated every 30–60 min during surgery (last dose at least 30 min before emergence)
Postoperative infusion	0.1–0.3 mg/kg bolus at induction, followed by 0.1–0.2 mg/kg/h infusion that can be continued for 24 up to 72 h; consider reducing does at 10 mg/h or less after 24 h

hypophosphatemia, hyperkalemia, and visual changes. Contraindications to the administration of magnesium sulphate include hypersensitivity, heart blocks, myocardial damage, hypomagnesemia, and hypercalcemia. It should be used with caution in patients with renal impairment, digitalized patients and those with myasthenia gravis or another neuromuscular disease. When administered, renal function should be monitored, as well as blood pressure, respiratory rate, and deep tendon reflexes.

Studies report a wide range in the dose administered to patients. Bolus ranges are usually between 30 and 50 mg/kg and can be continued by an infusion of 6–25 mg/kg/h until the end of the surgery. Continuous infusion in the postoperative period has also been described (up to 6 h). So far, no serious side effects have been mentioned. There is no recommended administration regime for use with magnesium sulphate, but it does seem that a bolus dose of 40–50 mg/kg efficiently decreases postoperative opioid consumption and pain scores [37].

3.1.3   Lidocaine

Lidocaine is a sodium channel blocker and has been used since the 1960s as an antiarrhythmic drug and local anesthetic. Intravenous lidocaine administration will affect peripheral and central nerve endings. Central sensitization resulting from tissue damage (somatic pain) may be minimized by intravenous lidocaine, due to its antihyperalgesic effects [38]. The analgesic effect of intravenous lidocaine administration may be a result of direct or indirect interaction with different receptors and nociceptive transmission pathways including: muscarinic antagonists, glycine inhibitors, a reduction in the production of excitatory amino acids, a reduction in the production of thromboxane A2, release of endogenous opioids, a reduction in neurokinins, and release of adenosine triphosphate [39]. Lidocaine infusion in the perioperative setting has been extensively studied because of its analgesic, antihyperalgesic, and anti-inflammatory effects [40].

Based on preclinical studies, we can say that lidocaine interacts with the inflammatory system, reducing the inflammatory response at very low concentrations (e.g., 0.1 μM lidocaine) when the exposure is for a long period of time (hours) [41].

It has been shown that lidocaine may decrease pain scores, reduce opioid consumption, increase the restoration of bowel movement, and even shorten the length of hospital stay for patients undergoing laparoscopic colon surgery [42–44].

However, a recent Cochrane meta-analysis assessed whether intravenous perioperative lidocaine, compared to placebo or no treatment, has a beneficial role in the postoperative period regarding gastrointestinal recovery, postoperative nausea, and opioid consumption [45].

Side effects of lidocaine intravenous administration should be made known and awareness raised (Table 3).

Concerning dose regimens, bolus doses of 1–3 mg/kg at induction have been recommended. Infusion doses vary from 1 to 5 mg/kg/h and are stopped either at skin closure or 48 h postoperatively. The lidocaine dose necessary to obtain analgesia in the perioperative period is thought to be between 1 and 2 mg/kg as an initial bolus, followed by continuous administration at 0.5–3 mg/kg/h [46].

Of note, the maximum dose of lidocaine to be administered to avoid local anesthetic systemic toxicity (LAST) is 5 mg/kg. It is likely that a continuous infusion of 2 mg/kg could be the most appropriate regimen for avoiding local anesthetics systemic toxicity (CNS toxicity occurs at lidocaine plasmatic concentrations >5 μg/ml, slightly above the therapeutic plasma level 2.5–3.5 μg/ml).

Table 3
Lidocaine side effects

Cardiovascular	Bradycardia, cardiac arrhythmias, circulatory shock, coronary artery vasospasm, edema, flushing, heart block, hypotension, local thrombophlebitis
Central nervous system	Agitation, anxiety, apprehension, coma, confusion, disorientation, dizziness, drowsiness, euphoria, hallucinations, hyperesthesia, hypoesthesia, intolerance to temperature, lethargy, loss of consciousness, metallic taste, nervousness, paresthesia, psychosis, seizure, slurred speech, twitching
Respiratory	Bronchospasm, dyspnea, respiratory depression
Neuromuscular and skeletal	Tremor, weakness
Gastrointestinal	Nausea and vomiting
Others	Tinnitus

3.1.4 Dexamethasone

Dexamethasone is a glucocorticoid agonist that acts by inhibiting leukocyte infiltration at the site of inflammation. It interferes with the function of the mediators of the inflammatory response, suppresses humoral immune responses, and reduces edema or scar tissue [47]. It is administered in the surgical setting to prevent postoperative nausea and vomiting [48]. Its intravenous administration permits the prolongation of peripheral nerve blocks [49]. It has been proven that a sole dose administered at the beginning of the surgical intervention of 0.1 mg/kg can lead to a reduction in opioid consumption, reduction of postoperative nausea and vomiting, reduction of fatigue and a better postoperative rehabilitation [50]. After administration, the plasmatic concentration peaks after 2–12 h, and it has a half-life of 36–72 h. It can suppress cortisol levels for up to 1 week [51]. A meta-analysis on whether a single dose of dexamethasone administered in the perioperative setting will lead to an increase in adverse effects has not found evidence that it increases the risk of postoperative wound infection or delayed wound healing. However, they did find that it significantly increased the glucose values for the first 12 h post-surgery [52]. Considerations should be made as to whether it is safe to be administered to diabetic patients. The dosage for the antiemetic effect is 50 μg/kg, whereas a reduction of postoperative analgesic requirements is obtained with a dose of 100 μg/kg. For the anti-inflammatory effect, it should be administered prior to incision. When administered in an awake patient it is associated with a painful perineal sensation.

3.1.5 Alpha-2 Adrenoceptor Agonists: Clonidine and Dexmedetomidine

Clonidine

Clonidine is an imidazoline derivative and acts as an α2-adrenergic agonist. Three alpha-2 receptors have been identified: α2A, α2B, and α2C with a different distribution. The α2A receptor is widely distributed throughout the CNS (locus coeruleus, brain stem nuclei, cerebral cortex, septum, hypothalamus, and hippocampus), in the kidneys, spleen, thymus, lung, and salivary glands. The α2B-receptor is located primarily in the periphery (kidney, liver, lung, and heart) with low levels of expression in the thalamic nuclei of the CNS. The α2C-receptor is primarily expressed in the brain, including the striatum, olfactory tubercle, hippocampus, and cerebral cortex. Low levels of the α2C-subtype are also found in the kidneys. The α2A and α2C receptors are located presynaptically and inhibit the release of noradrenaline from sympathetic nerves. Thus, their stimulation will lead to decreased sympathetic tone, blood pressure, and heart rate. Stimulation of centrally located α2A receptors will lead to sedation and analgesia; it also mediates components of the analgesic effect of nitrous oxide in the spinal cord. The stimulation of α2B-receptors leads to a constriction of vascular smooth muscle. Clonidine acts on all three receptor subtypes with similar potency.

Clonidine is used in multiple clinical situations including prophylaxis of vascular migraine headaches, treatment of severe dysmenorrhea, rapid detoxification in the management of opiate withdrawal, treatment of alcohol withdrawal used together with benzodiazepines, management of nicotine dependence, and treatment of attention-deficit hyperactivity disorder (ADHD). In the perioperative setting, it is used for its antihypertensive, analgesic, sedative, and anxiolytic effects. It can provide a maximum of 50% reduction in requirements of inhalation anesthetics to maintain 1 MAC.

Clonidine is rapidly absorbed after oral administration with a maximum plasma concentration achieved between 1.5 and 2 h [53]. It has a half-life of approximately 8–12 h [54]. Its elimination half-life is of 12–16 h, and its distribution half-life is more than 10 min. It is excreted mainly in the urine.

A recent systematic review and meta-analysis of 57 trials show that clonidine improves postoperative outcomes by improving pain control, reducing PONV, and improving hemodynamic and sympathetic stabilities. It does not seem to have consequences on renal function or awakening time and does not influence the cardiac outcome in the general population after noncardiac surgery. It is important to mention that nonfatal bradycardia/nonfatal cardiac arrest and hypotension have been described [55].

Used as a premedication, a dose of 2–4 µg/kg clonidine orally provides sedation, hypnosis, and anti-sialagogue effect [56]. Intravenous administration has been used with a dosage ranging from 1 to 4 µg/kg, infused over 10–15 min at different perioperative moments (60–90 min preoperatively, induction, and during maintenance).

Caution with its use should be taken for hemodynamically unstable patients, severe coronary insufficiency, recent myocardial infarction, cerebrovascular disease, chronic renal failure, cardiac conduction abnormalities, and elderly patients.

Dexmedetomidine

Dexmedetomidine is a specific and selective alpha-2 adrenoceptor agonist which acts via presynaptic and postsynaptic alpha-2-receptors. It is 8–10 times more selective towards the alpha-2-adrenoreceptor than clonidine [57]. It also seems to have higher alpha 2A and alpha 2C affinity than clonidine [58]. Its binding to the presynaptic alpha-2 receptor will inhibit norepinephrine release and therefore terminate the propagation of pain signals. By activation of the postsynaptic alpha-2 receptor, it will decrease the sympathetic activity, with attenuation of the neuroendocrine and hemodynamic responses to anesthesia and surgery. Its administration will lead to a reduction of anesthetic and opioid requirements, sedation, and analgesia. It preserves respiratory function and is used with a predictable decrease in mean arterial pressure and heart rate. It has sedative and anxiolytic effects and it is used in intensive care for ventilated patients that remain easily rousable and cooperative

Table 4
Suggested dose regimens for dexmedetomidine useful for periprocedural sedation [adapted from [63]]

Loading infusion of 1 µg/kg over 10 min; a reduced loading infusion of 0.5 µg/kg over 10 min is recommended for patients over 65 years
Maintenance dose is generally initiated at 0.6 µg/kg/h and titrated at the desired clinical effect between doses of 0.2 and 1.0 µg/kg/h
A reduction in the loading and maintenance dose is recommended in patients with hepatic impairment
A reduction in concomitant administered sedatives, anesthetics, hypnotics and opioids doses is required

during treatment. It offers reversible memory impairment and has been shown to provide a reduction of up to 90% in requirements of inhalational anesthesia to maintain a 1 MAC. Its most common side effects are hypotension, bradycardia, nausea, vomiting, and a dry mouth. A short hypertensive phase and subsequent hypotension are observed following initiation of infusion. Bradycardia and sinus arrest have occurred in young, healthy volunteers with high vagal tone or with different routes of administration, including rapid intravenous or bolus administration that were effectively treated with anticholinergic agents (atropine, glycopyrrolate). It is not recommended in patients with advanced heart block or severe ventricular dysfunction [59]. Cardiovascular adverse effects may be more pronounced in hypovolemic patients, in those with diabetes mellitus, chronic hypertension, in the elderly, and in those with high vagal tone.

It is metabolized by the liver and excreted mostly by the urine. It has a plasma half-life of 2–2 1/2 h, an elimination half-life of 2 h, and a distribution half-life of 5 min.

Intraoperative administration has been proven to reduce the requirements of other anesthetic agents. It can attenuate the hemodynamic response to intubation [60] and extubation [61] and can be continued in the postoperative period. It potentiates the effect of other anesthetic agents that can be administered intravenously, inhalational or by regional anesthesia. It has a significant opioid-sparing effect and is useful in intractable neuropathic pain [62]. Different doses are recommended when used for periprocedural sedation (Table 4).

4   Indications and Contraindications

Indications for OFA are multiple. Patients that can benefit from this technique are morbidly obese who have undergone gastric bypass, especially considering the postoperative respiratory disorders associated with opioids. Other populations include opioid-tolerant patients, those with chronic pain syndromes, or chronic respiratory impairment.

Indications for OFA include [64]:

- Narcotic history (acute and chronic opioid addiction).
- Opioid intolerance.
- Morbidly obese patients with obstructive sleep apnea.
- Hyperalgesia, history of chronic pain.
- Immune deficiency, oncologic surgery, inflammatory disease.
- Chronic obstructive pulmonary disease, asthma.

The OFA approach is contraindicated in patients expressing allergy either to the drug or to its excipient. Since OFA's drugs act on the sympathic system inducing bradycardia and/or hypotension, it is obvious that it should be used cautiously in patients with disorders of autonomic system, cerebrovascular disease, critical or acute coronary ischemia, arythmia particularly in extreme bradycardia, nonstabilized hypovolemic shock or polytrauma patients, interventions where controlled hypotension for minimal blood loss is necessary, and elderly patients with beta-blockers.

5   Protocols

While different OFA protocols have been reported in the literature, the publications rely mostly on case reports and different experiments with these drugs [65–69]. A study by Bakan et al. [65] compared OFA using dexmedetomidine and lidocaine, to opioid-based anesthesia induced with remifentanil for laparoscopic cholecystectomy. The authors showed a statistically significant reduction in fentanyl consumption at 2 h postoperatively, although this difference was not maintained at 6 h when fentanyl consumption was comparable between the two groups. They noted more hypotensive events within the opioid-based anesthesia group and hypertensive events within the opioid-free anesthesia protocol. The OFA had higher recovery times, but significantly lower pain scores and needed less analgesic rescue and treatment for nausea. In the opioid-free group, the patients ($n = 40$) received a loading dose of 0.6 µg/kg of dexmedetomidine, followed by infusion at a rate of 0.3 µg/kg/h and 1.5 mg/kg of lidocaine, then an infusion rate of 2 mg/kg/h. In the opioid-based anesthesia group, the patients ($n = 40$) received 2 µg/kg fentanyl at induction, followed by remifentanil infusion at a rate of 0.25 µg/kg/min. Lidocaine infusion was stopped 10 min before skin closure and skin incisions were infiltrated with bupivacaine. Dexmedetomidine and remifentanil infusion were terminated during skin closure. All patients received 8 mg Dexamethasone and 50 mg of dexketoprofen trometamol after anesthesia induction and 1 g of acetaminophen at the end of the surgery. All patients had a PCA pump set to deliver IV fentanyl.

A case report concerning a 40-year-old morbidly obese patient with a BMI of 50 kg/m^2 scheduled for a laparoscopic vertical sleeve gastrectomy was published by Aronsohn et al. [66]. No episodes of desaturation were noted in the postoperative period for this patient that suffered from severe OSA and was noncompliant with home CPAP; no opioids were administered in the perioperative period. She received a dose of 8 mg of dexamethasone at induction and 1 g of acetaminophen. The patient received an initial bolus of 5 mg/kg of ketamine, followed by a continuous infusion at 5 μg/kg/min. Surgeons injected the port sites with liposomal bupivacaine. Anesthesia was maintained with propofol infusion. During emergence, the patient received a dose of 4 mg ondansetron and 30 mg ketorolac. The patient was able to ambulate unassisted at 90 min postoperative. She was discharged from the hospital the following morning.

Landry et al. published a case report of OFA for an open aortic valve replacement [67]. As the patient had an allergy, manifested in the form of a diffuse skin rash to oxycodone and other drugs from the same class, it was advised to use as few opioids as possible perioperatively. The patient received 1 g of oral acetaminophen and 300 mg of gabapentin preoperatively. During induction, she received a 50 mg ketamine bolus, followed by an infusion of 5 μg/kg/min. Dexmedetomidine was administered as a bolus of 0.5 μg/kg during induction, followed by a 0.7 μg/kg/h infusion. She received 2 g of magnesium 30 min prior to sternotomy and 2 g after separation from cardiopulmonary bypass. A 2 mg/min infusion of lidocaine was started during induction and discontinued 1 h prior to injection of liposomal bupivacaine administered by the surgeon into the sternotomy site and intercostal nerves associated with bilateral ribs one to five (20 cc of 0.25% bupivacaine hydrochloride). On the first day postoperatively, the patient received a ketamine infusion 5 μg/kg/min that was decreased to 1 μg/kg/min after 1 h in the ICU. Dexmedetomidine at 0.7 μg/kg/h was discontinued after arrival to ICU. She received 1 g of Acetaminophen every 6 h and two doses of 15 mg of Ketorolac every 8 h. In the postoperative period from day 1 to discharge, 300 mg of gabapentin PO was given every 8 h and 1 g of Acetaminophen every 6 h for analgesia.

The authors demonstrated the feasibility of an OFA technique for adequate pain control for open cardiac surgery. Hontoir et al. investigated the effect of perioperative opioid-free anesthesia on postoperative patient comfort in 66 patients undergoing breast cancer surgery [68]. The study was made on two groups, one opioid group and the other opioid-free group. The opioid group received a target-controlled infusion of remifentanil. The opioid-free group received a loading dose of clonidine of 0.2 μg/kg. Both groups received a dose of 0.3 mg/kg ketamine and 1.5 mg/kg lidocaine. Upon surgical incision, patients in both groups received 1000 mg of acetaminophen and 75 mg of diclofenac. A total of three boluses of 0.2 mg/kg of ketamine were allowed in the opioid-free group if sevoflurane adjustments were not efficient at maintaining hemodynamic stability. At the end of the procedure, 4 mg of ondansetron was administered and a bolus of 0.03 mg/kg of piritramide was given at ski closure. The study showed that the patients in the opioid-free group were more sedated upon arrival in the post-anesthesia care unit but did not have a longer stay in the PACU. It also showed a statistically significant reduction in opioid consumption and an improved recovery during the first 24 postoperative hours.

Maevo Bello et al. studied the association of locoregional anesthesia with OFA in thoracic surgery [69]. Patients undergoing thoracic surgery are at high risk of postoperative pain. Considering opioid-related hyperalgesia that could lead to an increased postoperative analgesic consumption, this team investigated the effects of OFA on epidural ropivacaine requirements after thoracotomy. This observational, retrospective study included 50 patients in two groups of 25: an opioid-based anesthesia group (OBA) and an opioid-free anesthesia group (OFA). All patients received a thoracic epidural at T4/5 or T5/6 that was used intraoperatively with 0.2% ropivacaine at 4–6 ml/h, without any adjuvant or bolus. All patients received postoperative nausea and vomiting prophylaxis based on their Apfel score; they were given 4–8 mg Dexamethasone for one risk factor, to which droperidol was added (0.625 mg 30 min before the end of surgery) if two risk factors were present. Patients in the OBA group received total intravenous anesthesia with TCI propofol and remifentanil and depending on the anesthetist in charge, a single bolus dose of ketamine was administered at induction (25 mg). Patients in the OFA group received a preinduction single dose of IV clonidine (75–150 μg). They received a single IV dose of lidocaine (1.5 mg/kg) on arrival in the operating theatre. A bolus of 0.25–0.5 mg/kg of ketamine was administered followed by a continuous infusion at a rate of 0.25 mg/kg/h that was stopped at wound closure. TCI of propofol was used for induction and maintenance and neuromuscular blockade was achieved using rocuronium in both groups. All patients received 1 g of paracetamol intraoperatively, associated or not with nefopam (20 mg), tramadol (50 mg), or ketoprofen (100 mg). In the ICU patients received 1 g of paracetamol every 6 h in both groups and TEA was maintained via patient-controlled epidural anesthesia for 48 h using ropivacaine 2% with sufentanil (0.25 μg/ml) in both groups. The PCEA settings were 5 ml/h continuous infusion and a bolus of 5 ml with a 20-min lockout interval. The infusion was increased at 6 ml/h if there was an inadequate sensory block level and intravenous rescue analgesia was administered if the NRS score was superior to three using nefopam, tramadol, ketoprofen, or morphine patient-controlled analgesia. The authors found a significantly higher postoperative epidural ropivacaine consumption in the first 48 h in the OBA group (919,311 mg versus 693,270 mg, $P = 0.002$). The numerical rating scale at six and 24 h were also significantly lower in the OFA group (1[0–2] versus 3 [1–5], $P = 0.0005$ and 1[0–2] versus 3.5 [1–5]). In the post-anesthesia care unit, more patients in the OBA group required morphine for rescue analgesia (42% versus 4%, $P < 0.001$). During anesthesia, the OBA group required more vasopressor support as opposed to the OFA group, who were more hypertensive.

6   Perspectives

Although it seems logical that if patients are not given opioids during anesthesia, they will be less exposed to opioid-induced side effects, the hypothesis still needs to be proven in this era of evidence-based medicine. A shift in medical practice can only be made if a large body of evidence shows clear advantages of one technique over another. There are several ongoing clinical trials registered on clinicaltrials.gov studying opioid-free anesthesia in major spine surgery, bariatric surgery, cardiac surgery with cardiopulmonary bypass and laparoscopic colectomy. An article published in BMJ in 2018 announced a multicenter, double-blind, randomized, controlled clinical trial comparing opioid-free versus opioid anesthesia on postoperative opioid-related adverse events after major or intermediate noncardiac surgery. This POFA trial study is specifically studying OFA using dexmedetomidine [70]. More studies should be made on the interactions of the different drugs used with OFA. This will give clinicians a better understanding of the effects of their choices and will let them be better prepared for the patients' responses to these drugs. There is also a very large variation regarding the doses of the drugs used currently in OFA, so protocols need to be established.

7   Conclusions

Opioids are the most potent drugs used to control severe pain at this time. Their use in the perioperative period provides hemodynamic stability while establishing pain control and they remain one of the pillars of anesthesia and pain management. However, their administration has been proven to be linked to many side effects that will impact the quality of recovery, patient satisfaction, and hospitalization length. All of these factors, as well as problems related to opioid-induced hyperalgesia, opioid chronic use, and addiction, are now of great concern for modern society, not only because of the financial aspect.

In order to reduce opioid administration, research has investigated other drugs that target alternative receptors in an attempt to manage and control pain. These drugs also have a synergistic effect that will reduce further opioid requirements, while maintaining adequate pain control and hemodynamic stability. This multimodal approach for analgesia utilizes drugs such as ketamine, lidocaine, dexamethasone, acetaminophen, NSAIDs, and magnesium. In the same multimodal analgesia context, locoregional anesthesia is utilized. Neuraxial or a peripheral nerve block can be successfully employed to reduce opioid use in the preoperative period.

These different techniques, whether opioid-reducing opioid-free anesthesia combine adequate pain control and a reduction of postoperative nausea and vomiting, ileus, hyperalgesia, and postoperative delirium [71, 72].

The literature reports many different protocols using varying combinations of drugs. It seems that the those practicing opioid-reducing and opioid-free anesthesia techniques have developed their own personal protocols but have not studied the effects of their cocktail on large populations. At the same time, not enough data is available to determine the safety of different combinations of these drugs or the superiority of one protocol compared to another.

More studies are necessary to assess what the interactions between these drugs are. Another question that needs to be answered concerns indications for OFA. While some patient populations have been proven to benefit clearly from OFA (obesity, obstructive sleep apnea syndrome, or opioid dependence), we still need to assess whether there is an interest in OFA for all patients and in every surgical situation.

To this regard, clinical researchers must design studies with rigorous methodology in order to correctly assess the risks and benefits of OFA for patients in different surgical settings.

Ultimately, opioids are not the enemy but must be approached with an educated mind. The literature shows that opioid monotherapy is not the best option for pain management; a case-by-case approach is always desirable and should be implemented. Extreme attitudes towards opioid utilization, like opioid monotherapy versus opioid-free, could be dangerous.

References

1. Cohen MM (1969) The history of opium and opiates. Tex Med 65:76–85
2. Yim M, Parsa FD (2018) From the origins of the opioid use (and misuse) to the challenge of opioid-free pain management in surgery. Pain Treat. https://doi.org/10.5772/intechopen.82675
3. Janssen PA (1982) Potent new analgesics tailor-made for different purposes. Acta Anaesthesiol Scand 26:262–268
4. Duarte DF (2005) Opium and opioids: a brief history. Rev Bras Anestesiol 55(1):135–146

5. Chia YY et al (1999) Intraoperative high dose fentanyl induces postoperative fentanyl tolerance. Can J Anaesth 46:872–877

6. Guignard B et al (2000) Acute opioid tolerance: intraoperative remifentanil increases postoperative pain and morphine requirement. Anesthesiology 93:409–417

7. Cuomo A, Bimonte S, Forte CA et al (2019) Multimodal approaches and tailored therapies for pain management: the trolley analgesic model. J Pain Res 12:711–714. https://doi.org/10.2147/JPR.S178910

8. Elkassabany NM, Mariano ER (2019) Opioid-free anaesthesia – what would Inigo Montoya say? Anesth Edit 74(5):560–563. https://doi.org/10.1111/anae.14611

9. Kehlet H, Dahl JB (1993) The value of "multimodal" or "balanced analgesia" in postoperative pain treatment. Anesth Analg 77(5):1048–1056

10. Mulier J (2017) Opioid free general anesthesia: a paradigm shift? Rev Españ Anestesiol Reanim 64(8):427–430

11. Weinbroum A (2015) Role of anaesthetics and opioids in perioperative hyperalgesia. Eur J Anaesthesiol 32:230–231

12. Van Zee A (2009) The promotion and marketing of oxycontin: commercial triumph, public health tragedy. Am J Public Health 99(2):221–227

13. Joly V, Richebe P, Guignard B et al (2005) Remifentanil-induced postoperative hyperalgesia and its prevention with small-dose ketamine. Anesthesiology 103(1):147–155

14. Brummett CM, Waljee JF, Goesling J et al (2017) New persistent opioid use after minor and major surgical procedures in US adults. JAMA Surg 152(6):e170504. https://doi.org/10.1001/jamasurg.2017.0504

15. Shah A, Hayes CJ, Martin BC (2017) Factors influencing long-term opioid use among opioid naive patients: an examination of initial prescription characteristics and pain etiologies. J Pain 18(11):1374–1383. https://doi.org/10.1016/j.jpain.2017.06.010

16. Levy N, Mills P (2018) Controlled-release opioids cause harm and should be avoided in management of postoperative pain in opioid naïve patients. Br J Anaesth 122(6):e86–e90. https://doi.org/10.1016/j.bja.2018.09.005

17. Dumas EO (2018) Opioid tolerance development: a pharmacokinetic/pharmacodynamic perspective. AAPS J 4:537

18. Allouche S, Noble F, Marie N (2014) Opioid receptor desensitization: mechanisms and its link to tolerance. Front Pharmacol 5:280

19. Hayhurst CJ, Durieux ME (2016) Differential opioid tolerance and opioid-induced hyperalgesia: a clinical reality. Anesthesiology 2(124):483–488

20. Fletcher D, Martinez V (2014) Opioid-induced hyperalgesia in patient after surgery: a systematic review and a meta-analysis. Br J Anesth 112(6):991–1004

21. Afsharimani B, Baran J, Watanabe S (2014) Morphine and tumor growth and metastasis. Clin Exp Metastasis 31(2):149–158. https://doi.org/10.1007/s10585-013-9616-3

22. Gach K, Wyrębska A, Fichna J et al (2011) The role of morphine in regulation of cancer cell growth. Naunyn Schmiedeberg's Arch Pharmacol 384(3):221–230. https://doi.org/10.1007/s00210-011-0672-4

23. Gupta K, Kshirsagar S, Chang L (2002) Morphine stimulates angiogenesis by activating proangiogenic and survival-promoting signaling and promotes breast tumor growth. Cancer Res 62(3):4491–4498

24. Lennon FE, Mirzapoiazova T, Mambetsariev B (2012) Overexpression of the mu-opioid receptor in human non-small cell lung cancer promotes Akt and mTor activation, tumor growth, and metastasis. Anesthesiology 116(4):857–867

25. Mathew B, Lennon FE, Siegler J (2011) The novel role of the mu-opioid receptor in lung cancer progression: a laboratory investigation. Anesth Analg 112(3):558–567

26. Singleton PA, Mirzapoiazova T, Hasina R (2014) Increased mu-opioid receptor expression in metastatic lung cancer. Br J Anaesth 113(Suppl 1):i103–i108

27. Bimonte S, Barbieri A, Cascella M et al (2019) Naloxone counteracts the promoting tumor growth effects induced by morphine in an animal model of triple-negative breast cancer. In Vivo 33(3):821–825

28. Frauenknecht J, Kirkham KR, Jacot-Guillarmod A (2019) Analgesic impact of intra-operative opioid vs. opioid-free anaesthesia: a systematic review and meta-analysis. Anesthesia 74(5):651–662. https://doi.org/10.1111/anae.14582

29. Kawasaki Y, Kohno T, Zhuang ZY et al (2004) Ionotropic and metabotropic receptors, protein kinase A, protein kinase C, and Src contribute to C-fiber-induced ERK activation and cAMP response element-binding protein phosphorylation in dorsal horn neurons, leading to central sensitization. J Neurosci 24(38):8310–8321

30. Thompson T, Whiter F, Gallop K et al (2019) NMDA receptor antagonists and pain relief A

meta-analysis of experimental trials. Neurology 92(14):e1652–e1662. https://doi.org/10.1212/WNL.0000000000007238

31. Guillou N, Tanguy M, Seguin P et al (2003) The effects of small-dose ketamine on morphine consumption in surgical intensive care unit patients after major abdominal surgery. Anesth Analg 97(3):843–847

32. Yang Y, Cui Y, Sang K et al (2018) Ketamine blocks bursting in the lateral habenula to rapidly relieve depression. Nature 554 (7692):317–322

33. Kirby T (2015) Ketamine for depression: the highs and lows. Lancet Psychiatry 9:783–784. https://doi.org/10.1016/S2215-0366(15)00392-2

34. Gorlin AW, Rosenfeld DM, Ramakrishna H (2016) Intravenous sub-anesthetic ketamine for perioperative analgesia. J Anaesthesiol Clin Pharmacol 2:160–167. https://doi.org/10.4103/0970-9185.182085

35. Feria M, Abad F, Sanchez A et al (1993) Magnesium sulphate injected subcutaneously suppresses autotomy in peripherally deafferented rats. Pain 53:287–293

36. McCarthy RJ, Kroin JS, Tuman KJ et al (1998) Antinociceptive potentiation and attenuation of tolerance by intrathecal co-infusion of magnesium sulfate and morphine in rats. Anesth Analg 86:830–836

37. Albrecht E, Kirkham KR, Liu SS (2013) Perioperative intravenous administration of magnesium sulphate and postoperative pain: a meta-analysis. Anaesthesia 68(1):79–90. https://doi.org/10.1111/j.1365-2044.2012.07335.x

38. Kawamata M, Sugino S, Narimatsu E et al (2006) Effects of systemic administration of lidocaine and QX-314 on hyperexcitability of spinal dorsal horn neurons after incision in the rat. Pain 122:68–80

39. Abelson KS, Hoglund AU (2002) Intravenously administered lidocaine in therapeutic doses increases the intraspinal release of acetylcholine in rats. Neurosci Lett 317:93–96

40. Wright J, Durieux M, Groves D (2008) A brief review of innovative uses for local anesthetics. Curr Opin Anaesthesiol 21(5):651–656

41. Dunn LK, Durieux ME (2017) Perioperative use of intravenous lidocaine. Anesthesiology 126(4):729–737. https://doi.org/10.1097/ALN.0000000000001527

42. Koppert W, Weigand M, Neumann F et al (2004) Perioperative intravenous lidocaine has preventive effects on postoperative pain and morphine consumption after major abdominal surgery. Anesth Analg 98:1050–1055

43. Herroeder S, Pecher S, Schönherr ME et al (2007) Systemic lidocaine shortens length of hospital stay after colorectal surgery: a double-blinded, randomized, saline-controlled trial. Ann Surg 246:192–200

44. Tikuišis R, Miliauskas P, Samalavičius NE et al (2014) Intravenous lidocaine for postoperative pain relief after hand-assisted laparoscopic colon surgery: a randomized, saline-controlled clinical trial. Tech Coloproctol 18 (4):373–380. https://doi.org/10.1007/s10151-013-1065-0

45. Weibel S, Jelting Y, Pace NL et al (2018) Intravenous infusion of lidocaine starting at the time of surgery for reduction of pain and improvement of recovery after surgery. Cochrane Database Syst Rev 6:Art. No.:CD009642

46. Eipe N, Gupta S, Penning J (2016) Intravenous lidocaine for acute pain: an evidence-based clinical update. BJA Education 16 (9):292–298

47. Dexamethasone (DB01234) – dexamethasone- DrugBank. https://www.drugbank.ca/drugs/DB01234. Accessed 27 May 2019

48. De Oliveira GS Jr, Castro-Alves LJ, Ahmad S et al (2013) Dexamethasone to prevent postoperative nausea and vomiting: an updated meta-analysis of randomized controlled trials. Anesth Analg 116:58–74

49. Pehora C, Pearson AM, Kaushal A et al (2017) Dexamethasone as an adjuvant to peripheral nerve block. Cochrane Database Syst Rev 11: CD011770

50. De Oliveira GS Jr, Almeida MD, Benzon HT et al (2012) Perioperative single dose systemic dexamethasone for postoperative pain: a meta-analysis of randomized controlled trials. Anesthesiology 115:575–588. https://doi.org/10.1097/ALN.0b013e31822a24c2

51. Fauci AS, Dale DC, Balow JE (1976) Glucocorticosteroid therapy: mechanisms of action and clinical considerations. Ann Intern Med 84:304–315

52. Polderman JAW, Farhang-Razi V, van Dieren S et al (2019) Adverse side-effects of dexamethasone in surgical patients – an abridged. Cochrane systematic review. Anaesthesia. https://doi.org/10.1111/anae.14610

53. Davies DS, Wing LHM, Reid JL et al (1997) Pharmacokinetics and concentration-effect relationships of intravenous and oral clonidine. Clin Pharmacol Ther 21:593–601

54. Lowenthal DT, Matzek KM, MacGregor TR (1988) Clinical pharmacokinetics of clonidine. Clin Pharmacokinet 14:287–310

55. Sanchez Munoz MC, De Kock M, Forget P (2017) What is the place of clonidine in

anesthesia? Systematic review and meta-analyses of randomized controlled trials. J Clin Anesth 38:140–153. https://doi.org/10.1016/j.jclinane.2017.02.003

56. Carabine UA, Wright PM, Moore J (1991) Preanaesthetic medication with clonidine: a dose-response study. Br J Anaesth 67(1):79–83

57. Wagner DS, Brummett CM (2006) Dexmedetomidine: as safe as safe can be. Semin Anesth Perioper Med Pain 25:77–83

58. Fairbanks CA, Stone LS, Wilcox GL (2009) Pharmacological profiles of alpha 2 adrenergic receptor agonists identified using genetically altered mice and isobolographic analysis. Pharmacol Ther 123(2):224–238

59. Afsani N (2010) Clinical application of dexmedetomidine. S Afr J Anaesthesiol Analg 16:50–56

60. Scheinin B, Lindgren L, Randell T et al (1992) Dexmedetomidine attenuates sympathoadrenal responses to tracheal intubation and reduces the need for thiopentone and peroperative fentanyl. Br J Anaesth 68(2):126–131

61. Guler G, Akin A, Tosun Z et al (2005) Single-dose dexmedetomidine attenuates airway and circulatory reflexes during extubation. Acta Anaesthesiol Scand 49(8):1088–1091

62. Bekker A, Sturaitis MK (2005) Dexmedetomidine for neurological surgery. Neurosurgery 57 (1 Suppl):1–10. discussion 1–10

63. Scott-Warren VL, Sebastian J (2016) Dexmedetomidine: its use in intensive care medicine and anaesthesia. BJA Edu 16(7):242–246

64. Sultana A, Torres D, Schumann R (2017) Special indications for opioid free anaesthesia and analgesia, patient and procedure related: including obesity, sleep apnoea, chronic obstructive pulmonary disease, complex regional pain syndromes, opioid addiction and cancer surgery. Best Pract Res Clin Anaesthesiol 31(4):547–560. https://doi.org/10.1016/j.bpa.2017.11.002

65. Bakan M, Umutoglu T, Topuz U et al (2015) Opioid-free total intravenous anesthesia with propofol, dexmedetomidine and lidocaine infusions for laparoscopic cholecystectomy: a prospective, randomized, double-blinded study. Braz J Anesthesiol 65(3):191–199. https://doi.org/10.1016/j.bjane.2014.05.001

66. Aronsohn J, Orner G, Palleschi G et al (2019) Opioid-free total intravenous anesthesia with ketamine as part of an enhanced recovery protocol for bariatric surgery patients with sleep disordered breathing. J Clin Anesth 52:65–66

67. Landry E, Burns S, Pelletier MP et al (2018) A successful opioid- free anesthetic in a patient undergoing cardiac surgery. J Cardiothorac Vasc Anesth pii:S1053-0770(18)31092-9. https://doi.org/10.1053/j.jvca.2018.11.040

68. Hontoir S, Saxena S, Gatto P et al (2016) Opioid-free anesthesia: what about patient comfort? A prospective, randomized, controlled trial. Acta Anaesthesiol Belg 67(4):183–190

69. Bello M, Oger S, Bedon-Carte S (2019) Effect of opioid-free anaesthesia on postoperative epidural ropivacaine requirement after thoracic surgery: a retrospective unmatched case-control study. Anaesth Crit Care Pain Med pii:S2352-5568(18)30281-9. https://doi.org/10.1016/j.accpm.2019.01.013

70. Boloeil H, Lavoille B, Menard C et al (2018) POFA trial study protocol: a multicentre, double-bind,randomized, controlled clinical trial comparing opioid-free versus opioid anaesthesia on postoperative opioid-related adverse events after major or intermediate non-cardiac surgery. BMJ Open 8(6):e020873. https://doi.org/10.1136/bmjopen-2017-020873

71. Cascella M, Muzio MR, Bimonte S, Cuomo A, Jakobsson JG (2018) Postoperative delirium and postoperative cognitive dysfunction: updates in pathophysiology, potential translational approaches to clinical practice and further research perspectives. Minerva Anestesiol 84(2):246–260

72. Kim N, Matzon JL, Abboudi J et al (2016) A prospective evaluation of opioid utilization after upper-extremity surgical procedures. J Bone Joint Surg Am 98(20):e89

Chapter 12

New Insights into the Pharmacology of Dexmedetomidine and Open Issues for Neurosurgical Procedures

Mariantonietta Scafuro, Francesca Gargano, and Marco Fiore

Abstract

Patients undergoing neurosurgical procedures or requiring mechanical ventilation at the end of the neurosurgical procedure need analgosedation to reduce the anxiety and discomfort related to the intervention, as well as to minimize ventilator intolerance and desynchronizations. Dexmedetomidine is an alpha2-adrenergic agonist (C13H16HCl), as clonidine but more selective for alpha2-receptor. Dexmedetomidine has become increasingly popular for use in neurosurgical procedures and intensive care units (ICU) due to its proposed peculiarities for the management of systemic and cerebral hemodynamics, and the need for intraoperative cortical mapping.

Nowadays the approved therapeutic indication of the European Medical Agency for dexmedetomidine is the sedation of adult ICU patients, whereas the recommended use of the Food and Drug Administration, besides the sedation of adult ICU patients, is the sedation of non-ICU patients prior or during surgical and interventional procedures. However, in many studies, dexmedetomidine is used off-label.

This chapter would be a quick overview of the dexmedetomidine use in neurosurgical procedures, providing new insights into its pharmacology.

Key words Dexmedetomidine, Neurosurgery, Neuroanesthesia, Sedation, Intensive care unit, Chronic subdural hematoma, Awake craniotomy

1 Introduction

Dexmedetomidine (pharmacotherapeutic group: psycholeptics) is a particular alpha2-adrenergic agonist with sympatholytic, calming, anxiolytic, and narcotic-saving impacts with minimal unobtrusive respiratory depression [1]. The sedative and anxiolytic effects are due to the action of this medication on postsynaptic alpha2-adrenergic receptors (Fig. 1); in the locus coeruleus, as well, the sympathetic tone is decreased. Furthermore, the impact of the discharging of substance P in the dorsal horn of the spinal rope clarifies the drug's pain-relieving results; the subtypes of substance P receptors are present in sensory systems, veins, and vital organs (Fig. 1).

Marco Cascella (ed.), *General Anesthesia Research*, Neuromethods, vol. 150, https://doi.org/10.1007/978-1-4939-9891-3_12,
© Springer Science+Business Media, LLC, part of Springer Nature 2020

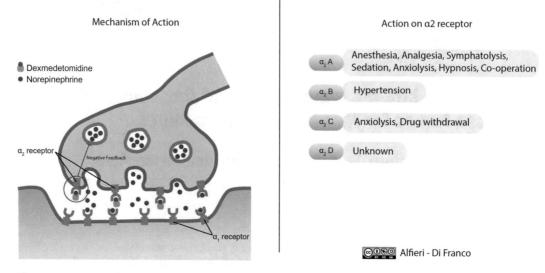

Fig. 1 Dexmedetomidine mechanism of action on α2-receptors

Dexmedetomidine is the dextrorotatory enantiomer of the monohydrochloride (+)-4-(2,3-dimethyl phenyl) ethyl-1H-imidazole. The affinity of dexmedetomidine for alpha2-receptors is eight times that of clonidine [2], and along these lines, dexmedetomidine has been utilized generally for patients treated in an intensive care unit (ICU).

The medical indications approved by the European Medical Agency (EMA) for dexmedetomidine is the sedation of patients treated in ICU requiring a sedation level not more profound than excitement in verbal stimulation [3], relating to Richmond Agitation-Sedation Scale (RASS) a score of 0 to −3 [4]. Actuality, the therapeutic indications of the Food and Drug Administration (FDA) incorporate sedation of ventilated patients during treatment in ICU (dexmedetomidine not to surpass 24 h), and the sedation of non-intubated patients before as well as during medical procedures [5].

2 Pharmacokinetics of Dexmedetomidine

2.1 Absorption

Concerning dexmedetomidine, numerous administration routes have been explored, although only the intravenous formulation is approved [3, 5]. Bioavailability of dexmedetomidine after oral administration is only 16% because of broad hepatic metabolization (hepatic first-pass) [6]. Then again, dexmedetomidine is absorbed either through the intranasal and buccal mucosae [7].

2.2 Distribution

The pharmacokinetics parameters of dexmedetomidine appear to be best portrayed by a two-compartment model, also in end-stage renal failure patients [8]; therefore, a dose redefinition is not required for patients with impaired renal function [9]. In the dosing range of 0.2–1.4 mcg/kg/h, dexmedetomidine shows straight pharmacokinetics, and no accumulation of the drug is seen when dexmedetomidine is given for as long as 14 days [9].

In ICU patients getting dexmedetomidine for 24 h, this medicine has a distribution volume at steady state >93 L [10]. Furthermore, dexmedetomidine displays a high plasma protein binding rate of 94% (albumin predominantly); in ICU patients with hypoalbuminemia, after long-term infusion, an expanded distribution volume at steady state is observed [11].

2.3 Metabolism

Dexmedetomidine is removed primarily through biotransformation by the liver; with a hepatic extraction proportion of 0.7 [12]. Direct N-glucuronidation by uridine 50-diphosphate-glucuronosyltransferase (UGT2B10, UGT1A4) represents about one-third of dexmedetomidine biotransformation. Furthermore, hydroxylation intervened by cytochrome P450 (CYP) proteins (essentially CYP2A6) is exhibited in liver microsomes [13]. The products of metabolism are hundredfold less powerful as alpha2-adrenoceptor agonist and therefore they are considered to be inactive.

2.4 Excretion

Less than 1% of dexmedetomidine is removed unaltered, 90–95% of the results of metabolism are discharged in the urine, and 4–10% in feces [13, 14]. The elimination half-life (t1/2) of 2.1–3.1 h, it is accounted in good health volunteers [14]. In ICU patients' similar qualities were found, with the elimination half-life going from 2.2 to 3.7 h and a clearance ranging from 0.53 to 0.80 ml/min/kg [15].

In patients with hypoalbuminemia, just a reduced half-life has been accounted. Therefore, clearance is not greatly influenced by hypoalbuminemia. This is in accordance with the "well-mixed" liver model, which expresses that for mixes with a high extraction proportion; the liver blood perfusion is the most significant factor administering hepatic clearance, and changes in plasma protein levels are relied upon not to result in modifying drug clearance [16].

The pharmacokinetic parameters of dexmedetomidine are shown in Tables 1 and 2.

3 Pharmacodynamics of Dexmedetomidine

Dexmedetomidine has a wide scope of pharmacological properties, mirroring the alpha2-receptors broad distribution all through the human body [17].

Table 1
Clinical pharmacokinetics of dexmedetomidine

	Author [ref.]	Journal (pub. year)	Main results: pro (P) et contra (C)
Absorption	Li [7]	Anaesthesia (2016)	P. Although only intravenous formulation is approved the drug is well absorbed through the intranasal and buccal mucosae
	Anttila [6]	Br J Clin Pharmacol (2003)	C. Bioavailability after oral administration, for extensive hepatic metabolism (first-pass effect) is only of 16%
Distribution	Zhong [8]	J Clin Pharm Ther (2017)	P. No dose adjustment is required for patients with end-stage renal failure
	Iirola [11]	Crit Care (2011)	C. In ICU patients with hypoalbuminemia, an increased volume of distribution at steady state is observed
Metabolism	Karol [13]	Best Pract Res Clin Anaesthesiol (2000)	P. Dexmedetomidine is eliminated mainly through biotransformation by the liver; with inactive products of metabolism
	Anttila [6]	Br J Clin Pharmacol (2003)	C. It should be used with caution in patients with hepatic impairment. A reduced maintenance dose may be considered
Extraction	Farag [14]	Curr Pharm Des (2012)	P. Less than 1% of dexmedetomidine is excreted unchanged, the 90–95% of the products of metabolism are excreted in the urine and 4–10% in feces
	Zhang [16]	J Clin Anesth (2015)	C. "Well-stirred" liver model: Liver blood flow governs hepatic clearance

Table 2
Pharmacokinetic parameters (mean $\pm$ standard deviation)

Pharmacokinetic parameters	Loading infusion (min)/total infusion duration (h)			
	10 min/12 h	10 min/24 h	10 min/24 h	35 min/24 h
	Dexmedetomidine Target concentration (ng/mL) and dose (μg/kg/h)			
	0.3/0.17	0.3/0.17	0.6/0.33	1.25/0.70
$t^{1/2}$	1.78 ± 0.30	2.22 ± 0.59	2.23 ± 0.21	2.50 ± 0.61
CL	46.3 ± 8.3	43.1 ± 6.5	35.3 ± 6.8	36.5 ± 7.5
Vss	88.7 ± 22.9	102.4 ± 20.3	93.6 ± 17.0	99.6 ± 17.8
Avg Css	0.27 ± 0.05	0.27 ± 0.05	0.67 ± 0.10	1.37 ± 0.20

Legend: terminal elimination half-life ($t1/2$) in hours; clearance (CL) estimated in L/h; steady-state volume of distribution (Vss) in liters; average steady-state concentration (Avg Css) in ng/mL
Inspired from Precedex™ Product Monograph—Hospira Healthcare Corporation

3.1 Sedative and Hypnotic Effects

Sedation with dexmedetomidine looks like common sleep and impersonates the profound recuperation rest that is seen after lack of sleep [18].

Dexmedetomidine's sedative and hypnotic effects are believed to be intervened through the enactment of focal presynaptic and postsynaptic alpha2-receptors in the locus coeruleus. It seems to impact endogenous sleep advancing pathways by diminishing the terminating of noradrenergic neurons in the brain stem, and by actuating endogenous non-rapid eye movement (non-REM) pathways; in this sense, dexmedetomidine produces a state intently looking like physiological stage 2 sleep. The exact mechanisms are not fully understood [19].

Portion subordinate sedation was seen in good health volunteers getting intravenous boluses of dexmedetomidine at the dose of 0.25–2 µg/kg/h, and the sedative impacts of dexmedetomidine have appeared both in healthy volunteers and in ICU patients [1].

A Ramsay sedation score (RSS) of 3 with an arousal sedation is generally achieved with a plasma concentration of dexmedetomidine of 0.2–0.3 ng/mL, and the sedation level reaches level at a plasma convergence of 0.7–1.25 ng/mL, comparing to a continuous infusion rate of 0.337–0.7 µg/kg/h. Unarousable profound sedation is thought to happen at plasma focuses above 1.9 ng/mL, although it is not easily obtained with dexmedetomidine [20].

3.2 Analgesic Effect

Alpha2-receptor, in the central nervous system and spinal cord, presumably intervenes in pain-relieving effects of dexmedetomidine. The transmission of neural impulses is smothered by hyperpolarization of interneurons and decrease of the arrival of nociceptive transmitters, for example, substance P and glutamate [21].

Studies examining the pain-relieving properties of dexmedetomidine found that a bringing about mild to profound sedation to obtain pain-relieving adequacy. At the point when administered as a sole medication in good health volunteers, dexmedetomidine at a dose up to 1.23 ng/mL did not give sufficient pain relief towards electrical or warmth stimulation [22]. Besides, in a study comparing pain relief and respiratory effects among dexmedetomidine and remifentanil, the plasma concentration of dexmedetomidine up to 2.4 ng/mL gave less pain relief than remifentanil. Taking everything into account, the pain-relieving effects of dexmedetomidine are as yet indistinct and may somewhat be inferable from an adjusted observation and a decreased anxiety [23].

3.3 Respiratory Effects

With plasma concentration up to 2.4 ng/mL, negligible respiratory depression is seen with dexmedetomidine, and the conservation of the ventilatory reaction to CO_2 is also noted [24]. In a study comparing remifentanil to dexmedetomidine in good health volunteers, no respiratory depression in the dexmedetomidine group was observed for plasma concentrations up to 2.4 ng/mL. The respiratory rate augmented proportionally with higher plasma concentrations, which made up for marginally diminished tidal volumes [25].

In any case, when dexmedetomidine is given in association with other anesthetic medications, an increased narcotic impact with an expanded danger of respiratory depression or apnea is accounted for, mainly depending on the combination of different drugs [26].

3.4 Cardiovascular Effects

Dexmedetomidine produces commonly a biphasic hemodynamic reaction: Bringing about low mean arterial pressure (MAP) at low plasma concentrations and high MAP at higher plasma concentrations. An intravenous (IV) bolus of dexmedetomidine results in a high (top) plasma concentration, with an increased MAP and a decreased heart rate (HR). During this stage, a stamped increment in fundamental vascular opposition has appeared.

The hemodynamic reaction is thought to begin from alpha2-receptor enactment in the vascular smooth muscles, causing rapid vasoconstriction along with hypertension. A rapid decrease joins this in HR, apparently brought about by the baroreceptor reflex. The primary hypertension is avoided if a loading dose is administered over a period of 10 min [27].

Following a couple of minutes, when dexmedetomidine plasma concentration decline, the vasoconstriction lessens, as dexmedetomidine likewise actuates alpha2-receptors in the vascular endothelial cells, bringing about vasodilation. Similarly, as with starting high plasma focuses after an IV bolus or quick stacking portion, higher dosages are related to dynamic increments in MAP. The hypertensive effects beat the hypotensive effects at a concentration range of 1.9–3.2 ng/mL [28]. Dexmedetomidine is not related with bounce back hypertension or tachycardia after infusion discontinuation; for example, insignificant changes in MAP and HR were seen following the sudden suspension of dexmedetomidine in critically ill patients [29].

Dexmedetomidine also showed to reduce the cerebral blood flow by about a third at 0.2–0.6 µg/kg/h in a dose-dependent manner. This property makes it particularly suitable for neurosurgical procedures [30].

4   Dexmedetomidine Use in Neurosurgical Methods

Dexmedetomidine is commonly used in different neurosurgical settings including chronic subdural hematoma, awake craniotomy, vertebroplasty/kyphoplasty, and laminotomy/discectomy, as well as for sedation and pain management in the postoperative course (Table 3).

Chronic subdural hematoma (CSDH) is a widespread clinical emergency experienced in neurosurgery. Monitored anesthesia care

Table 3

Dexmedetomidine use in neurosurgical procedures

Author [ref.]	Journal (pub. year)	Observation time	Study design (no. of patients enrolled)	Country, procedure	Main results: pros (P) and cons (C)
Wang [32]	Front Pharmacol (2016)	2014 (January) to 2015 (December)	RC (215)	China, CSDH	P. Few intraoperative movements, few rescue interventions, fast postoperative recovery, patient and surgeon satisfaction
Surve [33]	J Neurosurg Anesthesiol (2017)	2013 (May) to 2013 (December)	PC (76)	India, CSDH	P. Short operative time, short length of hospital stay, few hemodynamic fluctuations, few postoperative complications
Bishnoi [34]	J Neurosurg Anesthesiol (2016)	Not provided	PC (52)	India, CSDH	P. Few intraoperative movements, fast postoperative recovery, surgeon satisfaction
Suero Molina [38]	J Neurosurg (2018)	2009 (October) to 2015 (September)	RC (180)	Germany, AC	P. Few rescue interventions (opiates, vasoactive and antihypertensive drugs), short length of surgical and hospital stay
Elbakry [39]	Minerva Anestesiol (2017)	Not provided	PC (60)	Egypt, AC	P. Few incidences of nausea, vomiting, oxygen desaturation, and respiratory depression
Goettel [40]	Br J Anaesth (2016)	2012 (October) to 2014 (December)	RCT (50)	Canada, AC	P. Few respiratory adverse events
Garavaglia [41]	J Neurosurg Anesthesiol (2014)	2012 (March) to 2012 (December)	Case series (10)	Canada, AC	P. In association with scalp block, the procedure extends to patients who would traditionally not be considered
Lin [42]	Anesthesiology (2016)	2013 (April) to 2013 (August)	RCT (135)	China, AC	P. Few revealed neurologic dysfunctions
Lee [43]	J Int Med Res (2016)	Not provided	RCT (75)	South Korea, VP/KP	P. Few respiratory adverse events C. Less effective analgesia compared to remifentanil
Peng [44]	Clin Ther (2016)	Not provided	RCT (60)	China, Lam/Dis	P. More effective analgesia compared to midazolam

CSDH chronic subdural hematoma, *AC* awake craniotomy, *VP/KP* vertebroplasty/kyphoplasty, *Lam/Dis* laminotomy/discectomy

4.1 Chronic Subdural Hematoma

(MAC) is particular anesthesia management for therapeutic or diagnostic scopes performed under local anesthesia along with analgosedation [31] and probably is the favored anesthesia management among surgeons.

In a study by Wang et al. [32] the authors evaluated the neuroprotection and efficaciousness of dexmedetomidine compared to sufentanil in CSDH patients undergoing burr-hole surgery. The authors enrolled two-hundred and fifty patients divided into three groups: first group (67 patients, dexmedetomidine at dose of 0.5 μg/kg/h for 10 min), the second group (75 patients, dexmedetomidine infusion at dose of 1 μg/kg/h for 10 min), and a third group (73 patients, sufentanil at a dose of 0.3 μg/kg/h for 10 min). RSS score of all three groups was 3. The authors explored several outcomes: the onset time of anesthesia, the total amount of intraoperative patient movements, hemodynamic changes, the total dose of dexmedetomidine, time to initial dose and quantity of rescue benzodiazepine or opioid. Furthermore, the proportion of patients converted to other sedatives or other types of anesthetic, medical care, postsurgical recovery time, adverse events, and patients' and surgeons' satisfaction scores were recorded. Compared to sufentanil, dexmedetomidine at a dose of 1 μg/kg/h was related to fewer intraoperative movements, fewer rescue maneuvers, quicker postsurgical recovery, and higher patients' and surgeons' satisfaction scores [32]. Later on, Surveet al [33] compared dexmedetomidine sedation technique in combination with local anesthesia versus general anesthesia technique. The local anesthesia consisted of the infiltration of 0.5% bupivacaine (2 ml) and 2% lidocaine (2 ml) *plus* vasoconstrictive at every burr hole site. Seventy-six CSDH patients were enrolled and prospectively randomized into two groups: Dexmedetomidine group (38 patients, receiving dexmedetomidine at a dose of 1 μg/kg/h over 10 min followed by its infusion at a dose of 0.5 μg/kg/h) and General Anesthesia group (38 patients, of those four patients, were excluded, receiving general anesthesia). The dexmedetomidine sedation technique in combination with local anesthesia was related to considerably shorter surgical time, lower hemodynamic effects, fewer surgical complications, and shorter hospitalization [33].

More recently, Bishnoiet et al. [34] compared the effects of dexmedetomidine versus the fentanyl–midazolam combination. The authors enrolled 22 CSDH patients undergoing burr-hole surgery. The patients were randomized in a dexmedetomidine group and in a fentanyl–midazolam group. The dexmedetomidine was administered at a dose of 1 μg/kg/min over 10 min followed by the continuous infusion of dexmedetomidine at a dose of 0.03–0.07 μg/kg/h. Fentanyl at a dose of 0.5 μg/kg followed by a continuous infusion of fentanyl, 0.5–1.16 μg/kg/h, *plus* midazolam at a dose of 0.03 mg/kg over 10 min followed by continuous midazolam infusion of 0.03–0.07 mg/kg/h. The anesthetic agents were titrated to obtain an RSS score of 3. Intraoperative patient movement, surgical recovery time, and the surgeons' and patients' satisfaction scores were explored. Dexmedetomidine was associated with fewer intraoperative patient's movements, quicker surgical recovery, higher surgeons' satisfaction score, and patients' satisfaction score compared with fentanyl–midazolam association [34].

4.2 Awake Craniotomy

Awake craniotomy (AC) means the execution of at least a part of the neurosurgical procedure with an open skull and the patient awake [35]. Historically, AC procedures have been developed since the nineteenth century for removal epileptic foci under local anesthesia.

Nowadays thanks to recent monitoring techniques and the accessibility of newer anesthetic agents, indications of this operative modality have been extended to the execution of stereotactic biopsies, treatment of vascular lesions, resection of tumor lesions in the areas of language, and excision of supratentorial lesions located in different cortical areas. Commonly used anesthetic managements for AC are the MAC approach, and the asleep–awake–asleep (AAA) technique [36]. The MAC anesthesia focuses on conscious sedation that manages patients' pain and agitation, whereas maintaining the patient's ability to execute commands and patient's ability to protect the airway. Conversely, the AAA technique uses general anesthesia while invasive airway management, either laryngeal mask (LM) or tracheal tube (TT), before and once intraoperative mapping of the cortex [37]. Another approach for awake craniotomy is the Asleep–Awake (AA) technique which foresees a phase of general anesthesia followed by awakening for the mapping of the cortex after which subsequent sedation can be foreseen for the completion of the intervention. Suero Molina et al. [38] in a study enrolling one hundred eighty patients undergoing gliomas resection compared AC with dexmedetomidine (75 patients) and AAA with remifentanil and propofol (105 patients). In the AAA group, there was invasive airway management with an LM. The authors retrospectively evaluated the records of adverse events, the dose and frequency of drugs used, the operative time and the post-surgical hospitalization: In the dexmedetomidine group was significantly lower analgesic request, less use of vasoactive agents, shorter operative time and post-surgical hospitalization [38]. These findings suggest that dexmedetomidine is a despicable anesthetic agent in AC management. Furthermore, Elbakry et al. [39] recently evaluated the association of dexmedetomidine and propofol versus the association of remifentanil and propofol for epileptic patients undergoing AC surgery: 60 patients were randomly divided into two groups of 30 patients; the propofol–dexmedetomidine group and the propofol–remifentanil group. Sedation score, patients' satisfaction, surgeons' satisfaction, HR, MAP, and oxygen saturation were evaluated; adverse drug reactions, as well as respiratory depression, nausea, vomiting, and airways obstruction, were conjointly evaluated. Higher Sedation score was found within the propofol–remifentanil group compared to the propofol–dexmedetomidine group. There have been no important differences in patients' satisfaction scores between each group. The HR was lower within the propofol–dexmedetomidine group. The incidences of adverse events as nausea, vomiting, oxygen desaturation, and respiratory depression were statistically lower within the propofol–dexmedetomidine group. The authors concluded that propofol–dexmedetomidine association is as effective as propofol–remifentanil association; however, with fewer adverse effects for conscious sedation throughout AC [39]. In another randomized study comparing the association of remifentanil plus propofol versus dexmedetomidine, Goettel et al. [40] prospectively enrolled 50 patients (25 patients in the remifentanil–propofol group and 25 patients in the dexmedetomidine group) undergoing AC for surgical resection of a supratentorial tumor. The authors evaluated as outcomes the capability to perform intraoperative mapping of the cortex assessed on a numeric rating scale (NRS); the efficaciousness of sedation measured by the revised Observer's Assessment of Alertness/Sedation (OAA/S) scale; respiratory and hemodynamic changes; anxiety; pain; sedation; adverse drug events and patient's satisfaction. No significant differences regarding the capability to perform intraoperative mapping of the cortex and the level of sedation during the mapping were found; however, respiratory adverse events were less frequent in the dexmedetomidine group. The HR was lower within the dexmedetomidine group over time, although no treatment was necessary. In conclusion, dexmedetomidine was as effective as propofol–remifentanil association in performing intraoperative mapping of the cortex; however, with fewer adverse effects for conscious sedation throughout AC for resection of a supratentorial tumor [40]. Previously, Garavaglia et al. [41] described an anesthesiologic approach for AC consisting in dexmedetomidine analgosedation technique in combination with local infiltration of bupivacaine for the scalp nerve block. The authors enrolled ten patients, with a high risk of respiratory depression, undergoing AC for a resection of brain tumor requiring the intraoperative mapping of the cortex. The patients had a successful AC with a median operative time of 3.5 h (range 3–9 h), either in one case, with a prolonged surgery required due to the difficulty of surgical resection [41]. More recently, Lin et al. [42] enrolled 135 patients with supratentorial mass lesions (frontal parietal–temporal regions); the patients were randomized to four groups: "propofol," "midazolam," "fentanyl," and "dexmedetomidine" groups. The anesthetic agents were titrated to obtain sedation; however, totally cooperation (OAA/S score = 4); the Stroke Scale of the National Institutes of Health (NIHSS) was used to assess the neurological perform before and once sedation.

The primary outcome was the proportion of NIHSS-positive amendment in patients once sedation to OAA/S = 4. Of all patients enrolled, one-third had a neurological deficit at baseline. The proportion of NIHSS-positive amendment was 72% of changes in the midazolam group, 52% in the propofol group, 27% in the fentanyl group, and 23% in dexmedetomidine group: No statistical difference was observed in the midazolam group compared to propofol group, and no statistical difference was observed in the fentanyl group compared to the dexmedetomidine group. More focal neurologic deficits (mainly ataxia and limb motor weakness) due to the anesthetic agent, were observed in midazolam and propofol groups compered to fentanyl and dexmedetomidine groups. The population more susceptible to acquired focal neurologic deficits was patients with high-grade gliomas independently of the anesthetic agent used. In conclusion, at the same sedation level, less neurologic deficits were observed with fentanyl or dexmedetomidine use instead of midazolam or propofol use; patients with high-grade gliomas were more prone than those with low-grade gliomas to induced neurologic dysfunctions [42].

4.3   Vertebroplasty/ Kyphoplasty

In a randomized, prospective, double-blind study, Lee et al. [43] compared remifentanil versus dexmedetomidine for MAC management of less compressive vertebral fractures in 75 patients with an ASA class from 1 to 3 and aged more than 75 years. Thirty-seven patients received remifentanil infusion (remifentanil group) at a dose of 1–5 µg/kg/h and thirty-eight patients received dexmedetomidine (dexmedetomidine group) at a dose of 0.3–0.4 µg/kg followed by dexmedetomidine infusion at a dose of 0.2–1 µg/kg/h to obtain, during the procedure, an OAA/S scale inferior to 4. No significant differences were observed in the two groups about analgesia, recovery time, oxygen saturation, and researchers' satisfaction. However, in the dexmedetomidine group, the patients had a significantly lower HR and MAP with higher oxygen saturation and less respiratory depression. In conclusion, dexmedetomidine showed the same effectiveness with minor risk of respiratory depression [43].

4.4   Laminotomy/ Discectomy

In a controlled investigation, Peng et al. [44] enrolled 60 patients undergoing elective lumbar laminotomy and discectomy. The authors randomly assigned patients to receive either dexmedetomidine–fentanyl or midazolam–fentanyl for conscious sedation. Patient-controlled intravenous analgesia with fentanyl was used for postoperative pain management. Hemodynamic and respiratory changes, sedation scores, pain scores, fentanyl consumption, patient satisfaction, postoperative hospital stay, and adverse events were recorded. The HR was lower in the dexmedetomidine–fentanyl group compared to midazolam–fentanyl group; the intraoperative, postoperative, and total consumption of fentanyl were lower in the dexmedetomidine–fentanyl group; no significant differences were found for adverse events, postoperative hospital stay, or satisfaction between the two groups. The authors concluded that although awake lumbar disc surgery can be performed successfully under sedation with either midazolam–fentanyl combination or dexmedetomidine–fentanyl combination, the latter may be a better alternative for the opioid-sparing strategy [44].

4.5 Postoperative Care

Effective management of analgosedation in the ICU depends on the patient's needs, measured with subjective and/or objective tools, and the titration of the anesthetic drug to reach specific endpoints. Srivastava et al. [45] compared the efficacy of dexmedetomidine, propofol, and midazolam for sedation in neurosurgical patients during postoperative mechanical ventilation. Ninety patients, ASA class from I to III, undergoing neurosurgery and requiring postoperative ventilation were randomized: The patients were divided into three groups: The first group received dexmedetomidine (1 µg/kg over 15 min as a loading dose, followed by 0.4–0.7 µg/kg/h); the second group received propofol (1 mg/kg over 15 min followed by 1–3 mg/kg/h); and the third group received midazolam (0.04 mg/kg over 15 min followed by 0.08 mg/kg/h). In this study, HR, MAP, sedation level, fentanyl requirements, and ventilation and extubation times were assessed. An adequate sedation level was achieved with all drugs. Less fentanyl was required in the dexmedetomidine group for the postoperative pain control. There was a decrease in HR in patients in the dexmedetomidine group after the drug infusion, but no significant difference versus propofol group and midazolam. Extubation time was significantly lowest in the group receiving propofol [45].

5   New Clinical (Off-Label) Applications

Neither EMA [3] nor FDA [5] approved the use of dexmedetomidine in patients undergoing neurosurgery under General Anesthesia. However, there are several studies in the literature in which dexmedetomidine has been used in this anesthetic management. New dexmedetomidine use trends in neurosurgical procedures are illustrated in Table 4.

5.1 Craniotomy

Luo et al. [46] randomly enrolled 60 patients into two groups: 30 patients receiving intravenous saline solution and 30 patients receiving intravenous dexmedetomidine at a dose of 1 µg/kg 10 min before anesthesia induction, with a maintenance dose of 0.4 µg/Kg/h. MAP and HR were compared between the two groups before anesthesia induction (T1), before tracheal intubation (T2), immediately after tracheal intubation (T3), 1 min after tracheal intubation (T4), 3 min after TT (T5), 5 min after TT (T6).

Table 4
New dexmedetomidine use trends in neurosurgical procedures

Author [ref.]	Journal (pub. year)	Observation time span	Study design (no. of patients enrolled)	Country, procedure	Main results: pros (P) and cons (C)
Luo [46]	Clin Neurol Neurosurg (2016)	2013 (January) to 2015 (May)	RCT (60)	China, craniotomy	P. Increased hemodynamic stability, reduction of inflammatory, oxidative stress and brain injury markers
Kim [47]	Yonsei Med J (2016)	Not provided	RCT (64)	South Korea, craniotomy	P. Increased hemodynamic and respiratory stability
Tanskanen [48]	Br J Anaesth (2006)	Not provided	RCT (54)	Finland, craniotomy	P. Increased hemodynamic and respiratory stability
Kido [62]	Masui (2014)	Not provided	CR (1)	Japan, gasserian ganglion block	P. Patient comfort and had no communication difficulty

RCT randomized controlled trial, *CR* care report

The levels of serum interleukin-6 (IL-6), tumor necrosis factor-α (TNF-α), superoxide dismutase (SOD), malondialdehyde (MDA), neuron-specific enolase (NSE) and S100β were compared between the two groups before surgery, immediately post-surgery, and 24 h post-surgery. The authors found no significant differences in MAP and HR between the two groups at T1, T2, T5, and T6. The levels of MAP and HR in dexmedetomidine group were significantly lower than those in placebo group at T3 and T4. No significant differences of preoperative levels of serum TNF-α, IL-6, NSE, S100β, SOD, and MDA were found between the two groups. Serum levels of TNF-α, IL-6, NSE, S100β, and MDA were significantly reduced, whereas SOD was significantly lower in placebo group compared with those in dexmedetomidine group at surgery time and 24 h post-surgery [46]. In a contemporaneity study Kim et al. [47] compared the effects of dexmedetomidine and remifentanil on airways reflex and hemodynamic changes in patients undergoing craniotomy for clipping of unruptured cerebral aneurysm; in this prospective, randomized, double-blind study, the authors enrolled 74 patients who were divided into groups: the dexmedetomidine group (dexmedetomidine at a dose of 0.5 μg/kg over 5 min) and the remifentanil group (remifentanil at a concentration of 1.5 ng/mL, until extubation). Both the incidence and severity of cough and the hemodynamic changes were assessed during the recovery period. Hemodynamic changes, respiration rate, and sedation scale were measured after extubation and in the post-anesthetic care unit. The coughing (grade 2 and 3) at extubation was significantly reduced in the dexmedetomidine group (53.1%) compared to the remifentanil group (62.5%). At admission in the post-anesthetic care unit, both MAP and HR as well as HR at 10 min were significantly lower in the dexmedetomidine group. Respiratory rate was significantly lower in the remifentanil group at 2 and 5 min after extubation. The authors concluded that either a single bolus of dexmedetomidine or continuous infusion of remifentanil have the same efficacy in attenuating cough and hemodynamic changes in patients subjected to brain aneurysm clipping; however, dexmedetomidine presented better preservation of the cough reflex after extubation [47]. In a previous study Tanskan et al. [48] enrolled 54 patients undergoing elective surgery of supratentorial brain tumors; the authors randomly assigned to receive in a double-blind manner a continuous dexmedetomidine infusion (plasma target concentration 0.2 or 0.4 ng/ml) or placebo, beginning 20 min before anesthesia and continuing until the start of skin closure. Dexmedetomidine group received fentanyl 2 µg/kg at the induction of anesthesia and before the start of the operation, whereas the placebo group received an infusion of 4 µg/kg of fentanyl, respectively. Dexmedetomidine increased perioperative hemodynamic stability in patients undergoing brain tumor surgery. Compared with fentanyl, the extubation was faster, without respiratory depression, in the dexmedetomidine group [48].

5.2 Postoperative Care

Pain management for patients undergoing craniotomy remains challenging. Peng et al. [49] in a prospective study, randomly allocated into two equal groups 80 patients undergoing elective supratentorial craniotomy under sevoflurane/fentanyl anesthesia, to receive a continuous dexmedetomidine infusion of 0.5 µg/kg/h or placebo. In both groups, the administration began after induction and continued until the beginning of surgical stitching of the wound. Intravenous tramadol at a dose of 0.5 mg/kg was administered to achieve an assessment score of 4 or less of the 11-point numeric scale (NRS 11) in the post-anesthesia care unit and subsequently in the ward. Pain scores, tramadol consumption, sedation scores, postoperative nausea and vomiting (PONV) and adverse events were assessed in the first 24 h after surgery. Of the 80 patients, 76 were included in the analyses. Dexmedetomidine significantly reduced PONV scores, pain scores and postoperative tramadol consumption; fewer PONV events, requiring any treatment, were recorded in the dexmedetomidine group [49]. In a more recent randomized trial by Zhao et al. [50], the authors evaluated the efficacy and safety of dexmedetomidine for sedation and prophylactic analgesia in craniotomy patients who had delayed extubation after surgery. One-hundred and fifty patients with delayed extubation after craniotomy were randomized 1:1 and assigned either to the dexmedetomidine group (continuous dex-medetomidine infusion at a dose of 0.6 μg/kg/h) or to the control group (maintenance intravenous infusion of 0.9% sodium chloride). The following were identified as outcomes: the mean percentage of time under optimal sedation (SAS 3–4), the rate of patients who required rescue management with propofol–fentanyl, the total dose of propofol–fentanyl administered, the VAS, HR, MAP, and oxygen saturation. In the dexmedetomidine group, the percentage of time under optimal sedation was significantly higher than in the control group. Furthermore, in the dexmedetomidine group VAS was significantly lower than in the control group; HR and MAP were significantly lower in in patients undergoing continuous infusion of dexmedetomidine at the three specified time points (before TT, immediately after TT, and 30 min after TT). No significant difference in oxygen saturation was observed between the two groups. In the dexmedetomidine group, patients were more likely to develop bradycardia but were less likely to have tachycardia than in the control group. The author concluded that dexmedetomidine might be an effective prophylactic agent to induce sedation and analgesia in patients with delayed extubation after craniotomy [50].

Optimal anesthesia for craniotomies should eliminate the response to noxious stimuli, leaving patients sufficiently vigilant for early neurological assessment. Rajan et al. [51] compared postoperative blood pressure control, pain scores, and opioid requirement after anesthesia with dexmedetomidine versus remifentanil. There were two questions to which the authors wanted to answer: Whether the intraoperative administration of dexmedetomidine provides better control of postoperative blood pressure than remifentanil and whether patients treated with dexmedetomidine have less postoperative pain with an opioid-sparing management. Adults undergoing elective brain tumor excisions under balanced general anesthesia with endotracheal intubation were randomized to remifentanil (71 patients, infusion at a dose of 0.08–0.15 μg/kg/min) or dexmedetomidine (68 patients, infusion at a dose of 0.2–0.7 μg/kg/h). The values of MAP and pain scores were evaluated 15, 30, 45, 60, and 90 min after surgical procedure. Outcomes were assessed with joint hypothesis testing, evaluating noninferiority and superiority. The dexmedetomidine was more effective in controlling postoperative hemodynamics providing superior analgesia [51].

6   Discussion and Future Perspectives

Many anesthetics have been used for sedation, analgesia, and anxiolysis in neurosurgery procedures. When choosing an agent for neurosurgical sedation, multiple factors must be carefully

evaluated. Typical considerations include the indication for sedation, onset of action, duration of action, route of elimination, drug
interactions, and adverse effects. Goodwin et al. [52] suggested
that another (perhaps less commonly considered) criteria should be
taken into consideration: the potential impact of the sedative drug
on cognition, neurological sequelae and the probability that it
(appropriately used) reaches a state capable of providing a "cooperative sedation" [52–54], that seems to be the most crucial criterion
in neurosurgical procedures, in which the ability to perform intraoperative brain mapping is of paramount importance. Data emerging
from literature show that dexmedetomidine presents all these characteristics; furthermore, by anticonvulsive effects [55], stabilizing
hemodynamics, reducing inflammation, and inhibiting free radical
generation, it also plays an important role on brain protection
[44]. In addition, dexmedetomidine has neuroprotective effects
in various animal models of both ischemic and hemorrhagic brain
injury [56]. This aspect seems to be confirmed in patients with
traumatic brain injury (TBI) in a prospective controlled trial conducted by Wang et al. [57]. The authors examined the effect of
dexmedetomidine on cerebral blood flow (CBF) in critically ill
patients with or without TBI: 15 patients without TBI and
20 patients with TBI with a Glasgow Coma Scale score of 4–14
were enrolled. All patients received 1 μg/kg of dexmedetomidine
infused over 10 min, followed by a 0.4 μg/kg/h in continuous
infusion for 60 min. Blood pressure was maintained at the
pre-sedation level with dopamine in all patients. The CBF and
cerebral metabolic rate equivalent (CMRe) were measured before
sedation and 70 min after dexmedetomidine administration. Dexmedetomidine administration significantly decreased CBF in
patients without TBI, not altering the CMRe and CMRe/CBF
ratio. The percentage of CBF reduction was higher in the patients
without TBI compared to patients with TBI; however, CMRe and
CMRe/CBF were not significantly different in the TBI group
[57]. Schomer et al. [58] retrospectively studied the potential
benefit of adding dexmedetomidine infusion to standard sedative–analgesic infusions for the management of refractory intracranial hypertension. The primary objective of this study, that
represents the first clinical series regarding dexmedetomidine use
in neuro-critically ill patients for the treatment of refractory intracranial hypertension, was to determine the effect of dexmedetomidine on the need for rescue therapy [i.e., hyperosmolar boluses,
extraventricular drain (EVD) drainages] for refractory intracranial
hypertension. Secondary objectives included the number of intracranial pressure (ICP) excursions, bradycardic, hypotensive, and
compromised cerebral perfusion pressure episodes. This retrospective cohort study evaluated patients admitted to the neurosurgical
ICU who received dexmedetomidine for refractory intracranial

hypertension. Dexmedetomidine reduced the use of rescue therapies such as hyperosmolar boluses [58].

High concentrations of dexmedetomidine are significantly associated with an increase in systemic and pulmonary vascular resistance, resulting in systemic and pulmonary hypertension. This could be a limiting factor in the use of the drug, especially in patients with known heart problems, who can count on increasing their heart rate to provide an adequate cardiac output [27].

However, transesophageal echocardiographic evaluation in patients undergoing continuous infusion of dexmedetomidine during general anesthesia with the combination of propofol and remifentanil showed no alteration of cardiac function. The Stroke volume remains stable if the plasma concentration is <5 ng/mL and the Cardiac output was reduced cardiac output was due to lower heart rate [59].

Erdman et al. [60] in a retrospective, propensity-matched study, enrolled a total of 342 patients (105 dexmedetomidine and 237 propofol) with 190 matched (95 in each group) and compared the prevalence of severe hemodynamic effects in patients with a diagnosis of neurologic injury, receiving either dexmedetomidine or propofol. The authors identified as the primary outcome of the study severe hypotension (MAP <60 mmHg) associated to bradycardia (HR <50 beats/min): No difference was found in the primary composite result in both cohorts during sedative infusion [60].

An emerging therapeutic aspect is the perioperative infusion of dexmedetomidine, during general anesthesia, as an opioid-sparing strategy. Intraoperative dexmedetomidine infusion could be effective both for reducing pain and analgesic consumption [49] and postoperative MAP, providing superior analgesia after craniotomy [51] and can be safely used for both intubated and extubated patients following uneventful intracranial procedures [61].

Concerning administration strategies, the rate of infusion of dexmedetomidine is titrated during the procedures, to obtain the appropriate sedation level, and in our experience, about half of patients need to reach an infusion rate of 0.7 µg/kg/h. In our ongoing investigation during radiofrequency thermocoagulation, no mechanical ventilation was required for any patient. Ten patients experienced moderate bradycardia that did not require pharmacological treatment, in three patient it was necessary to infuse atropine 0.01 mg/kg and seven patients presented a hypotensive episode resolved by increasing the crystalloid infusion or using colloidal solutions. In no patient was it necessary to infuse additional doses of opioids and/or other analgesics.

Our experience suggests that dexmedetomidine off-label use could be implemented, as it is the case of the use of dexmedetomidine during radiofrequency thermocoagulation for pain treatment, where it could be a valid anesthetic option. In literature, there is only a Japanese case report: Kido et al. [62] described a case of successful procedural sedation using dexmedetomidine in a 68-year-old woman undergoing left gasserian ganglion block for intractable trigeminal neuralgia. During the entire duration of the intervention, the patient was comfortable and did not present verbal difficulties under sedation with dexmedetomidine [62].

Future studies are needed to evaluate the dexmedetomidine use in new pain management strategies.

References

1. Belleville JP, Ward DS, Bloor BC et al (1992) Effects of intravenous dexmedetomidine in humans. I. Sedation, ventilation, and metabolic rate. Anesthesiology 77:1125–1133

2. Virtanen R, Savola JM, Saano V et al (1988) Characterization of the selectivity, specificity and potency of medetomidine as an alpha 2-adrenoceptor agonist. Eur J Pharmacol 150:9–14

3. European Medicine Agency. Dexdor, dexmedetomidine. http://www.ema.europa.eu/ema/index.jsp?curl=pages/medicines/human/medicines/002268/human_med_001485.jsp&murl=menus/medicines/medicines.jsp&mid=WC0b01ac058001d124. Accessed 8 Jun 2019

4. Sessler CN, Mark S et al (2002) The Richmond agitation–sedation scale. Am J Respir Crit Care Med 166:1338–1344

5. Food and Drug Administration. Highlights on dexmedetomidine. https://www.accessdata.fda.gov/drugsatfda_docs/label/2013/02103 8s021lbl.pdf. Accessed 22 May 2019

6. Anttila M, Penttilä J, Helminen A et al (2003) Bioavailability of dexmedetomidine after extravascular doses in healthy subjects. Br J Clin Pharmacol 56:691–693

7. Li BL, Zhang N, Huang JX et al (2016) A comparison of intranasal dexmedetomidine for sedation in children administered either by atomiser or by drops. Anaesthesia 71:522–528

8. Zhong W, Zhang Y, Zhang MZ et al (2018) Pharmacokinetics of dexmedetomidine administered to patients with end-stage renal failure and secondary hyperparathyroidism undergoing general anaesthesia. J Clin Pharm Ther 43:414–421

9. https://ec.europa.eu/health/documents/community-register/2011/20110916108949/anx_10 8949_en.pdf

10. Valitalo PA, Ahtola-Satila T, Wighton A et al (2013) Population pharmacokinetics of dexmedetomidine in critically ill patients. Clin Drug Investig 33:579–587

11. Iirola T, Aantaa R, Laitio R et al (2011) Pharmacokinetics of prolonged infusion of high-dose dexmedetomidine in critically ill patients. Crit Care 15:R257

12. Dutta S, Lal R, Karol MD et al (2000) Influence of cardiac output on dexmedetomidine pharmacokinetics. J Pharm Sci 89:519–527

13. Karol MD, Maze M (2000) Pharmacokinetics and interaction pharmacodynamics of dexmedetomidine in humans. Best Pract Res Clin Anaesthesiol 14:261–269

14. Farag E, Argalious M, Abd-Elsayed A et al (2012) The use of dexmedetomidine in anesthesia and intensive care: a review. Curr Pharm Des 18:6257–6265

15. Venn RM, Karol MD, Grounds RM (2002) Pharmacokinetics of dexmedetomidine infusions for sedation of post-operative patients requiring intensive care. Br J Anaesth 88:669–675

16. Zhang T, Deng Y, He P et al (2015) Effects of mild hypoalbuminemia on the pharmacokinetics and pharmacodynamics of dexmedetomidine in patients after major abdominal or thoracic surgery. J Clin Anesth 27:632–637

17. Panzer O, Moitra V, Sladen RN (2011) Pharmacology of sedative analgesic agents: dexmedetomidine, remifentanil, ketamine, volatile anesthetics, and the role of peripheral mu antagonists. Anesthesiol Clin 29:587–605

18. Zhang Z, Ferretti V, Guntan I et al (2015) Neuronal ensembles sufficient for recovery sleep and the sedative actions of a2 adrenergic agonists. Nat Neurosci 18:553–561

19. Nelson LE, Lu J, Guo T et al (2003) The a2-adrenoceptor agonist dexmedetomidine converges on an endogenous sleep-promoting pathway to exert its sedative effects. Anesthesiology 98:428–436

20. European Medicines Agency (2015) Dexdor (dexmedetomidine): EU summary of product characteristics. http://www.ema.europa.eu/. Accessed 4 May 2015

21. Miller RD, Cohen NH, Eriksson LI et al (2015) Miller's anesthesia, 8th edn. Elsevier, Amsterdam, pp 854–859

22. Angst MS, Ramaswamy B, Davies MF et al (2004) Comparative analgesic and mental effects of increasing plasma concentrations of dexmedetomidine and alfentanil in humans. Anesthesiology 101:744–752

23. Weerink MAS, Struys MMRF, Hannivoort L et al (2017) Clinical pharmacokinetics and pharmacodynamics of dexmedetomidine. Clin Pharmacokinet 56:893–913

24. Alfieri A, Passavanti MB, Di Franco S et al (2019) Dexmedetomidine in the management of awake fiberoptic intubation. Open Anesth J 13:1–5

25. Hsu Y-W, Cortinez LI, Robertson KM et al (2004) Dexmedetomidine pharmacodynamics: part I: crossover comparison of the respiratory effects of dexmedetomidine and remifentanil in healthy volunteers. Anesthesiology 101:1066–1076

26. Ho AM-H (2005) Central apnoea after balanced general anaesthesia that included dexmedetomidine. Br J Anaesth 95:773–775

27. Ebert TJ, Hall JE, Barney JA et al (2000) The effects of increasing plasma concentrations of dexmedetomidine in humans. Anesthesiology 93:382–394

28. Talke P, Lobo E, Brown R (2003) Systemically administered alpha2-agonist-induced peripheral vasoconstriction in humans. Anesthesiology 99:65–70

29. Shehabi Y, Ruettimann U, Adamson H et al (2004) Dexmedetomidine infusion for more than 24 hours in critically ill patients: sedative and cardiovascular effects. Intensive Care Med 30:2188–2196

30. Prielipp RC, Wall MH, Tobin JR et al (2002) Dexmedetomidine-induced sedation in volunteers decreases regional and global cerebral blood flow. Anesth Analg 95:1052–1059

31. Das S, Ghosh S (2015) Monitored anesthesia care: an overview. J Anaesthesiol Clin Pharmacol 31:27–29

32. Wang W, Feng L, Bai F et al (2016) The safety and efficacy of dexmedetomidine vs. sufentanil in monitored anesthesia care during Burr-Hole surgery for chronic subdural hematoma: a retrospective clinical trial. Front Pharmacol 7:410

33. Surve RM, Bansal S, Reddy M et al (2017) Use of dexmedetomidine along with local infiltration versus general anesthesia for Burr Hole and evacuation of chronic subdural hematoma (CSDH). J Neurosurg Anesthesiol 29:274–280

34. Bishnoi V, Kumar B, Bhagat H et al (2016) Comparison of dexmedetomidine versus midazolam-fentanyl combination for monitored anesthesia care during burr-hole surgery for chronic subdural hematoma. J Neurosurg Anesthesiol 28:141–146

35. Meng L, McDonagh DL, Berger MS et al (2017) Anesthesia for awake craniotomy: a how-to guide for the occasional practitioner. Can J Anaesth 64:517–529

36. Carbone D, Lubrano G, Muzio MR et al (2019) Anesthetic management and psychological approaches for excision in awake craniotomy of lesions located within or near eloquent language areas. J Surg Forecast 2:1019

37. Eseonu CI, ReFaey K, Garcia O et al (2017) Awake craniotomy anesthesia: a comparison of the monitored anesthesia care and asleep-awake-asleep techniques. World Neurosurg 104:679–686

38. Suero Molina E, Schipmann S, Mueller I et al (2018) Conscious sedation with dexmedetomidine compared with asleep-awake-asleep craniotomies in glioma surgery: an analysis of 180 patients. J Neurosurg 129:1223–1230

39. Elbakry AE, Ibrahim E (2017) Propofol-dexmedetomidine versus propofol-remifentanil conscious sedation for awake craniotomy during epilepsy surgery. Minerva Anestesiol 83:1248–1254

40. Goettel N, Bharadwaj S, Venkatraghavan L et al (2016) Dexmedetomidine vs propofol-remifentanil conscious sedation for awake craniotomy: a prospective randomized controlled trial. Br J Anaesth 116:811–821

41. Garavaglia MM, Das S, Cusimano MD et al (2014) Anesthetic approach to high-risk patients and prolonged awake craniotomy using dexmedetomidine and scalp block. J Neurosurg Anesthesiol 26:226–233

42. Lin N, Han R, Zhou J, Gelb AW (2016) Mild sedation exacerbates or unmasks focal neurologic dysfunction in neurosurgical patients with supratentorial brain mass lesions in a drug-specific manner. Anesthesiology 124:598–607

43. Lee JM, Lee SK, Lee SJ et al (2016) Comparison of remifentanil with dexmedetomidine for monitored anaesthesia care in elderly patients during vertebroplasty and kyphoplasty. J Int Med Res 44:307–316

44. Peng K, Liu HY, Liu SL et al (2016) Dexmedetomidine-fentanyl compared with midazolam-fentanyl for conscious sedation in patients undergoing lumbar disc surgery. Clin Ther 38:192–201

45. Srivastava VK, Agrawal S, Kumar S et al (2014) Comparison of dexmedetomidine, propofol and midazolam for short-term sedation in postoperatively mechanically ventilated neurosurgical patients. J Clin Diagn Res 8: GC04–GC07

46. Luo X, Zheng X, Huang H (2016) Protective effects of dexmedetomidine on brain function of glioma patients undergoing craniotomy resection and its underlying mechanism. Clin Neurol Neurosurg 146:105–108

47. Kim H, Min KT, Lee JR et al (2016) Comparison of dexmedetomidine and remifentanil on airway reflex and hemodynamic changes during recovery after craniotomy. Yonsei Med J 57:980–986

48. Tanskanen PE, Kytta JV, Randell TT et al (2006) Dexmedetomidine as an anaesthetic adjuvant in patients undergoing intracranial tumour surgery: a double-blind, randomized and placebo-controlled study. Br J Anaesth 97:658–665

49. Peng K, Jin XH, Liu SL et al (2015) Effect of intraoperative dexmedetomidine on post-craniotomy pain. Clin Ther 37:1114–1121

50. Zhao LH, Shi ZH, Chen GQ et al (2017) Use of dexmedetomidine for prophylactic analgesia and sedation in patients with delayed extubation after craniotomy: a randomized controlled trial. J Neurosurg Anesthesiol 29:132–139

51. Rajan S, Hutcherson MT, Sessler DI et al (2016) The effects of dexmedetomidine and remifentanil on hemodynamic stability and analgesic requirement after craniotomy: a randomized controlled trial. J Neurosurg Anesthesiol 28:282–290

52. Goodwin H, Lewin JJ, Mirski MA (2012) 'Cooperative sedation': optimizing comfort while maximizing systemic and neurological function. Crit Care 16:217

53. Cascella M, Fusco R, Caliendo D et al (2017) Anesthetic dreaming, anesthesia awareness and patient satisfaction after deep sedation with propofol target controlled infusion: a prospective cohort study of patients undergoing day case breast surgery. Oncotarget 8:79248–79256

54. Cascella M, Bimonte S (2017) The role of general anesthetics and the mechanisms of hippocampal and extra-hippocampal dysfunctions in the genesis of postoperative cognitive dysfunction. Neural Regen Res 12:1780–1785

55. Fiore M, Torretta G, Passavanti MB et al (2019) Effectiveness of dexmedetomidine as adjunctive therapy, compared to the standard of care in the treatment of alcohol withdrawal syndrome: a systematic review protocol. JBI Database System Rev Implement Rep 2019. https://doi.org/10.11124/JBISRIR-2017-003949

56. Wang Y, Han R, Zuo Z (2016) Dexmedetomidine-induced neuroprotection: is it translational? Transl Perioper Pain Med 1:15–19

57. Wang X, Ji J, Fen L, Wang A (2013) Effects of dexmedetomidine on cerebral blood flow in critically ill patients with or without traumatic brain injury: a prospective controlled trial. Brain Inj 27:1617–1622

58. Schomer KJ, Sebat CM, Adams JY et al (2019) Dexmedetomidine for refractory intracranial hypertension. J Intensive Care Med 34:62–66

59. Lee SH, Choi YS, Hong GR, Oh YJ (2015) Echocardiographic evaluation of the effects of dexmedetomidine on cardiac function during total intravenous anaesthesia. Anaesthesia 70:1052–1059

60. Erdman MJ, Doepker BA, Gerlach AT et al (2014) A comparison of severe hemodynamic disturbances between dexmedetomidine and propofol for sedation in neurocritical care patients. Crit Care Med 42:1696–1702

61. Yokota H, Yokoyama K, Noguchi H et al (2011) Post-operative dexmedetomidine-based sedation after uneventful intracranial surgery for unruptured cerebral aneurysm: comparison with propofol-based sedation. Neurocrit Care 14:182–187

62. Kido H, Komasawa N, Fujiwara S et al (2014) Gasserian ganglion block for trigeminal neuralgia under dexmedetomidine sedation. Masui 63:901–913

Chapter 13

Transgenic Mouse Models, General Anesthetics, and Alzheimer Disease: Findings from Preclinical Studies

Sabrina Bimonte, Antonio Barbieri, Nagoth Joseph Amruthraj, Marco Cascella, Arturo Cuomo, and Claudio Arra

Abstract

Alzheimer's disease (AD) is a neurodegenerative pathology outlined by severe memory impairment and by a developing disruption of cognitive functions. Unfortunately, in contrast with many published data on this disorder, no successful pharmacological approaches and/or therapeutic drugs are available up to now. It has been postulated that the environmental factors (e.g., medications) through the interaction with genetic factors, are able to modulate AD. Growing shreds of evidence highlighted a presumable role of general anesthetics in progression of neurodegenerative processes, including AD. Interestingly, it has been hypothesized an association between anesthetic exposure and postoperative impairment of cognitive trajectory, in terms of postoperative cognitive dysfunction, and AD progression. Specifically, a hypothesis quite overtly postulates that especially inhaled anesthetics may have important effects on the progression of AD, leading to a severe cognitive impairment, although this is still debated. To shed light on this issue, and considering that it is not possible to perform studies in humans, many studies have been performed on AD transgenic animal models featuring neurofibrillary tangles, amyloid beta accumulation (plaque), and behavioral changes due to cognitive dysfunction. In this chapter, we highlight and describe these studies and the methodologies used. Findings emerged from these studies are inconclusive and indicate that anesthetics may promote, or not, the progression of AD, depending on the different experimental conditions, and probably anesthetics are not alone responsible for the cognitive impairment in the investigated animal models. Thus, further research in this paramount field is urgently needed.

Key words Alzheimer disease, Mild cognitive impairment, Inhaled anesthetics, Transgenic mice, Memory impairment, β-Amyloid protein, Tau protein, Neurofibrillary tangles

---

1 Introduction

The term dementia describes a mosaic of symptoms featured by a progressive impairment in mental abilities, including memory, attention, thinking, and judgment, as well as personality changes. These symptoms are associated with several clinical conditions such as infections, medication side effects, vitamin deficiency, depression, severe hypothyroidism, and alcoholism. While in these conditions (dementia-like symptoms) the memory impairment and other

Marco Cascella (ed.), *General Anesthesia Research*, Neuromethods, vol. 150, https://doi.org/10.1007/978-1-4939-9891-3_13,
© Springer Science+Business Media, LLC, part of Springer Nature 2020

described symptoms may ameliorate when the underlying disease is treated or addressed, in specific brain disorders featuring dementia, symptoms are permanent and worsen over time.

This mixed group of degenerative diseases encompasses the Alzheimer's disease (AD), the frontal temporal lobe degeneration, and the Lewy body dementia. Other types of progressive brain disease responsible for the occurrence of dementia symptoms are the vascular contributions to dementia and cognitive impairment, the mixed dementia (a combination of two or more types of dementia), the argyrophilic grain disease, the Creutzfeldt-Jakob disease, the Huntington's disease, the chronic traumatic encephalopathy, the Parkinson's disease (not always result in dementia), and the HIV-associated dementia.

Alzheimer's disease (AD) represents a devastating outline of dementia featured by memory impairments and dramatic cognitive dysfunctions. AD is the most common form of dementia, accounting for approximately two-thirds of all cases. In neuropathological terms, AD patients undergo a diffuse progressive injury of neurons and synapses which mainly involves the amygdala and the hippocampus. Although many have published data on the pathogenesis of this devastating neurodegenerative disorder, no successful pharmacological approaches and/or therapeutic drugs are available up to now. Indeed, current medications (e.g., cholinesterase inhibitors, and memantine) used in the early stages of the disease may only help lessen or stabilize cognitive symptoms for a limited time. In addition, approximately 50 million of persons globally live with dementia in 2017, and this number will get tripled in 2050. In the face of this alarming scenario, the search is on for potential new alternative strategies (e.g., natural compounds) [1–3].

Growing shreds of evidence highlighted a presumable association between general anesthetics medication exposition and accelerated progression of neurodegenerative processes in AD neuropathology [4, 5]. Interestingly, a link between anesthetic exposure and postoperative impairment of cognitive trajectory, in terms of postoperative cognitive dysfunction, and AD progression has been hypothesized [2, 6–18]. Investigations experimentally demonstrated that anesthetics alone may lead to an increment in the amyloid beta proteins (APP) production and aggregation [19–21] and are able to aggregate and to separate the tau protein [22–24].

To dissect this significant issue, many preclinical studies have been performed on AD transgenic (Tg) animal models [19–25]. In this chapter, we highlight and summarize these studies by focusing on the role of the inhaled anesthetics on the neuropathology of AD. Data emerged from these findings suggest that the anesthetics may contribute to accelerating AD's neuropathology but are not alone responsible for the cognitive impairment. Thus, is extremely urgent to conduct additional preclinical studies to ensure that clinicians can select the most appropriate anesthetic strategy for the safety of AD patients (to reduce the risk of worsening AD) and for those with a high risk of cognitive impairment such as people showing mild cognitive impairment (MCI). Indeed, clinically, dementia is preceded by this lesser degree of cognitive impairment which represents the stage between the typical cognitive decline of normal aging and the more severe decline of dementia. Symptoms of MCI may remain stable for years, progress to AD or another type of dementia (10–15%), or improve over time.

2 Alzheimer's Disease Pathogenesis: The Hypothesis of the Interplay Between Aβ Metabolism and Tau Pathology

The first step in the complex process of AD neuropathogenesis is represented by the formation of extracellular amyloid plaques (Aβ40 and Aβ42) and intraneuronal neurofibrillary tangles (NFTs), formed by tau proteins atypically hyperphosphorylated and put together into typical filaments [26]. These features are present in the hippocampal regions, which are involved in the storage of the memory, and in the temporal neocortex of the AD brain [27]. The aberrant depositions of the hyperphosphorylated tau protein and of the amyloid beta peptide (Aβ peptide) are responsible for the arrangement of amyloid plaques and NFTs leading to neurotoxicity and neuronal loss in the AD-affected brain [28]. A key role in AD pathogenesis is played by the accumulation of Aβ peptides, which, in turn, induces neurotoxicity, probably through the activation of the oxidative stress [29–32]. The degeneration and the death of neurons which represent the final step in AD neuropathogenesis are also caused by the microglia and astrocytes activation that are involved in the inflammatory reply to the deposition of Aβ (neuroinflammation) [33]. Microglia identified as the resident immune cells of the central nervous system (CNS), normally react to neuronal damage and discard the damaged cells by phagocytosis. However, these cells have adverse activities in different neuropathological conditions. The microglial activation (M1 phenotype) induces a complex cascade which involves the secretion of reactive oxygen species (ROS), reactive nitrogen species (NOS) and many pro-inflammatory molecules such as tumor necrosis factor-α (TNF-α), interleukin (IL) IL-6, and IL-1β. In turn, the oxidative stress and the neuroinflammation processes are responsible for an increased calcium influx, altered kinase and phosphatase activities, irregular regulation of transcription factors, and disruptions in ion channel and receptor gene expression. All these processes culminate in progressive impairment of the neurovascular working which leads to axonal demyelination, local hypoxia-ischemia, and white matter damages [29–33].

One of the mechanisms underlined the neuropathogenesis of AD is based on the axiom of the combined role of Aβ metabolism and tau pathology.

Accordingly, to the amyloid hypothesis, Aβ is produced thought consecutive reactions of proteolysis of amyloid precursor protein (APP) catalyzed by specific enzymes (α-secretase and β-secretase, an aspartyl protease β-site APP-cleaving enzyme) in different transmembrane domains [34–36]. After this process, Aβ is removed by the extracellular space and transferred into the blood and cerebrospinal fluid barrier (CSF) where is progressively degraded into metabolites with lower neurotoxicity by specific enzymes [37–40]. All these enzymes are able to regulate the Aβ catabolism and finally influenced the level of Aβ in the brain, whose alteration may contribute to AD development and progression, as previously demonstrated [41–43].

As mentioned before, also tau proteins contributed to AD pathogenesis. Normally localized in the axons, these proteins undergo to aberrant phosphorylation in AD and in more neurodegenerative diseases [44–47]. Tau phosphorylations are mediated by specific enzymes, predominantly the kinases [48, 49], while the phosphatases catalyze tau dephosphorylation in the brain [50]. Importantly, a correct balance in the activities of these enzymes is crucial for the maintenance of tau phosphorylation homeostasis. In absence of this balance, the NFTS will be formed leading to the neuronal cell death which represents the final step of AD neuropathogenesis [51] (Fig. 1).

3   Features of Alzheimer's Disease Transgenic Mouse Models

To dissect the molecular mechanisms hidden the AD pathogenesis and mechanisms of cognitive decline, several transgenic mouse (Tg) models have been generated. Specifically, the genetic approach was based on single or multiple mutations induced in the genes presenilin-1 (PSEN-1), presenilin-2 (PSEN-2), and APP which have been discovered in familial AD (FAD). Furthermore, because NTFs are also involved in AD pathogenesis, some researchers investigate on AD Tg models obtained by specific tau mutations.

These animal models are also used for dissecting the effects of general anesthetics, and particularly of the inhaled anesthetics, on behavioral changes and AD pathogenesis, as largely described through the studies of Tang et al. [19].

Basically, Tg mouse models have been developed by inducing mutations in PSEN1 and APP genes, in single or in combinations. For example, Struchler-Pierrat et al. [52] generated the APP23 mouse model by using the Thy-1 promoter which induced a mutation in the human APP gene. This model recapitulates the AD human condition, since the amyloid plaques, founded at 6 months

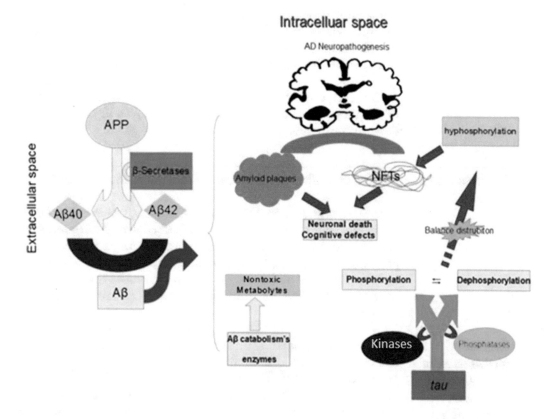

Fig. 1 Alzheimer's pathogenesis. The cartoon recapitulates the steps involved in Alzheimer's disease pathogenesis according to the hypothesis of an interplay between Aβ metabolism and tau pathology. Aβ is produced in the extracellular space by proteolysis of APP catalyzed by secretases. After this process, Aβ is removed by the extracellular space and transferred into the blood and cerebrospinal fluid barrier, where is progressively degraded into metabolites with lower neurotoxicity by specific enzymes. All these enzymes are able to regulate the Aβ catabolism and finally influenced the brain levels of Aβ plaques, whose alteration contribute to AD development and progression. Conjunctly, tau proteins are phosphorylated by kinases while the phosphates catalyze their dephosphorylation in the brain [50]. Importantly, a correct balance in the activities of these enzymes is crucial for the maintenance of tau phosphorylation homeostasis. In absence of this balance, the NFTs will be formed leading to the neuronal cell death and cognitive defects, as Aβ plaques, finally leading to onset of AD. *APP* amyloid precursor protein, *NTFs* neurofibrillary tangles, *Aβ40 and Aβ42* extracellular amyloid beta plaques

of age, are restricted in the cerebral vasculature, as attended in AD patients. Similar findings were founded in a classical AD Tg mouse model, the Tg2576 model, in which the APP gene is overexpressed due to a Swedish mutation (APPswe) [53]. Interestingly, combined APP and PSEN mutations have been used to generate AD Tg mouse model, to accelerate the formation of plaques (approximatively around 7 months) [25] and cognitive decline.

Regarding models of tau pathology, which are featured by the debut of NFTs, interesting results were obtained with the triple Tg AD mouse model (3×TgAD), obtained by insertion of three human mutations: in the genes PSEN1, APPswe, and Tau P301L. Plaques aroused between 10 and 12 months and NFTs appeared in the hippocampal region from 12 to 15 months, with a severe cognitive decline starting from the 24 months. Despite these models do not perfectly resemble the morphological features of AD (in human), they are considered useful tools to inquire the effects of the anesthetics on patients suffering from AD, or at high risk to develop the disease.

4   The Effects of Anesthetics on Alzheimer's Disease Transgenic Mouse Models

Several experiments on AD Tg mouse models, generated by single or multiple mutations as previously described, have been conducted in order to elucidate the effects of anesthetics exposure on AD progression [14, 16, 54–73]. Findings emerged from these studies indicate that anesthetics may or may not promote the progression of AD, depending on the different experimental conditions (Table 1).

4.1   The Effects of Inhaled Anesthetics on AD Progression: Shreds of Evidence from In Vivo Studies

The first important study on the topic was conducted, in 2008, by Bianchi et al. [54]. They investigated on the Tg2576 mouse, generated by inserting a double mutation into the human APP gene (APPswe). The authors based their study on the results obtained by Eckenhoff et al. [74] who demonstrated, in vivo, that the inhaled anesthetics increased the aggregation of Aβ [74]. Bianchi et al. [54] showed that exposure to halothane (0.8–1% in 30% oxygen for 5 days) but not isoflurane (0.9–1% in 30% oxygen for 5 days) increased amyloidopathy, by enhancing the deposition of plaques in AD-Tg2576 mice at 12 months of age. These results were deducted by quantifying immunohistochemically the Aβ plaque burden and the caspase-3 mediated apoptosis. On the contrary, no enhancement in cognitive decline, verified by the Morris Water Maze (MWM) test, was detected in these halothane-exposed AD-Tg2576 mice, while isoflurane exposure caused impairment in cognitive functions in wild-type mice, by acting probably on a different molecular pathway underlying AD neuropathogenesis. Taken together, these discoveries suggest that in that experimental setting the inhaled anesthetics influenced cognition and amyloidogenesis with different effects.

In order to elucidate these discrepancies, Perucho et al. [55] examined the effects of isoflurane (2% in 98% oxygen for 3 months), in Tg2576 and in wild-type mice at a presymptomatic stage (7-month-old). Data emerged from this study revealed that Tg2576 mice treated with isoflurane, differently from wild-type mice, manifested an increment in Aβ aggregates and in apoptosis especially in the hippocampal region. On the other hand, reduced autophagy and astroglial responses were detected in Tg2576

Table 1
The effects of inhaled anesthetic exposure in Alzheimer's transgenic mouse models

AD Tg mouse model, age	Type of anesthetic and dose	Time of anesthetic exposure	Effects	References
Tg2576 mice with *APPswe* (APP695) mutation; 12 months of age	Isoflurane (0.9–1 vol% in 30% oxygen) Halothane (0.8–1 vol% in 30% oxygen)	2 h/day for 5 days	Halothane induced amyloidopathy, no impairment of cognitive functions were observed	[54]
Tg2576 mice with *APPswe* (APP695) mutation;7 months of age	Isoflurane (2 vol% in 98% oxygen)	20–30 min/week for 3 months	Isoflurane increased apoptosis and amyloidopathy while reduced astroglia and autophagy and chaperone responses	[25]
3×TgAD mice with *APPswe, tauP301L, PS1M146V* mutations; 2, 4, and 6 months of age	Halothane 0.9–1.1 vol% in 30% oxygen); Isoflurane (0.9–1.1 vol% in 30% oxygen)	5 h per once a week for 4 weeks	Isoflurane increased phospho-tau protein at 3 months. Halothane improved cognitive functions	[56]
APPswe mice with *APPswe* mutation;11 months of age	Isoflurane (1 vol% in 30% oxygen); Trehalose (1% in fresh drinking water twice a week)	Three times for 1.5 h in 24 h	Isoflurane induced an increment in apoptosis in hippocampus and in the cerebral cortex and enhanced astrogliosis. Trehalose increased the expression levels of chaperones, reduced β tau plaques levels, amyloid peptide aggregates, and levels of phospho-tau	[57]
Tau.P301L mice, with *tauP301L* mutation; 8 months of age	Isoflurane		Isoflurane accelerated the tauopathy counteracted by hyperventilation or memantine	[58]
APP/PS1 mice with APPswe, PSEN1dE9 mutations;7 months of age	Isoflurane (1 vol% in 30% oxygen)	2 h/day for 5 days	Isoflurane enhanced spatial memory and the learning of aged mice, reduced	[59]

(continued)

Table 1
(continued)

AD Tg mouse model, age	Type of anesthetic and dose	Time of anesthetic exposure	Effects	References
			formations of the Aβ plaques and oligomers in the hippocampus	
APP23 mouse model, with APP mutation; young (4 months of age); old (14–16 months of age) mice	Isoflurane 1 MAC	2 h	Isoflurane improved cognitive performance and learning	[60]
Tg APP695 with APP mutation; 12 months of age	Isoflurane (1.5 vol% in 100% oxygen)	4 h	Isoflurane induced cognitive dysfunction and enhanced the phosphorylation of hippocampal tau protein at Ser262 site	[61]
APP/PS1 mice with APP and PS1 mutations; 10 months of age	Isoflurane (1.4 vol% in 100% oxygen) Zn	2 h	Isoflurane potentiated the memory impairment in zinc-deficient APP/PS1 transgenic mice.	[62]
APP,PSEN-1 transgenic mice with amyloid precursor protein swe, PSEN1dE9 mutations; 6 days of age	Sevoflurane(3 vol% or 2.1 vol% in 60% oxygen) IP3R2-APB (5 and 10 mg/kg, i.p.)	6 or 2 h	Sevoflurane induced caspase activation and apoptosis, impaired APP processing, and increased Aβ levels in the brain. Finally, IP3R-2-APB was able to mitigate/attenuate the devastating effect induced by sevoflurane	[63]

AD Alzheimer disease, *Tg* transgenic, *APP* amyloid precursor *protein*, *Swe* Swedish, *PSEN1* presenilin-1, *IP3R2-APB* inositol triphosphate receptor-antagonist 2-aminoethoxydiphenyl borate, *vol* volume, *i.p.* intraperitoneal injection treated mice. The results suggested that isoflurane was able to increase the amyloid disease only in AD Tg mouse models, but not on the control group. Moreover, a decrease in exploratory behavior was detected in Tg2576 isoflurane-treated mice.

A fascinating study was conducted by Tang et al. [56] on another Tg mouse model of AD (3×TgAD) in which APP (APPSwe), Presenilin-1 (PS-1M146V), and tauP301L were over-expressed. The authors studied the effects of the exposure to two different anesthetics, halothane (0.9–1.1% corresponding to 1.0 minimal alveolar concentration, MAC), and isoflurane (0.9–1.1% corresponding to 1.0 MAC), during the presymptomatic periods of AD (2, 4 and 6 months old). The behavior of mice was quantified by MWM test, while immunohistochemistry and biochemistry, were used to assess the accumulation and processing of Aβ, the processing of tau protein, the expression of synaptophysin, and inflammatory markers. Younger female mice treated with halothane showed an improvement in memory, which was not detected in isoflurane-treated mice. Importantly, the phospho-tau protein's levels in the hippocampus significantly increased in both groups of treatment particularly in mice with 6 months of exposure, although modifications amyloid, caspase, microglia or synaptophysin were not detected.

A different study was conducted by Perucho et al. [57] in APPswe mutant mice. Wild-type and APPswe mice were cotreated with trehalose (1% in fresh drinking water), a disaccharide which improves autophagy, and isoflurane (1% in 30% oxygen), at 11 months of age. Isoflurane induced an increment in apoptosis in the hippocampus and in the cerebral cortex. In addition, it provoked an enhancement in astrogliosis detected by immunohistochemistry and biochemistry assays. On the other side, these devastating effects were counteracted by the cotreatment with trehalose. Specifically, trehalose while increased the expression levels of chaperones was able to reduce β tau plaques levels, amyloid peptide aggregates, and phospho-tau levels. With these findings, it is presumably to assume that trehalose may be advised as a preventive strategy against pro-amyloidogenic effects induced (experimentally) by inhaled anesthetics.

Later on, Menuet et al. [58] demonstrated that isoflurane anesthesia accelerated the tauopathy and induced functional alterations in presymptomatic Tau.P301L mice, containing a mutation in the tau gene. Interestingly, these morphological findings were prevented by hyperventilation, or memantine, an antagonist of N-methyl-D-aspartate (NMDA). From these data, it can be deducted that NMDA-receptors are involved in the mechanisms underlying the isoflurane-induced changes, whereas their antagonists could be used for developing new safer protocols to adopt in AD patients suffering from tauopathy.

Interesting findings were reported by Su et al. [59]. The authors conducted research on APP/PS1 Tg-mice and wild-type subjected to isoflurane exposure (1% in 30% oxygen) during mid-life, which represents the presymptomatic phase of Aβ deposition. Surprisingly, isoflurane was able to enhance spatial memory and the learning (tested by using the MWM test) of both aged Tg and wild-type mice, used as control.

Moreover, reduced formations of the plaques and oligomers (detected by immunohistochemistry and Elisa test) in the hippocampal regions were detected in both group of treatments; however, the underlying mechanisms were not fully dissected. Finally, mice exposed to isoflurane maintained blood pressure and stable heart rates (detected respectively by blood gas analysis and hemodynamic recording) after and an initial decrement induced by isoflurane itself, thus avoiding the hypoxia. Thus, hemodynamics and behavioral changes were not correlated.

Several findings were described by Eckel et al. [60] in their evaluation on the effects of isoflurane-induced anesthesia conducted on behavioral performance and cognitive function in young and old AD Tg-mice, (APP23 mice). Accordingly, isoflurane (1 MAC, for 2 h) enhanced the cognitive performance, behavior flexibility and cognitive functions (assessed by modified hole-board test) in all experimental groups except for wild-type mice. These results indicate that in this animal model isoflurane exhibits a defensive effect on cognitive functions in AD.

In a subsequent investigation on AD Tg-mice (APP695) were evaluated the effects of isoflurane exposure (1.5 vol% in 100% oxygen for 4 h) on the hippocampal tau protein phosphorylation (determined by quantitative Western blotting), and on learning and memory behaviors (tested by MWM test) [61]. Importantly, isoflurane-induced dysfunctions at cognitive levels by improving the phosphorylation of hippocampal tau protein in a specific site (Ser262).

In a subsequent investigation, the effects of the isoflurane exposure on neuropathogenesis (determined by immunohistochemistry and western blotting analysis) and learning and memory function (detected by using the MWM test) were dissected in 10-month-old zinc-adequate, zinc-deficient, and zinc-treated APP/PS1 mice [62]. The principle is that because zinc (Zn) is involved in learning and memory processes, its absence, or reduction, may worsen AD pathology. Data emerged from this study, demonstrated that exposure of isoflurane in 10-month-old APP/PS1 mice with Zinc adequacy for 2 h induced transient neurotoxicity. On the other side, Zn deficiency reduced memory function and learning in 10-month-old APP/PS1 mice. Thus, in Zn-deficient APP/PS1 mice, exposure of isoflurane provoked severe neurotoxicity and dramatic memory and learning impairment. These outcomes suggested that Zn treatment may enhance memory function and learning of APP/PS1 mice by preventing the isoflurane-induced neurotoxicity. So AD patients having an exacerbated Zn deficiency should be more susceptible to isoflurane, and in the case of treatment with this volatile anesthetic, could worsen their pathological condition.

Different findings were reported by Lu et al. [63] who examined the effects of sevoflurane in newborn Tg mice with mutations in the APP and PSEN-1 genes. Sevoflurane activated the caspase and the apoptosis, impaired APP processing and enhanced brain Aβ levels in AD Tg mice more than in controls. Finally, inositol 1, 4, 5-trisphosphate receptor antagonist, 2-APB was able to mitigate the devastating effect induced by sevoflurane. Since sevoflurane is used mainly in pediatric patients, these outcomes suggest the needing of further studies, including clinical trials on humans, to dissect the potential neurotoxicity of sevoflurane, particularly in childhood.

4.2 The Effects of Propofol in the Onset and in the Progression of AD

Different findings were obtained from the studies conducted on the effects of propofol, an intravenous anesthetic with neuroprotective features [65–70], in AD Tg-mouse models (Table 2). Zhang et al. [71] tested the effects of propofol on the activation of caspase-3 induced by isoflurane in human neuroglioma cells, H4 wild type cells and stably transfected by APP, and in the cerebral tissue of APP/PSEN Tg-mice. Results from the in vitro studies showed that propofol reduced the activation of caspase-3 (determined by Western Blotting analysis) previously induced by isoflurane in H4-APP cells (treated with propofol 100 nM, with or without 2% isoflurane for 6 h) but not in H4 wild-type cells (propofol-treated as H4-APP cells). Moreover, propofol reduced the oligomerization of Aβ42, but not of Aβ40 induced by isoflurane in H4-APP cells. The authors also conducted an in vivo experiment (APP/PSEN Tg-mice) and obtained the same results. Basically, to attenuate the caspase-3 activation induced by isoflurane (1.4% isoflurane in 100% oxygen for 6 h), they treated the Tg and wild-type mice (6 days old) with propofol [200 mg/kg, intraperitoneal injection (i.p.)]. Results showed that propofol diminished the caspase-3 activation (determined by Western blots analysis) induced by isoflurane.

Later on, other studies were conducted on the effect of propofol on AD progression in 18 weeks old AD Tg-mice with mutations in APPswe and PSN1 genes. Shao et al. [16] demonstrated that chronic treatment with propofol (50 mg/kg/week, i.p. for 8 or 12 weeks) enhanced cognitive function (revealed by MWM test) by reducing the Aβ-induced mitochondria dysfunction (detected by Flow cytometric analysis) and caspase activation (detected by western blotting analyses). Similar results were obtained in human neuroglioma cells, H4, and mouse neuroblastoma cells, N2.

Table 2
The effects of propofol exposure in Alzheimer's transgenic mouse models

AD Tg mouse model, age	Type of anesthetic and dose	Time of anesthetic exposure	Effects	References
TgAD mice with APPswe, PSEN1dE9 mutations; 6 days of age	Isoflurane (1.4 vol % in 100% oxygen) Propopol (200 mg/kg, i.p.)	6 h	Propofol attenuated the activation of caspases-3 induced by propofol	[71]
Tg2576 mice with APPswe, PSEN1dE9 mutations; 18 months of age	Propofol (50 mg/kg, i.p.)	One a week for 4 weeks	Propofol enhanced cognitive function in AD Tg and WT mice by acting on the mitochondrial pathway	[16]
APP Tg2576 mice with APP mutation;15 months of age	Isoflurane (1%); Propofol, (26 mg/kg bolus followed by 2 mg/kg/min infusion); Diazepam (4 mg/kg); ketamine, (80 mg/kg); Pentobarbital, (males 50 mg/kg and females 40 mg/kg); fentanyl, (0.165 mg/kg)	Once a week for 90 min for 4 weeks	Repetitive use of analgesics and anesthetic had no effects on Cognitive Function, reduced deposition in the cortex. No significant effects were observed in Aβ fibril deposition	[14]
3×TgA with PS1M146V, APPswe, and tau P301L mutations; 11 months of age	Propofol (250 mg/kg, i.p.) with or without caecal ligation and excision	20–30 min	Propofol anesthesia, with or without surgical procedures, induced no changes in neuropathological markers of AD, and minimal changes in cognitive function	[16]
APP/PS1 with APPSWE/ PSENdE9 mutations; 6 months of age	Propofol (200 mg/kg, i.p.)	Three times at 6, 7 and 8 months of age	Propofol not exacerbated plaque deposition and loss of synapses in AD	[73]

AD Alzheimer disease, Tg transgenic, APP amyloid precursor protein, Swe Swedish, PSEN1 presenilin-1, i.p. intraperitoneal injection

A different experimentation was reported by Quiroga et al. [14] in Tg2576 AD mouse model subjected to repetitive exposure to isoflurane (1%), propofol (26 mg/kg bolus followed by 2 mg/kg/min infusion), diazepam (4 mg/kg), ketamine (80 mg/kg), pentobarbital (males 50 mg/kg and females 40 mg/kg); fentanyl (0.165 mg/kg). No behavioral differences were detected between Tg and wild-type mice, although a reduced deposition of Aβ (revealed by immunohistochemistry) in the frontal cortex and in the hippocampus was observed in exposed Tg-mice. Overall, the repetitive exposure did not provoke significant damage in Tg7526 mice.

Two recent studies highlighted the positive effects of propofol on neuropathology and cognition in AD Tg-mouse models. Specifically, Mardini et al. [72] proved that propofol anesthesia, associated or not with surgical procedures (caecal ligation and excision), was associated with better cognitive effects (detected by behavioral tests) in the aged brain of 11-month-old 3×TgAD mice. Interestingly, neuroinflammation (tested by biochemical tests) and tau aggregation and plaque formation (detected by immunohistochemistry) were unaltered in Tg propofol-treated mice with or without surgery. These findings strongly indicated that in preclinical settings anesthesia with propofol may improve cognitive outcomes in the aged AD brain compared with inhaled anesthetics.

Finally, Woodhouse et al. [73] studied the effects of repeated propofol exposure on synapses and plaque deposition in the APP/PS1 Tg-AD-mouse model. No increment in plaque deposition, or loss of synapse (determined by immunohistochemistry and western blotting analysis), was detected in APP/PS1 propofol-exposed mice compared to controls.

Overall, these results strongly suggest that differently from the inhaled anesthetic, propofol may attenuate AD progression.

5   Future Perspectives

In this chapter, we summarize the effects of anesthetics on AD progression. To try to recapitulate AD, different Tg mouse models have been engineered, with single or multiple mutations in genes involved in cognitive dysfunction, Aβ accumulation, and NFT formation. Despite the fact that these models did not precisely reproduce the human neurodegenerative disorder, they allowed the scientists to clarify the effects of anesthetics on the onset or the progression of AD. Depending on specific experimental conditions applied to different Tg mouse models, these studies underlined a promoting role for inhaled anesthetics on the acceleration of AD neuropathology. On the contrary, a mitigating role for intravenous anesthetics, as propofol, was found. These controversial results strongly suggest the need of future studies taking into account not only the cognitive and neuropathological differences in the models to be tested but also different types of anesthetics to be used. As the difficulty of carrying out translational studies is evident, preclinical research on this topic is of considerable importance. Particularly, these studies should address (1) new animal models designed on the basis of recent findings concerning AD pathogenesis and (2) methods to dissect the molecular mechanisms hidden in the effects of the use of these drugs on cognitive impairment and AD progression.

References

1. Cascella M, Bimonte S, Muzio MR et al (2017) The efficacy of Epigallocatechin-3-gallate (green tea) in the treatment of Alzheimer's disease: an overview of pre-clinical studies and translational perspectives in clinical practice. Infect Agents Cancer 12:36

2. Cascella M, Bimonte S (2017) The role of general anesthetics and the mechanisms of hippocampal and extra-hippocampal dysfunctions in the genesis of postoperative cognitive dysfunction. Neural Regen Res 12 (11):1780–1785

3. Cascella M, Bimonte S, Barbieri A et al (2018) Dissecting the potential roles of Nigella sativa and its constituent thymoquinone on the prevention and on the progression of Alzheimer's disease. Front Aging Neurosci 10:16. https://doi.org/10.3389/fnagi.2018.00016

4. Baranov D, Bickler PE, Crosby GJ et al (2009) First international workshop on anesthetics and Alzheimer's disease. Consensus statement: first international workshop on anesthetics and Alzheimer's disease. Anesth Analg 108 (5):1627–1630

5. Yang CW, Fuh JL (2015) Exposure to general anesthesia and the risk of dementia. J Pain Res 8:711–718

6. Arora SS, Gooch JL, Garcia PS (2014) Postoperative cognitive dysfunction, Alzheimer's disease, and anesthesia. Int J Neurosci 124 (4):236–242

7. Dong Y, Wu X, Xu Z et al (2012) Anesthetic isoflurane increases phosphorylated tau levels mediated by caspase activation and Abeta generation. PLoS One 7(6):e39386

8. Hudson AE, Hemmings HC et al (2011) Are anaesthetics toxic to the brain? Br J Anaesth 107(1):30–37

9. Kapila AK, Watts HR, Wang T et al (2014) The impact of surgery and anesthesia on postoperative cognitive decline and Alzheimer's disease development: biomarkers and preventive strategies. J Alzheimers Dis 41(1):1–13

10. Li HC, Chen YS, Chiu MJ et al (2014) Delirium, subsyndromal delirium, and cognitive changes in individuals undergoing elective coronary artery bypass graft surgery. J Cardiovasc Nurs 30(4):340–345

11. Liu H, Weng H (2014) Up-regulation of Alzheimer's disease-associated proteins may cause enflurane anesthesia induced cognitive decline in aged rats. Neurol Sci 35(2):185–189

12. Liu Y, Pan N, Ma Y et al (2013) Inhaled sevoflurane may promote progression of amnestic mild cognitive impairment: a prospective, randomized parallel-group study. Am J Med Sci 345(5):355–360

13. Papon MA, Whittington RA, El-Khoury NB et al (2011) Alzheimer's disease and anesthesia. Front Neurosci 4:272

14. Quiroga C, Chaparro RE, Karlnoski R et al (2014) Effects of repetitive exposure to anesthetics and analgesics in the Tg2576 mouse Alzheimer's model. Neurotox Res 26 (4):414–421

15. Reinsfelt B, Westerlind A, Blennow K et al (2013) Open-heart surgery increases cerebrospinal fluid levels of Alzheimer-associated amyloid beta. Acta Anaesthesiol Scand 57 (1):82–88

16. Shao H, Zhang Y, Dong Y et al (2014) Chronic treatment with anesthetic Propofol improves cognitive function and attenuates caspase activation in both aged and Alzheimer's disease transgenic mice. J Alzheimers Dis 41 (2):499–413

17. Sprung J, Jankowski CJ, Roberts RO et al (2013) Anesthesia and incident dementia: a population-based, nested, case-control study. Mayo Clin Proc 88(6):552–561

18. Xie Z, McAuliffe S, Swain CA et al (2013) Cerebrospinal fluid abeta to tau ratio and postoperative cognitive change. Ann Surg 258 (2):364–369

19. Jiang J, Jiang H (2015) Effect of the inhaled anesthetics isoflurane, sevoflurane and desflurane on the neuropathogenesis of Alzheimer's disease (review). Mol Med Rep 12(1):3–12

20. Eckenhoff G, Johansson JS, Wei H et al (2004) Inhaled anesthetic enhancement of amyloid-beta oligomerization and cytotoxicity. Anesthesiology 101:703–709

21. Carnini A, Lear JD, Eckenhoff RD (2007) Inhaled anesthetic modulation of amyloid β (1-40) assembly and growth. Curr Alzheimer Res 4:233–241

22. Lu SM, Gui B, Dong HQ et al (2015) Prophylactic lithium alleviates splenectomy-induced cognitive dysfunction possibly by inhibiting hippocampal TLR4 activation in aged rats. Brain Res Bull 114:31–41

23. Xu Z, Dong Y, Wang H et al (2014) Peripheral surgical wounding and age-dependent neuroinflammation in mice. PLoS One 9:e96752

24. Cibelli M, Fidalgo AR, Terrando N et al (2010) Role of interleukin-1β in postoperative cognitive dysfunction. Ann Neurol 68:360–368

25. Tang JX, Eckenhoff F (2012) Anesthetic effects in Alzheimer transgenic mouse models. Prog Neuro-Psychopharmacol Biol Psychiatry 47:167–171

26. Grundke-Iqbal I, Iqbal K, Quinlan M et al (1986) Microtubule-associated protein tau. A component of Alzheimer paired helical filaments. J Biol Chem 261:6084–6089

27. Taguchi K, Yamagata HD, Zhong WK et al (2015) Identification of hippocampus-related candidate genes for Alzheimer's disease. Ann Neurol 57:585–588

28. Chen G, Chen KS, Knox J et al (2000) A learning deficit related to age and betaamyloid plaques in a mouse model of Alzheimer's disease. Nature 408:975–979

29. Cotman CW, Su JH (1996) Mechanisms of neuronal death in Alzheimer's disease. Brain Pathol 6:493–506

30. Aslan M, Ozben T (2004) Reactive oxygen and nitrogen species in Alzheimer's disease. Curr Alzheimers Res 1:111–119

31. Butterfield DA, Griffin S, Munch G et al (2002) Amyloid beta-peptide and amyloid pathology are central to the oxidative stress and inflammatory cascades under which Alzheimer's disease brain exists. J Alzheimers Dis 4:193–201

32. Opazo C, Huang X, Cherny RA et al (2002) Metalloenzyme-like activity of Alzheimer's disease beta-amyloid. Cu-dependent catalytic conversion of dopamine, cholesterol, and biological reducing agents to neurotoxic H(2)O(2). J Biol Chem 277:40302–40308

33. Lustbader JW, Cirilli M, Lin XHW et al (2004) ABAD directly links Abeta to mitochondrial toxicity in Alzheimer's disease. Science 304:448–452

34. Gu Y, Misonou H, Sato T et al (2001) Distinct intramembrane cleavage of the beta-amyloid precursor protein family resembling gamma-secretase-like cleavage of Notch. J Biol Chem 276:35235–35238

35. Sastre M, Steiner H, Fuchs K et al (2001) Presenilin-dependent gamma-secretase processing of beta-amyloid precursor protein at a site corresponding to the S3 cleavage of Notch. EMBO Rep 2:835–841

36. Yu C, Kim SH, Ikeuchi T et al (2001) Characterization of a presenilin-mediated amyloid precursor protein carboxyl-terminal fragment gamma. Evidence for distinct mechanisms involved in gamma-secretase processing of the APP and Notch1 transmembrane domains. J Biol Chem 276:43756–43760

37. Miners JS, Baig S, Palmer J et al (2008) Abeta-degrading enzyme-degrading enzymes in Alzheimer's disease. Brain Pathol 18(240–252):2008

38. Bates KA, Verdile G, Li QX et al (2009) Clearance mechanisms of Alzheimer's amyloid-beta peptide: implications for therapeutic design and diagnostic tests. Mol Psychiatry 14:469–486

39. Eckman EA, Eckman CB (2005) Abeta-degrading enzymes: modulators of Alzheimer's disease pathogenesis and targets for therapeutic intervention. Biochem Soc Trans 33:1101–1105

40. Turner AJ, Tanzawa K (1997) Mammalian membrane metallopeptidases: NEP, ECE, KELL, and PEX. FASEB J 11:355–364

41. Cook DG, Leverenz JB, McMillan PJ et al (2003) Reduced hippocampal insulin-degrading enzyme in late-onset Alzheimer's disease is associated with the apolipoprotein E-epsilon4 allele. Am J Pathol 162:313–319

42. Ashe KH, Zahs KR (2010) Probing the biology of Alzheimer's disease in mice. Neuron 66:631–645

43. Sakono M, Zako T (2010) Amyloid oligomers: formation and toxicity of Abeta oligomers. FEBS J 277:1348–1358

44. Buée L, Bussière T, Buée-Scherrer V et al (2000) Tau protein isoforms, phosphorylation and role in neurodegenerative disorders. Brain Res Brain Res Rev 33:95–130

45. Sergeant N, Bretteville A, Hamdane M et al (2008) Biochemistry of tau in Alzheimer's disease and related neurological disorders. Expert Rev Proteomics 5:207–224

46. Ittner LM, Ke YD, Delerue F et al (2010) Dendritic function of tau mediates amyloid-beta toxicity in Alzheimer's disease mouse models. Cell 142:387–397

47. Sultan A, Nesslany F, Violet M et al (2011) Nuclear tau, a key player in neuronal DNA protection. J Biol Chem 286:4566–4575

48. Buée L, Troquier L, Burnouf S et al (2010) From tau phosphorylation to tau aggregation: What about neuronal death? Biochem Soc Trans 38:967–972

49. Iqbal K, Liu F, Gong CX et al (2010) Tau in Alzheimer disease and related tauopathies. Curr Alzheimer Res 7:656–664

50. Iqbal K, Grundke-Iqbal I (2008) Alzheimer neurofibrillary degeneration: significance, etiopathogenesis, therapeutics and prevention. J Cell Mol Med 12(1):38–55

51. Guillozet AL, Weintraub S, Mash DC et al (2003) Neurofibrillary tangles, amyloid, and memory in aging and mild cognitive impairment. Arch Neurol 60:729–736

52. Sturchler-Pierrat C, Abramowski D, Duke M et al (1997) Two amyloid precursor protein transgenic mouse models with Alzheimer disease-like pathology. Proc Natl Acad Sci U S A 94(24):13287–13292

53. Hsiao K, Chapman P, Nilsen S et al (1996) Correlative memory deficits, Abeta elevation, and amyloid plaques in transgenic mice. Science 274:99–102

54. Bianchi SL, Tran T, Liu C et al (2008) Brain and behavior changes in 12-month-old Tg2576 and nontransgenic mice exposed to anesthetics. Neurobiol Aging 29:1002–1010

55. Perucho J, Rubio I, Casarejos MJ et al (2010) Anesthesia with isoflurane increases amyloid pathology in mice models of Alzheimer's disease. J Alzheimers Dis 19:1245–1257

56. Tang JX, Mardini F, Caltagarone BM et al (2011) Anesthesia in presymptomatic Alzheimer's disease: a study using the triple-transgenic mouse model. Alzheimers Dement 7(5):521–531

57. Perucho J, Casarejos MJ, Gomez A et al (2012) Trehalose protects from aggravation of amyloid pathology induced by isoflurane anesthesia in APP(swe) mutant mice. Curr Alzheimer Res 9 (3):334–343

58. Menuet C, Borghgraef P, Voituron N et al (2012) Isoflurane anesthesia precipitates tauopathy and upper airways dysfunction in pre-symptomatic Tau.P301L mice: possible implication for neurodegenerative diseases. Neurobiol Dis 46(1):234–243

59. Su D, Zhao Y, Xu H et al (2012) Isoflurane exposure during mid-adulthood attenuates age-related spatial memory impairment in APP/PS1 transgenic mice. PLoS One 7(11): e50172

60. Eckel B, Ohl F, Starker L et al (2013) Effects of isoflurane-induced anaesthesia on cognitive performance in a mouse model of Alzheimer's disease: a randomised trial in transgenic APP23 mice. Eur J Anaesthesiol 30(10):605–611

61. Li C, Liu S, Xing Y et al (2014) The role of hippocampal tau protein phosphorylation in isoflurane-induced cognitive dysfunction in transgenic APP695 mice. Anesth Analg 119 (2):413–419

62. Feng C, Liu Y, Yuan Y et al (2016) Isoflurane anesthesia exacerbates learning and memory impairment in zinc-deficient APP/PS1 transgenic mice. Neuropharmacology 111:119–129

63. Lu Y, Wu X, Dong Y et al (2010) Anesthetic sevoflurane causes neurotoxicity differently in neonatal naive and Alzheimer disease transgenic mice. Anesthesiology 112:1404–1141

64. Jevtovic-Todorovic V, Kirby CO et al (1997) Isoflurane and propofol block neurotoxicity caused by MK-801 in the rat posterior cingulate/retrosplenial cortex. J Cereb Blood Flow Metab 17:168–174

65. Jevtovic-Todorovic V, Wozniak DF, Powell S et al (2001) Propofol and sodium thiopental protect against MK-801-induced neuronal necrosis in the posterior cingulate/retrosplenial cortex. Brain Res 913:185–189

66. Rossaint J, Rossaint R, Weis J et al (2009) Propofol: neuroprotection in an in vitro model of traumatic brain injury. Crit Care 13: R61

67. Wu GJ, Chen WF, Hung HC et al (2011) Effects of propofol on proliferation and anti-apoptosis of neuroblastoma SH-SY5Y cell line: New insights into neuroprotection. Brain Res 1384:42–50

68. Acquaviva R, Campisi A, Murabito P et al (2004) Propofol attenuates peroxynitrite-mediated DNA damage and apoptosis in cultured astrocytes: an alternative protective mechanism. Anesthesiology 101:1363–1371

69. Bayona NA, Gelb AW, Jiang Z et al (2004) Propofol neuroprotection in cerebral ischemia and its effects on low-molecular-weight antioxidants and skilled motor tasks. Anesthesiology 100:1151–1159

70. Velly LJ, Guillet BA, Masmejean FM et al (2003) Neuroprotective effects of propofol in a model of ischemic cortical cell cultures: role of glutamate and its transporters. Anesthesiology 99:368–375

71. Zhang Y, Zhen Y, Dong Y et al (2011) Anesthetic propofol attenuates the isoflurane-

induced caspase-3 activation and Aβ oligomerization. PLoS One 6(11):e27019

72. Mardini F, Tang JX, Li JC et al (2017) Effects of propofol and surgery on neuropathology and cognition in the 3xTgAD Alzheimer transgenic mouse model. Br J Anaesth 119 (3):472–480

73. Woodhouse A, Fernandez-Martos CM et al (2018) Repeat propofol anesthesia does not

exacerbate plaque deposition or synapse loss in APP/PS1 Alzheimer's disease mice. BMC Anesthesiol 18(1):47

74. Eckenhoff RG, Johansson JS, Wei H et al (2012) Inhaled anesthetic enhancement of amyloid-beta oligomerization and cytotoxicity. Anesthesiology 101(3):703–709

Chapter 14

The Biochemical Basis of Delirium

Matthew Umholtz and Nader D. Nader

Abstract

Delirium is an important clinical diagnosis that is common in the post-operative period and in critically ill patients. It is associated with an increase in morbidity, mortality and resource utilization. Though the pathogenesis of delirium has been increasingly recognized for its importance and has been an intensively studied in recent years, the biochemical mechanism for its development is still debated. This review scrutinized a number of studies in order to better characterize the biochemical basis for delirium, with particular focus paid to the interactions of the cholinergic system, the cholinergic anti-inflammatory pathway, the immune system and neuroinflammation. Despite the clinical impact of delirium, evidence-based protocols for the prevention and treatment are still lacking. Several previous trials have attempted to prevent or treat delirium by modulation of the cholinergic system with acetylcholinesterase inhibitors, the results of which have been largely ambiguous at best. As the biochemical basis of delirium becomes more clearly defined, future research into therapeutics based on immune modulation and treatment of neuroinflammation may prove to be very promising.

Key words Postoperative delirium, Postoperative cognitive dysfunction, Neuroinflammation, Cholinergic system, Anesthetic complications

1 Introduction

Delirium is a common acute neuropsychiatric disorder characterized by impairments in attention, consciousness, cognition, and behavioral disturbances. Delirium in hospitalized patients is usually the result of some precipitating stressful stimulus such as trauma, major surgery, or severe illness. This dysfunctional response to a stressful event then results in acute failure of the central nervous system (CNS). For more details on clinical features of postoperative delirium (POD) and postoperative cognitive dysfunction (POCD) see the other chapter on the topic in this book.

The precise mechanism by which this dysfunction occurs remains elusive. Meanwhile, the dominant theories by which delirium occurs relate to deficiency or dysfunction of neurotransmitter systems (e.g., involving the cholinergic system), an aberrant stress response, and neuroinflammation [1, 2].

Marco Cascella (ed.), *General Anesthesia Research*, Neuromethods, vol. 150, https://doi.org/10.1007/978-1-4939-9891-3_14,

2   Neurotransmitter Systems Impairment

One of the leading hypotheses for the biochemical mechanism of delirium is deficiency or dysfunction of the cholinergic system [3]. Cholinergic inputs in the basal forebrain have a critical role in many higher functions including memory, attention, arousal, and sensory processing [4]. The cholinergic deficiency hypothesis was originally derived from observations that anticholinergic substances could induce delirium in some patients. Risk of delirium and its severity are correlated with increased exposure to medications that have known anticholinergic activity [5, 6]. Additionally, cholinergic neurons located in basal section of forebrain degenerate extensively in Alzheimer's disease (AD), which shares many similarities with delirium and these patients are known to be at increased risk of developing acute delirium [7–9].

Impaired synthesis may play a leading role in acetylcholine deficiency. Acetylcholine (ACH) is derived from acetyl coenzyme A and choline with choline acetyltransferase acting as the key enzyme in its synthesis. Its breakdown is mediated by ACH esterase into the inactive metabolites choline and acetate. Precursor substrate deficiency is a potential cause of impaired ACH synthesis [10]. Acetyl CoA is produced by the breakdown of carbohydrates during glycolysis. Acetyl CoA then enters the citric acid cycle where its oxidation results in energy production. Thus, hypoglycemia or severe malnutrition may result in cholinergic deficiency by way of decreased production of acetyl Co-A. In one experimental study, severe hypoglycemia induced by high dose insulin, decreased the synthesis of ACH in rat brain (as measured by the incorporation of [2h4] choline in to ACH) [10, 11]. Interestingly, pretreatment with a cholinesterase inhibitor, physostigmine, prior to induced hypoglycemia decreased the number of rat deaths at 3 h [11].

ACH acts at both nicotinic receptors (nACR) and muscarinic receptors (mACR or M receptors) in the CNS. Of these two receptors, muscarinic receptors are more widely distributed and probably play a more important role in the development of delirium and cognitive dysfunction [4, 12]. The most widely expressed subtype in the CNS is the M1 receptor and is involved in perception attention and cognitive functioning. Toxins and drugs with anticholinergic effect primarily act by blocking the postsynaptic M1 muscarinic receptors and produce symptoms such as hallucinations, confusion and other cognitive deficits. Some drugs have anticholinergic activity by blocking presynaptic release of ACH. Cannabinoids inhibit the presynaptic release of multiple neurotransmitters, including ACH through the binding of specific cannabinoid receptors [13]. Opiates also act presynaptically through opiate receptors to inhibit acetylcholine release [14]. Both opiates and cannabinoids act through G-protein coupled receptors to inhibit calcium channel opening. The blocking of calcium channels prevents nerve depolarization, which affects neurotransmitter release (ACH).

Inhibition of ACH function at nACRs in the CNS may play a role in the development of delirium and cognitive dysfunction. nACRs in the brain have been implicated in several important functions such as memory, learning, arousal and reward. Some anesthetic drugs such as isoflurane, propofol and nitrous oxide (N_2O) inhibit nACRs [15, 16]. Though primarily thought to produce their effects by inhibition of GABA receptors, blockade of nACRs may contribute to their hypnotic and amnestic properties [17]. However, it is also quite plausible that this blockade may in part explain the occasional undesirable side effect emergence delirium.

Thiamine deficiency is known to cause Wernicke encephalopathy, which is characterized by opthalmoplegia, ataxia, and confusion. These deficits are thought to be the result of ACH deficiency due to altered acetyl Co-A metabolism in the CNS resulting in cholinergic deficits [18]. In support of the hypothesis that cholinergic deficiency plays a key role in the development in the deficits seen in thiamine deficiency, acetylcholinesterase inhibitors are able to reduce symptoms and mortality in experiment animals with induced thiamine deficiency [19]. Thiamine plays a key role as a component of several enzymes involved in carbohydrate metabolism. Particularly important are alpha ketaglutarate dehydrogenase, which is the rate limiting step of the citric acid cycle and pyruvate dehydrogenase. The decrease of these two key enzymes is the main cause of energy and acetyl Co-A depletion in thiamine deficiency [20, 21].

The destruction and loss of cholinergic neurons in the basal forebrain of patients with advanced AD has been well documented. AD shares many similarities with other causes of delirium and cognitive dysfunction [22]. These patients also demonstrate an increased risk of acute worsening of delirium and cognitive dysfunction at times of stress such as acute illness or major surgery [23]. Analysis of cholinergic markers from the brains of patients with Alzheimer's shows decreased choline acyltransferase activity, decreased high affinity choline uptake activity and deceased ACH activity. In these sites of decreased cholinergic activity there is also decreased activity of pyruvate dehydrogenase and alpha ketoglutarate dehydrogenase. The decreased enzymatic function may prove to be particularly deleterious to cholinergic neurons as they result in decreased acetyl CoA and choline, which are important not just for energy production and synthesis of ACH but also for the production of structural lipids. Additionally, these patients also exhibit decreased glucose uptake and oxidative metabolism, most likely due to reduction of glucose transport in the brains of patients with AD [22].

Hypoxia is a potentially important cause of impaired ACH deficiency [24]. ACH synthesis is dependent on aerobic metabolism and the citric acid cycle for the production of acetyl CoA and thus is particularly susceptible to disturbances in cerebral metabolism. Histotoxic hypoxia induced in rat brain by cyanide toxin can reduce the incorporation of [2H4] choline into ACH [24]. Likewise, anemic hypoxia induced by infusion of sodium nitrite ($NaNO_2$) (to produce methemoglobinemia) impairs ACH synthesis along the same pathway. Interestingly, this impaired synthesis occurs even with normal ATP/ADP values, suggesting that even mild cyanide toxicity and low levels of methemoglobinemia can significantly affect ACH levels [11].

Mild hypoxia impairs ACH synthesis in rat brain, as in one study evaluating the synthesis of ACH, control rats breathing 30% oxygen ($PaO_2 = 120$ mmHg) were compared to rats breathing moderate hypoxic mixture of 15% oxygen ($PaO_2 = 57$ mmHg) and severe hypoxic mixture at oxygen concentrations of 10% ($PaO_2 = 42$ mmHg). Synthesis of ACH was measure by analysis of U-14C glucose. The synthesis of ACH was decreased by 35% in rats breathing 15% oxygen and by 54% in rats breathing 10% oxygen, respectively. Similarly, ACH synthesis as measured with [1-2H2, 2-2H2] choline, was decreased by 50% with 15% Oxygen, and by 68% with 10% oxygen, respectively. The rats breathing hypoxic mixtures had increased brain lactates and increased cortical blood flow compared to rats breathing 30% oxygen [24]. These results suggest that even milder forms of hypoxia may impair ACH synthesis and possibly contribute to delirium and cognitive dysfunction. These findings may help to explain the delirium that accompanies patients who have suffered from cerebral vascular accidents, as well as, their increased sensitivity to the psychogenic effects of anticholinergic drugs.

Interestingly, in the same study, rats breathing 70% N_2O decreased ACH synthesis by 45% and 53% (as measured by [U-14C] glucose and [1-2H2, 2-2H2] choline, respectively [24]). In other studies, N_2O has been shown to be neurotoxic to rat brains in high concentration in an age-dependent manner. However, these affects have not been shown in human studies [25]. In a study of elderly patients undergoing noncardiac surgery, the use of N_2O did not increase the risk of POD/POCD when compared to patients who did not receive N_2O [26]. To date, there is no link between N_2O use during anesthesia and increased incidence of POD or POCD. The reason for these deleterious effects in rats and the apparent discrepancy in human studies is not entirely clear.

The cholinergic system has important interactions with other neurotransmitter systems, dysfunction of which also have been associated with delirium. In particular, dopamine excess has been associated with hyperactive delirium and psychosis

[3, 27]. Dopamine acts via D1 receptors in the hippocampus and prefrontal cortex which plays an important role in short term memory and executive functions [28]. Dopamine excess has been linked with a concomitant decrease in acetylcholine. This inverse relationship may be one of the key mechanisms in the pathogenesis of delirium. Hypoxia is a well-documented cause of acute delirium. Hypoxia causes a dopamine surge in the CNS that then results in a decrease of acetylcholine release from neurons, resulting in delirium [29]. This relationship between dopamine levels and ACH is further evidenced by the efficacy of neuroleptics (particularly Haloperidol) that act as dopamine antagonists in the treatment of acute delirium. Furthermore, dompaminergic medications (such as Levodopa) and drugs that influence the intrasynaptic levels of dopamine (cocaine, amphetamine) have been associated with delirium in a dose dependent manner [30, 31].

Serotonin (5-HT) is another important neurotransmitter that affects acetylcholine levels; however, the mechanism by which this occurs appears to be quite complex. 5-HT has the potential to increase or decrease acetylcholine levels, and these effects are likely mediated by serotonin receptor subtype and also the location in the brain in which the receptor is stimulated. For example, stimulation of rat hippocampal 5-HT release with fenfluramine further stimulates the release of ACH by way of HT3 receptors. This exaggerated acetylcholine release can be inhibited by 5-HT3 receptor blockers [32]. Likewise, stimulation of ACH release from rat hippocampus by way of direct serotonin infusion can be blocked by 5-HT1A receptor antagonists, suggesting that 5-HT actions at 5-HT1A receptors may play an important role in ACH levels [33]. In the frontal cortex of rat brain, ACH release can be stimulated by binding of 5-HT4 receptor agonists [34]. Conversely, there is evidence that 5-HT blocks the release of ACH in rat striatum [35, 36]. The production of 5-HT is dependent on the precursor tryptophan (TRP). It is hypothesized that decreased TRP levels can decrease 5-HT which ultimately leads to delirium. One recent study found that low levels of preoperative TRP levels were associated with an increased incidence of postoperative delirium in high risk individuals [37]. However, another recent study found no correlation between plasma 5-HT levels and incidence of delirium in critically ill patients [38]. The effect of 5-HT on ACH in the CNS and the role it may play in delirium risk certainly deserves further investigation.

Please add details on other systems involved as it is well known that dopamine excess, gamma-amino-butyric acid (GABA) and serotonin alterations may contribute to the genesis of delirium.

3   The Inflammatory Response

The inflammatory response that accompanies acute illness or major surgery is thought to be one of the main factors for the cause of POD, POCD, as well as, delirium that often accompanies trauma or serious illness [39, 40]. The mechanism by which inflammation contributes to delirium is likely multifactorial. Tissue concentrations of TNF in the hippocampus are one of the hallmarks of depressive mood [41]. The activation of the inflammatory cascade is associated with release of proinflammatory cytokines into the blood stream as well as recruitment of immune cells [42–44]. A healthy immune response is crucial for proper wound healing, repair of tissue damage as well as to combat infection without detriment to the host's own cells and tissues. The inflammatory response can be deleterious, resulting in severe damage or death unless kept in check by counter-regulatory mechanisms [45]. Thus, counterregulatory anti-inflammatory mechanisms must be simultaneously activated during the inflammatory response to counterbalance its potentially harmful actions. The main components of this anti-inflammatory response are the release of anti-inflammatory cytokines (i.e., IL-10), stress hormones and interestingly, a neuroinhibitory response [43, 46]. The afferent neural circuit of the nervous system response are peripheral nerves which sense and transmit signals to the CNS regarding infection or injury. These signals then trigger an efferent pathway from the CNS via the vagus nerve through the release of acetylcholine [47]. This release of ACH attenuates the production of TNF, Il-6, and Il-1β by immune cells expressing the nACH subunit alpha-7 [40, 42, 48, 49]. Direct electrical stimulation of the vagus nerve significantly inhibits the production of cytokines by innate immune cells in organs innervated by the vagus nerve (spleen, liver, GI tract, and heart) [50]. This vagal nerve stimulation significantly downregulates the production of TNF, IL-1, IL-6, and IL-8 but does not alter the production of the anti-inflammatory cytokine IL-10. The attenuation of these proinflammatory cytokines maintains the proinflammatory response within an adequate range [42]. This mediation of the immune response protects the organism from organ damage and death during syndromes of excessive cytokine release such as trauma, sepsis, endotoxemia, and hemorrhagic shock. This neuroinhibitory response likely represents an evolutionary ancient adaptation, dating back to invertebrate nematode worms which possess a primitive nervous and immune system [51].

The discovery of this neural pathway has raised the hypothesis that the progression of some autoimmune diseases may be related to the loss of neuroinhibitory activity [52]. In one study of nonobese diabetic mouse model, neurons surrounding the insulin producing beta cells are destroyed prior to the beta cells themselves.

This has raised the hypothesis that the loss of a neuroinhibitory pathway leads to later immune related destruction of beta cells [53]. Another related study showed that in mice subjected to collagen-induced arthritis, there is destruction of neurons in the spleen prior to the onset of clinical symptoms of the disease. Further studies have then showed evidence that stimulating the cholinergic anti-inflammatory pathway in these mice then reversed the arthritis [54, 55]. These results suggest that this neural circuit may prove to be crucial in preventing excessive immune mediated damage by inhibiting cytokine release as the result of autoimmunity, shock, trauma and severe illness.

A deregulated neuroimmune response combined with a dysfunctional cholinergic system may prove to be a plausible cause behind delirium physiology. Patients suffering from postoperative delirium have lower levels of cholinesterase activity even before surgery [40]. Additionally, low levels of cholinesterase activity are associated with elevated proinflammatory cytokines during acute illness [56, 57]. This suggests that low levels of cholinesterase activity represent a dysfunctional or degenerative cholinergic system which then may predispose or contribute to a high risk of delirium either postoperatively or during acute illness [40]. The decrease of cholinesterase activity is particularly intriguing, as these findings have also been well documented on necropsy of deceased patients suffering from AD. Though these results do not provide a conclusion how acetylcholinesterase levels correlate with peripheral levels of ACH or the responsiveness of the vagal nerve to inflammation, these results do suggest that baseline acetylcholinesterase levels can be a candidate biomarker of an elevated inflammatory response postoperatively and a risk factor for POD/POCD [58].

As mentioned previously, the stress of surgery, trauma or acute illness triggers an inflammatory response by the immune system characterized by the release of both proinflammatory and anti-inflammatory cytokines. However, in patients suffering from POD, levels of C-reactive protein are found to be significantly higher versus patients who are not exhibiting delirium [59, 60]. Additionally, these patients also show a gross imbalance of proinflammatory (IL-6, IL-8, IL-1βa, TNF-α) to anti-inflammatory cytokines (IL-10). This imbalance towards a greater proinflammatory ratio of cytokines likely is a key mechanism behind the delirium seen postoperatively or during acute stress [39].

Though it seems plausible that dysfunction of the cholinergic system has a critical role in a further dysfunctional inflammatory response to an acute stress, the mechanism by which this inflammatory response leads delirium and cognitive dysfunction needs to be explored. As mentioned previously, in patients suffering from delirium, an aberrant stress response leads to an excessive proinflammatory state. In the CNS, microglia activation results in the excessive release of proinflammatory cytokines [58, 61]. The result of this excessive inflammatory response is neuroinflammation within the central nervous system. In particular, the hippocampus may be particularly susceptible to neuroinflammatory insults as it is known to be critically important to cognition and highly vulnerable to aging [41, 62]. Additionally, as AD shares some similarities with delirium, it is worth noting that in patients with severe AD, degeneration of the cortex is a common feature with the hippocampus appearing to be particularly vulnerable to the devastating effects of the disease [63, 64].

One key mechanism behind hippocampal memory dysfunction in delirium relates to the effects of the proinflammatory cytokine IL-1β [65]. One interesting study evaluating fear conditioning in rats, administration of agents that induce IL-1β affected the memory consolidation of memories that depend on hippocampus function but had no effect of hippocampal independent memory pathways (auditory cue fear conditioning) [48]. Furthermore, the memory processing dysfunction associated with IL-1β is prevented by blocking the actions of this cytokine. Elderly patients with hip fracture are among the highest risk group for POD/POCD. A recent study of elderly patients with acute hip fracture presenting for surgery, patients that develop POD have been found to have elevated IL-1β in their cerebral spinal fluid when compared to patients who did not go on to develop delirium [66]. In another recent study using a mouse model of orthopedic surgery, mice that underwent surgery of the tibia under general anesthesia, memory impairment was associated with increased plasma cytokines, increased hippocampal microgliosis, as well as increased expression of IL-1β in the hippocampus [67].

TNF is one of the key triggers for the inflammatory cascade by acting upstream of IL-1β and other proinflammatory cytokines. In analysis of cytokine levels after major surgery in a mouse model, TNFα was one of the first proinflammatory cytokines that can be detected in plasma (detected at approximately 30 min postoperatively). In contrast, IL-1β was not detected until approximately 6 h after surgery [68]. Preoperative administration of anti-TNF effectively reduced IL-1β at 6 and 24 h after surgery. However, if anti-TNF administration was delayed until 1-h postoperatively, its administration had no effect on levels of proinflammatory cytokines. To relate the inflammatory process to cognitive changes in a rat model, contextual fear response was analyzed (noxious foot shock stimulation) postoperatively. Fear response was significantly decreased postoperatively representing a form of cognitive dysfunction and hippocampal dependent memory impairment. Preoperative treatment with anti-TNF significantly ameliorated this cognitive impairment [68].

Preoperative administration of anti-TNF effectively reduced brain microgliosis after an orthopedic surgery in mice [68]. Microglia are the innate immune cells of the CNS. They function as macrophages and represent the main form of immune defense in the CNS. At baseline, they remain in a quiescent, resting state, but are activated by proinflammatory cytokines, cell necrosis factors, lipopolysaccharide and several other factors. Upon activation, microglia transform their cell bodies from small, thin cell bodies with long branching processes, to a large amoeboid shape capable of phagocytosis (features described as microgliosis) [69]. This activation of microglia is accompanied by the production of numerous potentially toxic mediators, as well as further potentiation and release of proinflammatory cytokines. Preoperative treatment of mice with anti-TNF significantly prevents microgliosis postoperatively and decreases the systemic cytokine response [68]. Preoperative treatment with anti-TNF also decreased memory impairment postoperatively (as measured by contextual fear response), and decreases hippocampal inflammation.

Advanced age is among the major preoperative risk factors for the development of POD and POCD. There is evidence that microglia in the geriatric population are shifted more towards the inflammatory phenotype [58, 70]. This relative shift in phenotype is termed microglial priming and may reflect an increase in inflammation associated with aging. Also, this priming of the microglia may be a key mechanism to explain the susceptibility of geriatric patients to POD or POCD [58].

MicroRNAs (miRNAs) are noncoding RNA strands that are able to influence gene expression. MiRNAs have mainly a negative regulatory role and act by influencing gene expression by degrading mRNAs or inhibiting the translation of target genes [71]. Recent research has increasingly implicated the deregulation of miRNAs in a variety of disease processes, including AD and cognitive dysfunction. There is evidence that upregulation of inflammation by miRNA-146 is linked to the development and progression of AD [72]. It is therefore conceivable that miRNA dysregulation may influence upregulation of inflammation and influence POCD. In particular, decreased expression of miRNA-572 has been implicated as an early marker in POCD. It is hypothesized that in early POCD miRNA-572 may be down-regulated in order to upregulate its downstream target genes. In a rat model of POCD, injection of a mi-RNA572 inhibitor into the lateral ventricle improved cognitive function as it was measured by way of a water maze test [73].

The precise mechanism behind the pathophysiology of delirium and cognitive dysfunction that is associated with major surgery or severe illness remains incompletely understood. There are likely a myriad of factors that may ultimately prove to be involved in what is clearly a complex form of pathology. Since there is currently no proven modality to effectively prevent or treat delirium and cognitive dysfunction, hopefully future research into the mechanism and biochemical basis of delirium will lead to possible treatments or preventative strategies. Deficiency or dysfunction of the cholinergic system is likely one key aspect of delirium physiology. Despite the similarities between delirium and AD, the proven treatments for AD have not proven to do be reliable to prevent or treat postoperative delirium. There was initially great optimism that acetylcholine esterase inhibitors, due to their limited efficacy in treating AD, may have some benefit in treating or preventing POD and POCD. Thus far, the results have been mixed. While one recent report showed that rivastigmine patch was able to prevent POD in high-risk patients presenting for hip fracture surgery, another study showed no benefit in cardiac surgery for prevention with preoperative oral rivastigmine [74, 75]. Experimental evidence in animal studies, however, has showed that administration of acetylcholinesterase inhibitors suppresses systemic inflammation and enhances survival of septic animals [76]. However, a recent study of critically ill patients treated with rivastigmine as an adjunct to haloperidol for delirium actually had an increased risk of mortality than patients treated with haloperidol alone [77]. These mixed results demonstrate the complexity of the cholinergic system and its crucial role not just in cognition but also in immunity.

4   Further Research Perspectives

Future research into modulation of the immune response to surgery, trauma and critical illness may prove to be beneficial in the prevention or treatment of delirium and cognitive dysfunction. In particular, cytokine selective therapy is an intriguing modality that may prove useful. Anti-TNF has had an acceptable safety profile and has shown benefit in treating inflammatory disorders such rheumatoid arthritis, Crohn's disease, and ankylosing spondylitis. The evidence suggests that anti-TNF may be of benefit in preventing or treating POD/POCD. Furthermore, the relationship between neuroinflammation and miRNA is an exciting new field of study for further research to explore. Modulating the neuroinflammatory response with agonist or antagonist miRNAs is a promising approach to treating delirium and cognitive dysfunction.

The clinical impact for patients and cost to society due to delirium and cognitive dysfunction cannot be understated. Delirium and cognitive dysfunction is associated with long-term disability and increased mortality. In the geriatric hospitalized population, the incidence of delirium occurs in 20–60% and contributes to nearly seven billion dollars in Medicare hospital costs annually [1, 78]. Further research into the mechanism of delirium and cognitive dysfunction may lead to modalities to prevent or treat these disabling disorders.

References

1. Inouye SK (2006) Delirium in older persons. N Engl J Med 354:1157–1165

2. Inouye SK, Ferrucci L (2006) Elucidating the pathophysiology of delirium and the interrelationship of delirium and dementia. J Gerontol A Biol Sci Med Sci 61:1277–1280

3. Hshieh TT, Fong TG, Marcantonio ER et al (2008) Cholinergic deficiency hypothesis in delirium: a synthesis of current evidence. J Gerontol A Biol Sci Med Sci 63:764–772

4. Benarroch EE (2010) Acetylcholine in the cerebral cortex: effects and clinical implications. Neurology 75:659–665

5. Han L, McCusker J, Cole M et al (2001) Use of medications with anticholinergic effect predicts clinical severity of delirium symptoms in older medical inpatients. Arch Intern Med 161:1099–1105

6. Tune LE, Damlouji NF, Holland A et al (1981) Association of postoperative delirium with raised serum levels of anticholinergic drugs. Lancet 2:651–653

7. Bartus RT, Dean RL III, Beer B et al (1982) The cholinergic hypothesis of geriatric memory dysfunction. Science 217:408–414

8. Deiner S, Silverstein JH (2009) Postoperative delirium and cognitive dysfunction. Br J Anaesth 103(Suppl 1):i41–i46

9. Siedlecki KL, Stern Y, Reuben A et al (2009) Construct validity of cognitive reserve in a multiethnic cohort: the Northern Manhattan Study. J Int Neuropsychol Soc 15:558–569

10. Blass JP, Gibson GE (1999) Cerebrometabolic aspects of delirium in relationship to dementia. Dement Geriatr Cogn Disord 10:335–338

11. Gibson GE, Blass JP (1976) Impaired synthesis of acetylcholine in brain accompanying mild hypoxia and hypoglycemia. J Neurochem 27:37–42

12. Pratico C, Quattrone D, Lucanto T et al (2005) Drugs of anesthesia acting on central cholinergic system may cause post-operative cognitive dysfunction and delirium. Med Hypotheses 65:972–982

13. Mackie K (2008) Cannabinoid receptors: where they are and what they do. J Neuroendocrinol 20(Suppl 1):10–14

14. Kearns IR, Morton RA, Bulters DO et al (2001) Opioid receptor regulation of muscarinic acetylcholine receptor-mediated synaptic responses in the hippocampus. Neuropharmacology 41:565–573

15. Flood P, Ramirez-Latorre J, Role L (1997) Alpha 4 beta 2 neuronal nicotinic acetylcholine receptors in the central nervous system are inhibited by isoflurane and propofol, but alpha 7-type nicotinic acetylcholine receptors are unaffected. Anesthesiology 86:859–865

16. Yamakura T, Harris RA (2000) Effects of gaseous anesthetics nitrous oxide and xenon on ligand-gated ion channels. Comparison with isoflurane and ethanol. Anesthesiology 93:1095–1101

17. Westphalen RI, Hemmings HC Jr (2003) Effects of isoflurane and propofol on glutamate and GABA transporters in isolated cortical nerve terminals. Anesthesiology 98:364–372

18. Nakagawasai O (2005) Behavioral and neurochemical alterations following thiamine deficiency in rodents: relationship to functions of cholinergic neurons. Yakugaku Zasshi 125:549–554

19. Barclay LL, Gibson GE, Blass JP (1981) Impairment of behavior and acetylcholine metabolism in thiamine deficiency. J Pharmacol Exp Ther 217:537–543

20. Butterworth RF, Giguere JF, Besnard AM (1985) Activities of thiamine-dependent enzymes in two experimental models of thiamine-deficiency encephalopathy: 1. The pyruvate dehydrogenase complex. Neurochem Res 10:1417–1428

21. Butterworth RF, Giguere JF, Besnard AM (1986) Activities of thiamine-dependent enzymes in two experimental models of thiamine-deficiency encephalopathy. 2. Aalpha-Ketoglutarate dehydrogenase. Neurochem Res 11:567–577

22. Szutowicz A, Tomaszewicz M, Bielarczyk H (1996) Disturbances of acetyl-CoA, energy and acetylcholine metabolism in some encephalopathies. Acta Neurobiol Exp (Wars) 56:323–339

23. Robinson TN, Raeburn CD, Tran ZV et al (2009) Postoperative delirium in the elderly: risk factors and outcomes. Ann Surg 249:173–178

24. Gibson GE, Duffy TE (1981) Impaired synthesis of acetylcholine by mild hypoxic hypoxia or nitrous oxide. J Neurochem 36:28–33

25. Jevtovic-Todorovic V, Todorovic SM, Mennerick S et al (1998) Nitrous oxide (laughing gas) is an NMDA antagonist, neuroprotectant and neurotoxin. Nat Med 4:460–463

26. Leung JM, Sands LP, Vaurio LE et al (2006) Nitrous oxide does not change the incidence of postoperative delirium or cognitive decline in elderly surgical patients. Br J Anaesth 96:754–760

27. Tost H, Alam T, Meyer-Lindenberg A (2010) Dopamine and psychosis: theory, pathomechansims and Intermediate Phenotypes. Neurosci Biobehav Rev 34(5):689–700

28. Seamans JK, Floresco SB, Phillips AG (1998) D1 receptor modulation of hippocampal-prefrontal cortical circuits integrating spatial memory with executive functions in the rat. J Neurosci 18:1613–1621

29. Broderick PA, Gibson GE (1998) Dopamine and serotonin in rat striatum during in vivo hypoxic-hypoxia. Metab Brain Dis 4:143–153

30. Alagiakrishnan K, Wiens CA (2004) An approach to drug induced delirium in the elderly. Post Grad Med J 80:388–393

31. Takeuchi A, Ahern TL, Henderson SO (2011) Excited delirium. West J Emerg Med 12 (1):77–83

32. Console S, Bertorelli R, Russi G et al (1994) Serotonergic facilitation of acetylcholine release in vivo from rat dorsal hippocampus via serotonin 5-HT3 Receptors. J Neurochem 62(6):2254–2261

33. Izumi J, Washizuka M, Miura N et al (1994) Hippocampal serotonin 5-HT1A receptor enhances acetylcholine release in conscious rats. J Neurochem 62(5):1804–1808

34. Consolo S, Arnaboldi S, Giorgi S et al (1994) 5-HT4 receptor stimulation facilitates acetylcholine release in rat frontal cortex. Neuroreport 5(10):1230–1232

35. Gillet G, Ammor S, Fillion G (1985) Serotonin inhibits acetylcholine release from rat striatum slices: evidence for a presynaptic receptor mediated effect. J Neurochem 45 (6):1687–1691

36. Jackson D, Stachowiak MK, Bruno JP et al (1988) Inhibition of striatal acetylcholine release by endogenous serotonin. Brain Res 457(2):259–266

37. Robinson TN, Raeburn CD, Angles EM et al (2008) Low tryptophan levels are associated with post-operative delirium in the elderly. Am J Surg 196(5):670–674

38. Tomas CD, Salluh J, Soares M et al (2015) Baseline acetylcholinesterase activity and serotonin plasma levels are not associated with delirium in critically ill patients. Rev Bras Ter Intens 27(2):170–177

39. Cerejeira J, Firmino H, Vaz-Serra A et al (2010) The neuroinflammatory hypothesis of delirium. Acta Neuropathol 119:737–754

40. Cerejeira J, Nogueira V, Luis P et al (2012) The cholinergic system and inflammation: common pathways in delirium pathophysiology. J Am Geriatr Soc 60:669–675

41. Fasick V, Spengler RN, Samankan S et al (2015) The hippocampus and TNF: common links between chronic pain and depression. Neurosci Biobehav Rev 53:139–159

42. Parrish WR, Rosas-Ballina M, Gallowitsch-Puerta M et al (2008) Modulation of TNF release by choline requires alpha7 subunit nicotinic acetylcholine receptor-mediated signaling. Mol Med 14:567–574

43. Porhomayon J, Kolesnikov S, Nader ND (2014) The impact of stress hormones on post-traumatic stress disorders symptoms and memory in cardiac surgery patients. J Cardiovasc Thorac Res 6:79–84

44. Tracey KJ (2009) Reflex control of immunity. Nat Rev Immunol 9:418–428

45. Pol RA, van Leeuwen BL, Izaks GJ et al (2014) C-reactive protein predicts postoperative delirium following vascular surgery. Ann Vasc Surg 28:1923–1930

46. de Waal Malefyt R, Abrams J, Bennett B et al (1991) Interleukin 10(IL-10) inhibits cytokine synthesis by human monocytes: an autoregulatory role of IL-10 produced by monocytes. J Exp Med 174:1209–1220

47. Blalock JE, Smith EM (2007) Conceptual development of the immune system as a sixth sense. Brain Behav Immun 21:23–33

48. Pugh RC, Fleshner M, Watkins LR et al (2001) The immune system and memory consolidation: a role for the cytokine IL-1beta. Neurosci Biobehav Rev 25:29–41

49. Roytblat L, Talmor D, Rachinsky M et al (1998) Ketamine attenuates the interleukin-6 response after cardiopulmonary bypass. Anesth Analg 87:266–271

50. Borovikova LV, Ivanova S, Zhang M et al (2000) Vagus nerve stimulation attenuates the systemic inflammatory response to endotoxin. Nature 405:458–462

51. Andersson U, Tracey KJ (2012) Neural reflexes in inflammation and immunity. J Exp Med 209:1057–1068

52. Saravia F, Homo-Delarche F (2003) Is innervation an early target in autoimmune diabetes? Trends Immunol 24:574–579

53. Straub RH, Rauch L, Fassold A et al (2008) Neuronally released sympathetic neurotransmitters stimulate splenic interferon-gamma secretion from T cells in early type II collagen-induced arthritis. Arthritis Rheum 58:3450–3460

54. van Maanen MA, Lebre MC, van der Poll T et al (2009) Stimulation of nicotinic acetylcholine receptors attenuates collagen-induced arthritis in mice. Arthritis Rheum 60:114–122

55. Zhang P, Han D, Tang T et al (2008) Inhibition of the development of collagen-induced arthritis in Wistar rats through vagus nerve suspension: a 3-month observation. Inflamm Res 57:322–328

56. Abou-Hatab K, Nixon LS, O'Mahony MS et al (1999) Plasma esterases in cystic fibrosis: the impact of a respiratory exacerbation and its treatment. Eur J Clin Pharmacol 54:937–941

57. Hubbard RE, O'Mahony MS, Calver BL et al (2008) Plasma esterases and inflammation in ageing and frailty. Eur J Clin Pharmacol 64:895–900

58. Locatelli FM, Kawano T (2017) Postoperative cognitive dysfunction: preclinical highlights and perspectives on preventative strategies. Curr Topics Anesthesiol. In: Erbay RH (ed) InTech, London. https://doi.org/10.5772/66574. https://www.intechopen.com/books/current-topics-in-anesthesiology/postoperative-cognitive-dysfunction-preclinical-highlights-and-perspectives-on-preventive-strategies

59. Macdonald A, Adamis D, Treloar A et al (2007) C-reactive protein levels predict the incidence of delirium and recovery from it. Age Ageing 36:222–225

60. White S, Eeles E, O'Mahony S et al (2008) Delirium and C-reactive protein. Age Ageing 37:123–124. author reply 124

61. Wilson CJ, Finch CE, Cohen HJ (2002) Cytokines and cognition—the case for a head-to-toe inflammatory paradigm. J Am Geriatr Soc 50:2041–2056

62. Bartsch T, Wulff P (2015) The hippocampus in aging and disease: From plasticity to vulnerability. Neuroscience 309:1–16

63. Ball MJ (1977) Neuronal loss, neurofibrillary tangles and granulovacuolar degeneration in the hippocampus with ageing and dementia. A quantitative study. Acta Neuropathol 37:111–118

64. West MJ, Coleman PD, Flood DG et al (1994) Differences in the pattern of hippocampal neuronal loss in normal ageing and Alzheimer's disease. Lancet 344:769–772

65. Murray CA, Lynch MA (1998) Evidence that increased hippocampal expression of the cytokine interleukin-1 beta is a common trigger for age- and stress-induced impairments in long-term potentiation. J Neurosci 18:2974–2981

66. Cape E, Hall RJ, van Munster BC et al (2014) Cerebrospinal fluid markers of neuroinflammation in delirium: a role for interleukin-1beta in delirium after hip fracture. J Psychosom Res 77:219–225

67. Cibelli M, Fidalgo AR, Terrando N et al (2010) Role of interleukin-1beta in postoperative cognitive dysfunction. Ann Neurol 68:360–368

68. Terrando N, Monaco C, Ma D et al (2010) Tumor necrosis factor-alpha triggers a cytokine cascade yielding postoperative cognitive decline. Proc Natl Acad Sci U S A 107:20518–20522

69. Aloisi F (2001) Immune function of microglia. Glia 36:165–179

70. Xanthos DN, Sandkuhler J (2014) Neurogenic neuroinflammation: inflammatory CNS reactions in response to neuronal activity. Nat Rev Neurosci 15:43–53

71. Murchison EP, Hannon GJ (2004) miRNAs on the move: miRNA biogenesis and the RNAi machinery. Curr Opin Cell Biol 16:223–229

72. Wang G, Huang Y, Wang LL et al (2016) MicroRNA-146a suppresses ROCK1 allowing hyperphosphorylation of tau in Alzheimer's disease. Sci Rep 6:26697

73. Yu X, Liu S, Li J et al (2015) MicroRNA-572 improves early post-operative cognitive dysfunction by down-regulating neural cell adhesion molecule 1. PLoS One 10:e0118511

74. Gamberini M, Bolliger D, Lurati Buse GA et al (2009) Rivastigmine for the prevention of postoperative delirium in elderly patients undergoing elective cardiac surgery—a randomized controlled trial. Crit Care Med 37:1762–1768

75. Youn YC, Shin HW, Choi BS et al (2016) Rivastigmine patch reduces the incidence of postoperative delirium in older patients with cognitive impairment. Int J Geriatr Psychiatry 32(10):1079–1084

76. Wang H, Liao H, Ochani M et al (2004) Cholinergic agonists inhibit HMGB1 release and improve survival in experimental sepsis. Nat Med 10:1216–1221

77. van Eijk MM, Roes KC, Honing ML et al (2010) Effect of rivastigmine as an adjunct to usual care with haloperidol on duration of delirium and mortality in critically ill patients: a multicentre, double-blind, placebo-controlled randomised trial. Lancet 376:1829–1837

78. Hevesi ZG, Hammel LL (2012) Geriatric dDisorders. In: Stoelting RK, Hines RL, Marschall KE (eds) Stoelting's anesthesia and co-existing disease. Saunders/Elsevier, Philadelphia, PA, p 674

Chapter 15

Postoperative Delirium and Postoperative Cognitive Dysfunction

Matthew Umholtz and Nader D. Nader

Abstract

Postoperative delirium (POD) and postoperative cognitive dysfunction (POCD) are post-surgical phenomena associated with an increase in morbidity, mortality, and resource utilization. This review scrutinizes a number of studies in order to better characterize POD/POCD, with particular focus paid to the conditions' etiology, associated risk factors, prevention, and management. Despite their clinical importance, evidence-based protocols for the prevention and treatment of POD and POCD are still lacking. In order to ensure improved safety and cost-effective management of POD/POCD future research should focus on screening protocols to identify at risk patients, in addition to formulating a standardized treatment regimen for patients most at risk.

Key words Postoperative delirium, Postoperative cognitive dysfunction, Anesthesia recovery, Anesthetic complications, Perioperative complications, Cognitive impairment

1 Introduction

Postoperative delirium (POD) and cognitive dysfunction (POCD) have long been recognized as potential complications of anesthesia and surgery, particularly in the geriatric population. Both are discrete entities that are associated with debilitating morbidity and mortality. Far from a benign occurrence, POD and POCD are important predictors of persistent decline and early mortality of the patients, as well as a significant burden on health care cost, length of hospital admission, cost per day of hospital stay as well as an increased incidence of admission to long-term assisted living after hospital discharge [1].

Marco Cascella (ed.), General Anesthesia Research, Neuromethods, vol. 150, https://doi.org/10.1007/978-1-4939-9891-3_15,
© Springer Science+Business Media, LLC, part of Springer Nature 2020

2   Postoperative Delirium

Delirium has been well described in the Diagnostic and Statistical Manual of Mental Disorders fifth edition [2]. Its key characteristic is an acute change of mental status from the patient's baseline. There is usually an acute onset and a fluctuating clinical course. Patients will typically have exhibit decreased attention, which may be observed by patients being easily distracted or having difficulty following a conversation. Patients will frequently display disorganized thinking. Language may be disorganized, incoherent or illogical. Patients may present behaviors that range from hypoactive to hyperactive or mixed psychomotor impairments. Frequently there is memory impairment. Emergence delirium should not be confused with true postoperative delirium as POD is not relegated only to the time the patients is emerging from anesthesia. Indeed, most patients with POD exhibit a lucid period after emergence only to exhibit POD between postoperative days 1–3 [1].

Several methods have been devised to diagnose patients with POD. The most commonly used method used in research studies has been the Confusion Assessment Method (CAM) [3]. There are also a CAM tool for intensive care unit (CAM-ICU), and a short CAM which contains only items 1–4 of the original CAM. The Delirium Rating Scale Revised-98 is a 16-item clinician-rated scale with 13 severity items and 3 diagnostic items. It appears to be a comprehensive instrument especially for monitoring patients over a given period [4]. More detailed is the Delirium Symptom Interview which is a structured interview with 107 items of which 63 are interview questions and the remanding ones observations [5]. The NEECHAM Confusion Scale is a screening scale which can be used by nurses to rate the patient's behavior while providing routine care to patients. The scale takes 10 min to complete. It has high inter-rater reliability, good validity, and high sensitivity (95%) and specificity [6].

3   Emergence Delirium

Delirium that occurs as the patient is emerging from a general anesthetic is termed emergence delirium (ED). ED, also referred to as emergence agitation, is temporally related to the period of time in which the patient is emerging from general anesthesia and typically resolves within the 1 h after recovery from anesthesia (in contrast to POD which follows an initial lucid period after an anesthetic and may last several days). Although it is more common in the pediatric population, all age ranges can potentially exhibit the condition. There is also an increased incidence in the geriatric population compared to younger adults. Though there is a range of potential psychomotor behaviors seen in emergence delirium, most often these patients exhibit agitation, confusion, and violent behavior.

Risk factors for emergence delirium include the age of patient. Patients under the age of 40 and over the age of 64 are at increased risk [7]. There are conflicting reports on the effect of anesthetic drugs on the incidence of ED. Some studies have implicated benzodiazepines in the development of ED while others have shown benzodiazepines to have a protective effect [8–10]. Ketamine has been implicated in increasing the risk of ED in some studies [11]. Inhalation anesthetics may play a role in the development of ED, with sevoflurane and desflurane being particularly problematic in comparison to older agents with a higher blood solubility (and thus a slower emergence from general anesthesia). It is hypothesized that the relatively rapid emergence from general anesthesia with new generation of volatile anesthetics increases the susceptibility to ED [10–12]. Certain surgical procedures show higher incidences of ED with breast surgery and abdominal surgery showing the highest risk. Patients with a history of posttraumatic stress disorder (PTSD) are known to have an increased risk of developing ED after a general anesthetic [13]. Typically, as long as the patient is protected from harming themselves, emergence delirium usually resolves without sequelae.

4 Postoperative Cognitive Dysfunction

The potential relationship between POD and POCD has been the subject of not insubstantial controversy over the past several years. Although POD and POCD are strongly correlated phenomena, they represent two distinct clinical entities along the continuum of cognitive impairment after anesthesia and surgery. POCD can manifest after the occurrence of POD and its incidence seems to depend on the duration of POD. However, a decline in cognition abilities temporally associated with surgery could manifest itself also without a clearly identifiable occurrence of POD; thus, POCD can occur also in patients who did not experience POD. In contrast to POD, POCD is characterized by long term deterioration in mental abilities postoperatively. It is more difficult to define and diagnose because, in contrast to POD, which requires the detection of symptoms, the diagnosis of POCD requires sophisticated baseline neuropsychological testing both before and after surgery [14, 15]. Moreover, no universally accepted diagnostic criteria for POCD exist and it has been not listed in the latest versions of the International Classification of Diseases (ICD-11) and the Diagnostic and Statistical Manual of Mental Disorders. The manifestations for POCD are usually subtle and the spectrum of cognitive abilities that may be affected are diverse. Most psychometric tests analyzing cognitive function focus on memory and intellectual tasks that measure abilities involving learning, verbal fluency, perception, attention, executive function and abstract thinking. Cases of post-operative cognitive dysfunction are identified by conducting pre-operative neurophysiologic tests that test abilities regarding learning, memory, verbal abilities, perception, attention, executive function and abstract thinking. Unfortunately, there has not been a standardized methodology used in clinical practice or during studies to date to identify POCD and to evaluate its prevalence. Hence, it is somewhat more difficult to identify the presence and thus the incidence of POCD. Regardless, POCD seems to be a separate, though similar entity to POD in which the temporal relationship with the surgical event is not fully elucidated. Nevertheless, the risk factors for both appear to be very similar, as well as, the temporal relationship to occur in the postoperative time period after an initial transient lucid period.

The incidence of POD ranges from 5% to 15% with certain high-risk groups such as hip fracture patients ranging as high as an average of 35% [16]. The incidence of POCD is more difficult to describe given the definition of POCD used and the variety of tests that can be used to establish the diagnosis as well as their statistical evaluation. In some studies, the incidence of POCD has been has high as 40% in patients over age 60 at hospital discharge, and in approximately 10% of these patients the POCD persists for at least 3 months postoperatively [15]. Nevertheless, POCD has been described in young adults, adults as well as in the geriatric population. Similar to POD, POCD is more common in the geriatric population, as well as having a more prolonged course in this population. POCD is usually transient, but in a small subset of patients (approximately 1%) it can persist for years. Whether these persistent cases of POCD reflect permanent POCD remains controversial [17–19].

5   Risk Factors

The cause of delirium is multifactorial. Risk factors include advanced age, baseline dementia, or previous history of POD, history of hearing loss or visual impairment, baseline cognitive dysfunction or dementia. Metabolic and physiologic disturbances have been implicated as well. Hypoxemia, hypercarbia, and hypoglycemia are potentially life-threatening causes of delirium which should be recognized and treated without delay.

POCD has long been recognized as a potential complication of cardiac surgery. The cause for increased prevalence in cardiac surgery is likely multifactorial and related to surgical, anesthetic and patient factors. POCD had previously been thought to be a complication   related   physiologic   disturbance   related   to cardiopulmonary bypass (CPB) [20–22]. However, randomized controlled trials have shown the incidence of POCD to be similar whether cardiac surgery is performed with cardiopulmonary bypass or on a beating heart ("off pump") [23]. Markers of inflammation are also similar in patients randomized to "on pump" versus "off pump" cardiac surgery [24]. Additionally, when cardiopulmonary bypass is utilized, there does not appear to be an advantage in using pulsatile versus nonpulsatile flow in relation to prevention of POD/POCD [25]. The stress of open-heart surgery places a significant systemic inflammatory burden on the body. This inflammatory response causes the upregulation of macrophages, neutrophils, cytokines and free radical production. This inflammatory process can disrupt the blood brain barrier and increase the susceptibility of the brain to ischemic insults. This acute cerebral inflammatory process is likely an exacerbation of a chronic inflammatory state caused by the atherosclerosis process as well as aging. This acute cerebral inflammatory state likely causes neuron dysfunction and neuronal loss and is the most likely mechanism to explain the high incidence of POD/POCD in this population [26].

Hypothermia is frequently used during CPB as it may be an important modality useful for neuroprotection, however it may also contribute to POD/POCD seen in cardiac surgery [27]. Hypothermia decreases the cerebral metabolic rate, attenuates the neuroinflammatory response to CPB, decreases the production of free reactive species and excitatory neurotransmitter release (which may play a role in neuronal death). However, during rapid rewarming prior to the commencement of CPB may disrupt cerebral autoregulatory mechanisms and lead to cerebral edema [28]. The resulting cerebral edema may cause increased intracranial pressure, impaired cerebral metabolism and lead to increased risk of POD/POCD.

Metabolic syndrome is characterized as a cluster of conditions (hypertension, hyperglycemia, hyperlipidemia, excess body fat) that increase the risk of heart disease and atherosclerosis, is also a proinflammatory state that increases the risk of POD/POCD [29, 30]. This high inflammatory state predisposes geriatric patients to cognitive decline at baseline, as well as, an increased risk of POD/POCD [31]. This chronic inflammatory state compounded onto an acute exacerbation of inflammation is the likely mechanism for the increased incidence of POD/POCD after cardiac surgery. This acute rise in inflammation contributes to neuronal loss and dysfunction.

Hyperglycemia is an important modifiable risk factor for postoperative delirium. Hyperglycemia has long been recognized as a cause of a proinflammatory state, thus implying a possible mechanism for its association with delirium. Intraoperative hyperglycemia (>200 mg/dL) has been linked with POCD in patients undergoing open heart surgery [32]. Interestingly, tight glucose control (with intraoperative glucose between 80 and 100 mg/dL) may be deleterious for several reasons, one of which is potentially increasing the risk of POD when compared to conventional intraoperative glucose control [33].

Inadequate oxygenation, whether occurring acutely by perioperative hypoxia or secondary to anemia, both seem to be significant risk factors for the development of delirium. However, chronically impaired oxygenation may also have a role. Patients with obstructive sleep apnea are at increased risk of developing POD/POCD though the mechanism by which this occurs is not currently known [34]. Obstructive sleep apnea is an impairment of oxygenation during sleep that is most commonly caused by excessive tissue in the pharynx that causes obstruction during sleep, hypoxia, and hypercarbia, which leads to frequent arousals and impaired rest. This impaired oxygenation may represent an important cause of delirium in these patients. Consistent with this hypothesis is the phenomenon of increased risk of POCD after cardiac surgery in patients with cerebral oxygen desaturation (occurring with normal systemic oxygen saturation) [35].

In the geriatric patient, acute illness or exacerbation of chronic illness may precipitate delirium [36, 37]. The risk of POD delirium in this population is as high as 10%, but can be much higher depending on the surgical procedure and patient population. The highest risk surgery is hip surgery with an incidence of approximately 35%, probably reflecting the frailty and overall decline represented by hip fracture patients. There have also been reports of increased incidence related to cardiac, thoracic, vascular, and emergency surgery. In the elderly population even relatively minor surgeries such as cataract removal have been associated with POD, thus reflecting the particular susceptibility of this population.

Hyponatremia and azotemia have all been implicated to be factors that can cause or exacerbate delirium [36]. Infection, particularly urinary tract infections and respiratory infections are well-documented causes of delirium in both surgical and medical patients. Elevated leukocyte count preoperatively has been found to be a risk factor for the development of POD postoperatively, probably reflecting the risk of even occult infection or inflammation on developing delirium [38].

Advanced age has clearly been shown to be a prominent risk factor for developing postoperative delirium, as well as, developing delirium during acute illness. Poor preoperative functional status along with a higher American Society of Anesthesiologist class have both been shown to be independent risk factors for the development of postoperative delirium [39]. Lower preoperative albumin and high preoperative bilirubin have been associated with POD thus raising the possibility that nutritional strategies or improvement of liver function may help prevent or possibly treat POD.

Lifestyle modification may have important implications for prevention of POD. A history of smoking has been shown to increase the risk of POCD. In one study of patients undergoing abdominal aortic aneurysm surgery, the risk of POD was correlated with increasing pack/year history [40]. Potential mechanisms for smoking increasing risk of delirium are microvascular damage and increased atherosclerosis in cerebral vasculature. Also, nicotine withdrawal may have similar biochemical similarities with delirium [41]. Nicotine withdrawal is thought to be a relative acetylcholine deficiency in the central nervous system. Chronic exposure to nicotine causes an up regulation and desensitization of acetylcholine receptors [42]. During withdrawal from nicotine the unoccupied state of these receptors is believed to be a key mechanism of withdrawal symptoms. Nicotine withdrawal and POD share some similar characteristics such as restlessness, irritability, and confusion. Nicotine replacement in the form of transdermal patches may have therapeutic benefit to prevent POD in patients that are unable to quit smoking preoperatively.

Alcohol abuse has been identified as another risk factor for postoperative delirium and postoperative cognitive dysfunction. Chronic alcohol abuse causes atrophy of the frontal lobes, as well as, lower metabolism over the frontal cortex [43, 44]. These chronic changes lead to impaired executive function, impaired memory and in severe cases a condition indistinguishable from dementia not related to alcohol abuse. Alcoholics display greater neurocognitive impairment preoperatively compared to non-alcoholics and display greater levels of neurocognitive impairments in comparison to non-alcoholics [45, 46]. However, abstinence from alcohol use for up to 5 weeks has shown no protective benefit in chronic alcohol abusers suggesting that the neurocognitive impairments likely are chronic, not easily reversible and possibly permanent [47].

6   Pathophysiology

The etiology of POD and POCD remains unclear. Several mechanisms have been postulated. General anesthetics in aged rats have shown to negatively affect spatial memory for up to 2 weeks after exposure [48]. In addition, laboratory studies have shown enhancement of beta-amyloid oligomerization after exposure to inhaled volatile anesthetics [49]. Thus, it is tempting to speculate that the mechanism of POD and POCD are primarily related to the effect of the anesthetic on the CNS. Interestingly, regional anesthesia has not been shown to have a protective effect when given in place of a general anesthetic for surgery [50, 51]. However, major surgery does appear to be a significant culprit in the development of POD/POCD while ambulatory surgery is less of a risk [52]. This suggests that a primary cause of POD/POCD may be related to the increased inflammatory activity associated with major surgery [53]. In support of the role that inflammation plays in the development of delirium, elevated C-reactive protein has been associated with development of delirium while low levels of CRP predict the resolution of delirium [54, 55]. This points to a potential mechanism of delirium which involves leukocyte migration into the central nervous system and a breakdown of the blood–brain barrier [56].

Another potential risk factor for delirium as it relates to major surgical procedures is the association with increased blood loss and the development of POD. Postoperative delirium has found to be associated with increased intraoperative blood loss, increased postoperative blood transfusions and a postoperative hematocrit less than 30% [57]. Blood transfusion is a well-documented trigger and amplifier of systemic inflammation, as well as, a cause of increased oxidative damage [58]. Conversely, lower hemoglobin levels also correlate with increasing risk of POD/POCD [59]. At this time there is not a clear clinical recommendation that can be made regarding perioperative anemia management in its relationship to POD/POCD. However, it is increasingly clear that surgeries with increased risk for bleeding and the need for transfusion will likely significantly increase the risk of POD/POCD, especially in the geriatric population or patients with other risk factors for POD/POCD.

7   Prevention and Treatment

Given the prevalence and increased morbidity and mortality of POD/POCD, recent studies have focused on prevention and treatment of POD. Anti-inflammatory pharmacology shows promise in treating and possibly preventing POD. Hydroxy methyl glutaryl coenzyme-A reductase inhibitors (statins) have shown promise to have neuroprotective effects in both animal and human studies [60]. These neuroprotective effects are likely derived from statins immunomodulatory, anti-inflammatory and antithrombotic properties [61]. In one prospective observational study, statins given preoperatively were found to decrease POD in patients undergoing cardiac surgery with cardiopulmonary bypass [62]. Interestingly, another retrospective study found an increased association with preoperative statin use and increased POD after elective surgery in elderly patients [63]. Clearly, more research is needed to identify the role statins may play in the development of POD.

Treatment of POD should begin with identification of underlying medical conditions which may be contributing causes, optimization of environment and pain control and in refractory cases, pharmacologic treatment. In situations where POD places the patient or caregivers at risk of physical harm or when POD is interfering with routine postoperative care, pharmacologic treatment should be considered. Haloperidol, a typical antipsychotic that acts as a dopamine receptor type-2 (D2) antagonist remains one of the mainstays in the treatment of delirium, regardless of the cause. Recent studies have shown possible benefit from prophylactic use of haloperidol in preventing postoperative delirium [64]. When administered for treatment, a dose of 0.5–1 mg IV every 10–15 min can help control behavior. Larger doses may be used in severely agitated and combative patients, but may cause undesirable side effects, particularly over sedation. QT prolongation is an ever-present concern which can be monitored by EKG [65]. Extrapyramidal side effects have been reported and may be permanent [66]. Neuroleptic malignant syndrome has been reported in patients as a side effect of Haloperidol and manifests with hyperpyrexia, rigidity, cognitive changes, autonomic instability and elevated creatinine kinase levels [67]. In the surgical population patients with these symptoms may be confused to be exhibiting symptoms of malignant hyperthermia. In patients in whom side effects from haloperidol are of particular concern, atypical antipsychotic medications can be considered as they have a more favorable safety profile. In, particular, risperidone has shown promise in treating postoperative delirium, and in one study, was able to prevent POD in cardiac patients undergoing CPB [68].

The relative deficiency in acetylcholine remains as a leading hypothesis to explain the susceptibility to dementia and cognitive decline in the elderly population [69]. As such, several studies have investigated the role that acetylcholine esterase inhibitors may play in the prevention and treatment of POD. The beneficial effects of anticholinesterase inhibitors in the treatment of dementia have suggested potential usefulness in the treatment of POD. Thus far the results have been mixed, however, a recent study involving rivastigmine administered transdermal via a patch was able to decrease the incidence of POD in hip fracture patients [70]. Meanwhile, another prospective study failed to show benefit of oral rivastigmine in preventing POD in patients undergoing cardiac surgery [71].

The aging brain likely has increased sensitivity towards the adverse effects in the central nervous system (CNS) of drugs commonly used in the perioperative period. Virtually any drug with CNS activity has been implicated in causing delirium. Antihistamines such as diphenhydramine are often given as antiemetic medications, however these drugs often have significant anticholinergic effects that can cause agitation, confusion and disorientation [72]. Even small amounts of anticholinergic drugs such as atropine and scopolamine can cause delirium in the geriatric population (the quaternary amine glycopyrrolate which does not cross the blood brain barrier is a better choice in this population) [73]. If delirium caused by anticholinergic medications is suspected, the cholinesterase inhibitor physostigmine can be used as an antidote.

Effective management of the surgical pain is another important modifiable strategy to prevent POD. Increased postoperative pain scores during the first 3 postoperative days increases the likelihood of POD (after controlling for other known preoperative risk factors) [74]. Inadequate pain appears to be an important cause of POD. Opioids remain the mainstay for treating postoperative pain. There is inadequate evidence whether opioids in and of themselves contribute to increased risk or severity of POD. Since the importance of adequate pain control after surgery to prevent POD cannot be underestimated, opioids should be used to obtain adequate analgesia, postoperatively. Meperidine is the only opioid that has consistently shown an increased association with delirium (probably secondary to its anticholinergic properties) [75]. There does not seem to be any clear advantage among the more commonly used perioperative opioids (hydromorphone, fentanyl, and morphine) in preventing delirium. Because of the hemodynamic stability of opioids, high-dose opioids are common practice in cardiothoracic surgery, at least prior to "fast-tracking" of cardiac patients. Low doses of fentanyl for open-heart surgery (10 μg/kg) versus higher doses (50 μg/kg) does not prevent POCD in this high-risk group of patients but may aid in earlier extubation times postoperatively [76].

The literature regarding benzodiazepines and the role they may play in the development of POD or POCD in the elderly population has been mixed. Benzodiazepines have variable pharmacodynamics and pharmacodynamics and their cognitive effects can be more exaggerated in the elderly. Their effects may appear paradoxical in geriatric patients as these patients sometimes exhibit agitation and disinhibition. Metabolism of some benzodiazepines (such as diazepam) can be slow, as well as having several active metabolites. In elderly patients, the metabolites benzodiazepines, particularly the metabolites of longer acting drugs such as diazepam can be detected postoperatively even one week after surgery. However, the detection of these metabolites does not seem to correlate with POCD. Though the role that benzodiazepines may play in the development of POCD remains controversial, there does seem to be a stronger correlation in their role in increasing risk of POD [77]. This underscores the need to tailor treatment in this age group to smaller doses of this class of drug, or even avoiding them all together in the geriatric population. Longer acting benzodiazepines as well as increased doses of benzodiazepines have a stronger association with POD than shorter acting agents or low dose exposures [78]. Benzodiazepines may be necessary in certain subgroups of patients (active seizures, prevention of delirium tremens in known alcoholics and in benzodiazepine dependency) however, attempting to limit their use in geriatric patients in the perioperative period may prove prudent [75].

Dexmedetomidine is a highly selective alpha-2 agonist that may prevent POD in both cardiac and noncardiac surgery. When compared to propofol in elderly cardiac patients, dexmedetomidine had less incidence of POD as well as shorter duration and severity of POD [79]. An interesting study conducted in China compared low dose (0.1 μg/kg/h) intravenous dexmedetomidine infusions versus placebo in both intubated and non-intubated noncardiac postsurgical patients, the results of which showed a significant decrease in the incidence of POD (9% dexmedetomidine group versus 23% in the placebo) [80]. The dexmedetomidine group also reported better scores on a subjective sleep quality scale. Dexmedetomidine may prove to be particularly advantageous in the perioperative setting given its analgesic, sympatholytic and anxiolytic properties as well as its qualities as a sedative with minimal respiratory depressant compromise. Using dexmedetomidine for postoperative sedation may also decrease requirements for benzodiazepine use postoperatively. Dexmedetomidine may have important functions perioperatively as an anti-inflammatory agent. Dexmedetomidine given after surgery decreases important inflammatory markers such as interleukins (IL-6 and IL-8), as well as tumor necrosis factor alpha (TNF-α), which have been linked to increased risk of POCD [81, 82]. Dexmedetomidine decreases the risk of POCD in the geriatric population with its anti-inflammatory actions as the likely mechanism.

Ketamine is another important anesthetic agent that may have anti-inflammatory actions that can prevent POD/POCD. In one study of cardiac patients undergoing surgery requiring CPB, including Ketamine as one of the induction agents significantly decreased the risk of POD versus a control group (3% of the patients receiving Ketamine were diagnosed with POD versus 31% in the control group) [83]. The group receiving ketamine was also found to have lower C-reactive protein levels, thus giving support to the theory that the mechanism of protection is as an anti-inflammatory agent. Ketamine is also known to inhibit IL-6, a potent inflammatory cytokine [84].

Adequate research in the identification, prevention, and treatment of POD/POCD is of upmost importance for the benefit of surgical patients and for society. With an increasing segment of the population over the age of 65, the incidence of POD and POCD will likely become an increasingly common problem. POD and POCD are costly. They increase hospital length of stay, the utilization of resources, and increase the need for prolonged rehabilitation after discharge from the hospital. Patients with POD and POCD are at increased risk of mortality and are at risk for long-term cognitive and physical morbidity.

8   Future Research Perspectives

Future research may lead us into a clearer way to optimize care for these patients. As the mechanism of POD/POCD becomes more clearly defined, specific modalities to identify these patients perioperatively may be useful. Ideally, we could identify the patients at the highest risk of developing POD/POCD and implement specific strategies to prevent the progression of POD/POCD. Specifically, future studies focusing on detection of certain biomarkers of inflammation show promise in identifying patients at increased risk of POD/POCD. Also, specific modalities acting as anti-inflammatory agents may prove useful in the prevention and treatment of POD and the progression to POCD.

References

1. Deiner S, Silverstein JH (2009) Postoperative delirium and cognitive dysfunction. Br J Anaesth 103(Suppl 1):i41–i46
2. American Psychiatric Association (2013) Diagnostic and statistical manual of mental disorders (DSM-V), 5th edn. American Psychiatric Publishing, Washington, DC
3. Inouye SK, van Dyck CH, Alessi CA et al (1990) Clarifying confusion: the confusion assessment method. A new method for detection of delirium. Ann Intern Med 113:941–948
4. Albert MS, Leukoff SE, Reilly C (1992) The delirium symptom interview: an interview for the detection of delirium symptoms in hospital patients. J Geriatr Psychiatry Neurol 5 (1):14–21
5. Grover S, Kate N (2012) Assessment scales for delirium: a review. World J Psychiatry 2 (4):58–70
6. Neelon VJ, Champagne MT, Carlson JR et al (1996) The NEECHAM Confusion Scale: construction, validation and clinical testing. Nurs Res 45(6):324–330
7. Radtke FM, Franck M, Hagemann L et al (2010) Risk factors for inadequate emergence after anesthesia: emergence delirium and hypoactive emergence. Minerva Anestesiol 76:394–403
8. Lepouse C, Lautner CA, Liu L et al (2006) Emergence delirium in adults in the post-anaesthesia care unit. Br J Anaesth 96:747–753
9. Schor JD, Levkoff SE, Lipsitz LA et al (1992) Risk factors for delirium in hospitalized elderly. JAMA 267:827–831
10. Viswanath O, Kerner B, Jean YK et al (2015) Emergence delirium: a narrative review. J Anesthesiol Clin Sci 4:xx–xx
11. Lilburn JK, Dundee JW, Nair SG et al (1978) Ketamine sequelae. Evaluation of the ability of various premedicants to attenuate its psychic actions. Anaesthesia 33:307–311
12. Aono J, Ueda W, Mamiya K et al (1997) Greater incidence of delirium during recovery from sevoflurane anesthesia in preschool boys. Anesthesiology 87:1298–1300
13. Umholtz M, Cilnyk J, Wang CK et al (2016) Postanesthesia emergence in patients with post-traumatic stress disorder. J Clin Anesth 34:3–10
14. Rasmussen LS (1998) Defining postoperative cognitive dysfunction. Eur J Anaesthesiol 15:761–764
15. Silverstein JH, Steinmetz J, Reichenberg A et al (2007) Postoperative cognitive dysfunction in patients with preoperative cognitive impairment: which domains are most vulnerable? Anesthesiology 106:431–435
16. Rooke AG (2013) Anesthesia for the older patient. Wolters Kluwer Health/Lippincott Williams & Wilkins, xxi, Philadelphia, PA, p 1767
17. Abildstrom H, Rasmussen LS, Rentowl P et al (2000) Cognitive dysfunction 1-2 years after non-cardiac surgery in the elderly. ISPOCD group. International study of post-operative cognitive dysfunction. Acta Anaesthesiol Scand 44:1246–1251
18. McDonagh DL, Mathew JP, White WD, et al; Neurologic Outcome Research G (2010) Cognitive function after major noncardiac surgery, apolipoprotein E4 genotype, and biomarkers of brain injury. Anesthesiology 112:852–859
19. Newman MF, Kirchner JL, Phillips-Bute B, et al; Neurologic Outcome Research G, The Cardiothoracic   Anesthesiology   Research

Endeavors I (2001) Longitudinal assessment of neurocognitive function after coronary-artery bypass surgery. N Engl J Med 344:395–402

20. Day JR, Taylor KM (2005) The systemic inflammatory response syndrome and cardiopulmonary bypass. Int J Surg 3:129–140

21. Gao L, Taha R, Gauvin D et al (2005) Postoperative cognitive dysfunction after cardiac surgery. Chest 128:3664–3670

22. Roach GW, Kanchuger M, Mangano CM et al (1996) Adverse cerebral outcomes after coronary bypass surgery. Multicenter Study of Perioperative Ischemia Research Group and the Ischemia Research and Education Foundation Investigators. N Engl J Med 335:1857–1863

23. Kozora E, Kongs S, Collins JF et al (2010) Cognitive outcomes after on- versus off-pump coronary artery bypass surgery. Ann Thorac Surg 90:1134–1141

24. Parolari A, Camera M, Alamanni F et al (2007) Systemic inflammation after on-pump and off-pump coronary bypass surgery: a one-month follow-up. Ann Thorac Surg 84:823–828

25. Ozturk S, Sacar M, Baltalarli A et al (2016) Effect of the type of cardiopulmonary bypass pump flow on postoperative cognitive function in patients undergoing isolated coronary artery surgery. Anatol J Cardiol 16:875–880

26. van Harten AE, Scheeren TW, Absalom AR (2012) A review of postoperative cognitive dysfunction and neuroinflammation associated with cardiac surgery and anaesthesia. Anaesthesia 67:280–293

27. Miang Ying Tan A, Amoako D (2013) Postoperative cognitive dysfunction after cardiac surgery. Cont Edu Anaesth Crit Care Pain 13:218–223

28. Joshi B, Brady K, Lee J et al (2010) Impaired autoregulation of cerebral blood flow during rewarming from hypothermic cardiopulmonary bypass and its potential association with stroke. Anesth Analg 110:321–328

29. Hudetz JA, Patterson KM, Amole O et al (2011) Postoperative cognitive dysfunction after noncardiac surgery: effects of metabolic syndrome. J Anesth 25:337–344

30. Reijmer YD, van den Berg E, Dekker JM et al (2012) Development of vascular risk factors over 15 years in relation to cognition: the Hoorn Study. J Am Geriatr Soc 60:1426–1433

31. Yaffe K, Kanaya A, Lindquist K et al (2004) The metabolic syndrome, inflammation, and risk of cognitive decline. JAMA 292:2237–2242

32. Puskas F, Grocott HP, White WD et al (2007) Intraoperative hyperglycemia and cognitive decline after CABG. Ann Thorac Surg 84:1467–1473

33. Saager L, Duncan AE, Yared JP et al (2015) Intraoperative tight glucose control using hyperinsulinemic normoglycemia increases delirium after cardiac surgery. Anesthesiology 122:1214–1223

34. Roggenbach J, Klamann M, von Haken R et al (2014) Sleep-disordered breathing is a risk factor for delirium after cardiac surgery: a prospective cohort study. Crit Care 18:477

35. Slater JP, Guarino T, Stack J et al (2009) Cerebral oxygen desaturation predicts cognitive decline and longer hospital stay after cardiac surgery. Ann Thorac Surg 87:36–44. discussion 44–35

36. Aldemir M, Ozen S, Kara IH et al (2001) Predisposing factors for delirium in the surgical intensive care unit. Crit Care 5:265–270

37. Jitapunkul S, Pillay I, Ebrahim S (1992) Delirium in newly admitted elderly patients: a prospective study. Q J Med 83:307–314

38. Galanakis P, Bickel H, Gradinger R et al (2001) Acute confusional state in the elderly following hip surgery: incidence, risk factors and complications. Int J Geriatr Psychiatry 16:349–355

39. Raats JW, van Eijsden WA, Crolla RM et al (2015) Risk factors and outcomes for postoperative delirium after major surgery in elderly patients. PLoS One 10:e0136071

40. Benoit AG, Campbell BI, Tanner JR et al (2005) Risk factors and prevalence of perioperative cognitive dysfunction in abdominal aneurysm patients. J Vasc Surg 42:884–890

41. Hsieh SJ, Shum M, Lee AN et al (2013) Cigarette smoking as a risk factor for delirium in hospitalized and intensive care unit patients. A systematic review. Ann Am Thorac Soc 10:496–503

42. Benowitz NL (2009) Pharmacology of nicotine: addiction, smoking-induced disease, and therapeutics. Annu Rev Pharmacol Toxicol 49:57–71

43. Kril JJ, Halliday GM, Svoboda MD, Cartwright H et al (1997) The cerebral cortex is damaged in chronic alcoholics. Neuroscience 79:983–998

44. Kubota M, Nakazaki S, Hirai S et al (2001) Alcohol consumption and frontal lobe shrinkage: study of 1432 non-alcoholic subjects. J Neurol Neurosurg Psychiatry 71:104–106

45. Dao-Castellana MH, Samson Y, Legault F et al (1998) Frontal dysfunction in neurologically normal chronic alcoholic subjects: metabolic and neuropsychological findings. Psychol Med 28:1039–1048

46. Victor M (1993) Persistent altered mentation due to ethanol. Neurol Clin 11:639–661

47. Hudetz JA, Iqbal Z, Gandhi SD et al (2007) Postoperative cognitive dysfunction in older patients with a history of alcohol abuse. Anesthesiology 106:423–430

48. Culley DJ, Baxter M, Yukhananov R et al (2003) The memory effects of general anesthesia persist for weeks in young and aged rats. Anesth Analg 96:1004–1009

49. Eckenhoff RG, Johansson JS, Wei H et al (2004) Inhaled anesthetic enhancement of amyloid-beta oligomerization and cytotoxicity. Anesthesiology 101:703–709

50. Rasmussen LS, Johnson T, Kuipers HM et al (2003) Does anaesthesia cause postoperative cognitive dysfunction? A randomised study of regional versus general anaesthesia in 438 elderly patients. Acta Anaesthesiol Scand 47:260–266

51. Williams-Russo P, Sharrock NE, Mattis S, Szatrowski TP, Charlson ME (1995) Cognitive effects after epidural vs general anesthesia in older adults. A randomized trial. JAMA 274:44–50

52. Moller JT, Cluitmans P, Rasmussen LS et al (1998) Long-term postoperative cognitive dysfunction in the elderly ISPOCD1 study. ISPOCD Investigators International Study of Post-operative Cognitive Dysfunction. Lancet 351:857–861

53. Qiao Y, Feng H, Zhao T et al (2015) Postoperative cognitive dysfunction after inhalational anesthesia in elderly patients undergoing major surgery: the influence of anesthetic technique, cerebral injury and systemic inflammation. BMC Anesthesiol 15:154

54. Pol RA, van Leeuwen BL, Izaks GJ et al (2014) C-reactive protein predicts postoperative delirium following vascular surgery. Ann Vasc Surg 28:1923–1930

55. Speciale S, Bellelli G, Guerini F et al (2008) C-reactive protein levels and delirium in a rehabilitation ward. Age Ageing 37:122–123

56. Hevesi ZG, Hammel LL (2012) Geriatric disorders. In: Stoelting RK, Hines RL, Marschall KE (eds) Stoelting's anesthesia and co-existing disease. Saunders/Elsevier, Philadelphia', PA, p 674

57. Marcantonio ER, Goldman L, Orav EJ et al (1998) The association of intraoperative factors with the development of postoperative delirium. Am J Med 105:380–384

58. Rosa SD, Bristot Mde L, Topanotti MF et al (2011) Effect of red blood cell transfusion on parameters of inflammation and oxidative stress in critically ill patients. Rev Bras Ter Intens 23:30–35

59. Chen YL, Lin HC, Lin KH et al (2015) Low hemoglobin level is associated with the development of delirium after hepatectomy for hepatocellular carcinoma patients. PLoS One 10:e0119199

60. Patel A, Pisklakov SV (2012) Statins as potentially neuroprotective agents: a review. J Anesth Clin Res 3:1000251

61. De Loecker I, Preiser JC (2012) Statins in the critically ill. Ann Intensive Care 2:19

62. Katznelson R, Djaiani GN, Borger MA et al (2009) Preoperative use of statins is associated with reduced early delirium rates after cardiac surgery. Anesthesiology 110:67–73

63. Redelmeier DA, Thiruchelvam D, Daneman N (2008) Delirium after elective surgery among elderly patients taking statins. CMAJ 179:645–652

64. Schrader SL, Wellik KE, Demaerschalk BM et al (2008) Adjunctive haloperidol prophylaxis reduces postoperative delirium severity and duration in at-risk elderly patients. Neurologist 14:134–137

65. Hatta K, Takahashi T, Nakamura H et al (2001) The association between intravenous haloperidol and prolonged QT interval. J Clin Psychopharmacol 21:257–261

66. Caroff SN, Hurford I, Lybrand J et al (2011) Movement disorders induced by antipsychotic drugs: implications of the CATIE schizophrenia trial. Neurol Clin 29:127–148. viii

67. Levenson JL (1985) Neuroleptic malignant syndrome. Am J Psychiatry 142:1137–1145

68. Prakanrattana U, Prapaitrakool S (2007) Efficacy of risperidone for prevention of postoperative delirium in cardiac surgery. Anaesth Intensive Care 35:714–719

69. Coyle JT, Price DL, DeLong MR (1983) Alzheimer's disease: a disorder of cortical cholinergic innervation. Science 219:1184–1190

70. Youn YC, Shin HW, Choi BS et al (2017) Rivastigmine patch reduces the incidence of postoperative delirium in older patients with cognitive impairment. Int J Geriatr Psychiatry 32(10): 1079–1084

71. Gamberini M, Bolliger D, Lurati Buse GA et al (2009) Rivastigmine for the prevention of postoperative delirium in elderly patients undergoing elective cardiac surgery—a randomized controlled trial. Crit Care Med 37:1762–1768

72. Agostini JV, Leo-Summers LS, Inouye SK (2001) Cognitive and other adverse effects of diphenhydramine use in hospitalized older patients. Arch Intern Med 161:2091–2097

73. Seo SW, Suh MK, Chin J, Na DL (2009) Mental confusion associated with scopolamine patch in elderly with mild cognitive impairment (MCI). Arch Gerontol Geriatr 49:204–207

74. Lynch EP, Lazor MA, Gellis JE et al (1998) The impact of postoperative pain on the development of postoperative delirium. Anesth Analg 86:781–785

75. Marcantonio ER, Juarez G, Goldman L et al (1994) The relationship of postoperative delirium with psychoactive medications. JAMA 272:1518–1522

76. Silbert BS, Scott DA, Evered LA et al (2006) A comparison of the effect of high- and low-dose fentanyl on the incidence of postoperative cognitive dysfunction after coronary artery bypass surgery in the elderly. Anesthesiology 104:1137–1145

77. Rasmussen LS, Steentoft A, Rasmussen H, Kristensen PA, Moller JT (1999) Benzodiazepines and postoperative cognitive dysfunction in the elderly. ISPOCD Group. International Study of Postoperative Cognitive Dysfunction. Br J Anaesth 83:585–589

78. Fredman B, Lahav M, Zohar E et al (1999) The effect of midazolam premedication on mental and psychomotor recovery in geriatric patients undergoing brief surgical procedures. Anesth Analg 89:1161–1166

79. Maldonado JR, Wysong A, van der Starre PJ et al (2009) Dexmedetomidine and the reduction of postoperative delirium after cardiac surgery. Psychosomatics 50:206–217

80. Su X, Meng ZT, Wu XH et al (2016) Dexmedetomidine for prevention of delirium in elderly patients after non-cardiac surgery: a randomised, double-blind, placebo-controlled trial. Lancet 388:1893–1902

81. Li B, Li Y, Tian S et al (2015) Anti-inflammatory effects of perioperative dexmedetomidine administered as an adjunct to general anesthesia: a meta-analysis. Sci Rep 5:12342

82. Peng L, Xu L, Ouyang W (2013) Role of peripheral inflammatory markers in postoperative cognitive dysfunction (POCD): a meta-analysis. PLoS One 8:e79624

83. Hudetz JA, Patterson KM, Iqbal Z et al (2009) Ketamine attenuates delirium after cardiac surgery with cardiopulmonary bypass. J Cardiothorac Vasc Anesth 23:651–657

84. Roytblat L, Talmor D, Rachinsky M, Duvdenani A (1998) Ketamine attenuates the interleukin-6 response after cardiopulmonary bypass. Anesth Analg 87:266–271

INDEX

Marco Cascella (ed.), *General Anesthesia Research*, Neuromethods, vol. 150, https://doi.org/10.1007/978-1-4939-9891-3,
© Springer Science+Business Media, LLC, part of Springer Nature 2020

Printed in the United States
By Bookmasters